Plant and Vegetation

Volume 11

Series editor
M.J.A.Werger, Utrecht, The Netherlands

Plant and Vegetation is a new Springer series comprising a series of books that present current knowledge and new perspectives on world vegetation. Examining the ecology of plants and vegetation at all scales – from plant to landscape – and covering key issues such as globalization, invasive species, climate change and the dynamics of plant biodiversity, this book series draws together a wide range of material of interest to plant ecologists, vegetation scientists, and geographers around the world. The series provides a valuable resource for both graduate students and researchers in environmental and biological sciences, as well as for landscape planners and policy makers involved in land-use and restoration projects at local, regional and international levels.

More information about this series at http://www.springer.com/series/7549

Cindy Q. Tang

The Subtropical Vegetation of Southwestern China

Plant Distribution, Diversity and Ecology

Springer

Cindy Q. Tang
Institute of Ecology and Geobotany
Yunnan University
Kunming, Yunnan, China

ISSN 1875-1318 ISSN 1875-1326 (electronic)
Plant and Vegetation
ISBN 978-94-017-9740-5 ISBN 978-94-017-9741-2 (eBook)
DOI 10.1007/978-94-017-9741-2

Library of Congress Control Number: 2015930082

Springer Dordrecht Heidelberg New York London

Printed on acid-free paper

Springer Science+Business Media B.V. Dordrecht is part of Springer Science+Business Media (www.springer.com)

The woods are lovely, dark and deep,
But I have promises to keep,
And miles to go before I sleep,
And miles to go before I sleep.

–Robert Frost, 1923

Preface

Over recent decades, a number of superb studies have been carried out on the vegetation of China, but up to now, no Chinese geographic zone as a whole has received the kind of quantitative, systematically rigorous analysis that today's vegetation science demands. Southwestern China lags in knowledge regarding the dynamics, succession, ecology, and diversity mechanisms of its subtropical vegetation. This book is born of a desire to provide valid scientific information on the patterns and processes of vegetation change across a broad range of spatial and temporal scales in the region. I hope that it will serve this primary purpose. It is perhaps fitting that the first such treatment should be devoted to southwestern China, whose exceptional interest and importance will become evident in the pages that follow. As a plant ecologist and an educator of graduate students in the fields of ecology and geobotany, I am hopeful that it will also provide readers with some understanding of how vegetation dynamics evolve, how plants persist, and how plant diversity has changed over time and space in this specific geographic region.

The arrangement of chapters in this book is intended to promote an integration of ecological perspectives on China's marvelous subtropical vegetation. Following an introduction (Chap. 1), each chapter concentrates on the Yunnan area, except that Chap. 7 also focuses on the Yangtze River valley near the boundaries separating Chongqing, Sichuan, and Guizhou.

Without the help and support of others, this book would not have been realized. I am greatly indebted to my field coinvestigators through whose efforts the present studies have become possible: Long-Yuan He and Yongchuan Yang. I am very grateful to the staff of nature reserves and to local villagers who accompanied me in the field; through their efforts, I have been able to visit extremely remote areas. Their dependable good humor added tremendously to the pleasure of my field journeys, and their resourcefulness under difficult conditions was remarkable. I feel very grateful for the technical expertise of Long-Yuan He, who helped me to make the maps for the book. I especially express my gratitude to Professor Marinus J. A. Werger for his continuous encouragement through all stages of the production and for invaluable editing and critical comments on the manuscript, to Professor Masahiko Ohsawa, whose broad and profound knowledge of the vegetation of East

Asia has inspired me in the study of southwestern China, for his comments and suggestions on the manuscript, and to my colleagues and graduate students who have given me limitless assistance in obtaining research material. I particularly value and acknowledge the help of Thomas Montgomery, my husband, for his ongoing inspiration, support for my field research, and polishing and editing of the English text, and of Ian Tang Montgomery, my son, for his patience and understanding of my absence from home during my numerous long journeys in the field.

Kunming, Yunnan, China
February 2015

Cindy Q. Tang

Contents

Chapter 1
Introduction

Abstract The region known as southwestern China (SW China, 21°08′32″-34°20′ N, 97°30′-110°10′ E) includes Yunnan, Sichuan, Guizhou provinces and Chongqing Municipality. Topographically, it lies between the first altitudinal step, the Tibetan Plateau, and the third step, the middle-lower Yangtze Plain, and stretches over a vast area (1,135,567 km^2). The region includes the Hengduan Mountains, the Yunnan-Guizhou Plateau, and the Sichuan Basin. The Hengduan Mountain Range, situated at the juncture of Tibet, Sichuan and Yunnan in the southeast of the Tibetan Plateau, has an average elevation of 2,000–6,000 m. Its highest peak, Gonggashan, reaches 7,556 m above sea level. The climate is dominated by the Asian monsoon system. It falls into latitudes classified as subtropical, except for the small tropical area of southernmost Yunnan. With that exception, altitudinal climates range from subtropical to temperate, alpine and frigid.

The crust movements of various parts of SW China have undergone tortuous and complicated evolutionary processes over the long development of geological history. The most striking geological characteristic of the region is a massive crust movement strongly linked with the forming of the Hengduan Mountains – an extension of the eastern Himalayas.

The soils vary widely, including red earth, yellow, yellowish brown, brown, purple, and mountain gray-brown types.

In the Early Tertiary period, the areas now including eastern Yunnan, Sichuan, Chongqing and Guizhou bore subtropical dry and semi-dry sparse woodlands, while most areas now comprising Yunnan had a warm-rainy winter and hot-dry summer climate supporting subtropical evergreen broad-leaved forests. In the Late Tertiary period, the vegetation of SW China became a subtropical mixed evergreen and deciduous broad-leaved forest, except in southern Yunnan, home to a subtropical evergreen broad-leaved forest. In the Quaternary, vegetation changes were taking place at about 10,000 years BP (years before present), and by 7,000 years BP the pattern of the mixed evergreen and deciduous forest had become more complex. Parts of this region had not been covered by the ice sheet during the Quaternary. In general, since the Holocene, the natural vegetation types of SW China have been similar to those of the present time.

The rich flora and complicated plant geographic patterns of SW China result from a combination of elements from the Sino-Japanese, Sino-Himalayan, African-Mediterranean, and Indo-Malesian floristic regions. SW China is one of the most botanically diverse terrestrial regions on earth, with a high level of endemism.

C.Q. Tang, *The Subtropical Vegetation of Southwestern China*,
Plant and Vegetation 11, DOI 10.1007/978-94-017-9741-2_1

The vegetation includes tropical rain forest, subtropical evergreen broad-leaved forest (the vast majority), warm-temperate deciduous broad-leaved forest, temperate coniferous and broad-leaved mixed forest, cold-temperate coniferous forest, alpine scrub community, meadow, and wetland, as well as thorny shrubland and savanna in hot-dry valleys. There are also precious remnants: Tertiary relict forests in specific habitats such as scree slopes, limestone areas and stream banks. The subtropical vegetation of SW China offers an unparalleled diversity of plant life and has far-reaching importance for the entire world.

In the last few centuries human activities have increasingly changed the subtropical vegetation. Man's activities destroyed most of the original primary evergreen broad-leaved forests dominated by species of *Cyclobalanopsis, Castanopsis, Lithocarpus, Machilus, Phoebe, Cinnamomum, Schima, Michelia* etc. and they became restricted to remote mountain areas, nature reserves, steep slopes, and temple grounds. Complex mosaics of stands of varying ages, composition and structure intermingle with agricultural fields or mono-species plantations. So, SW China is truly a land of plant diversity, vegetation dynamics, and succession in time and space.

1.1 Environmental Setting

1.1.1 Physiography

The topography of China can be described as a series of steps decreasing in altitude from the Tibetan Plateau (Qinghai-Tibet Plateau), with a mean altitude of 4,000 m (highest elevation Mt. Everest, 8,848 m), which makes up one-third of the country, to the Xinjiang-Inner Mongolia, Loess, and Yunnan-Guizhou Plateau, followed by a descent eastward to a mean altitude of 2,000 m, and then a further descent to a mean altitude of 200–500 m, reaching finally to the sea (Liu and Ding 1984). These topographic variations make the terrain of China into a series of three altitudinal steps. Southwestern China, as referred to in this text, includes Yunnan, Sichuan, Guizhou provinces and Chongqing Municipality. This region (21°08′32″-34°20′ N, 97°30′-110°10′ E), lying between the first step, the Tibetan Plateau, and the third step, the middle-lower Yangtze Plain, stretches over an area of 1,135,567 km^2 (Fig. 1.1). It comprises a landscape of mountain chains interspersed with basins and alluvial plains. It includes the Hengduan Mountains, the Yunnan-Guizhou Plateau, and the Sichuan Basin (Fig. 1.2). The Hengduan Mountains are a series of parallel north-south ranges forming the south-eastern part of the Tibetan Plateau, Western Yunnan and Western Sichuan (Latitude 22°-32°05′ N, Longitude 97°-103° E). Most parts of the region are characterized by a series of parallel north-south high mountain ranges alternating with deep gorges and rivers. The north-south orientation of the Hengduan Mountains intersects the predominant east-west mountain system of eastern Asia (Fig. 1.3). The Chinese name for this region is "Hengduanshan" which

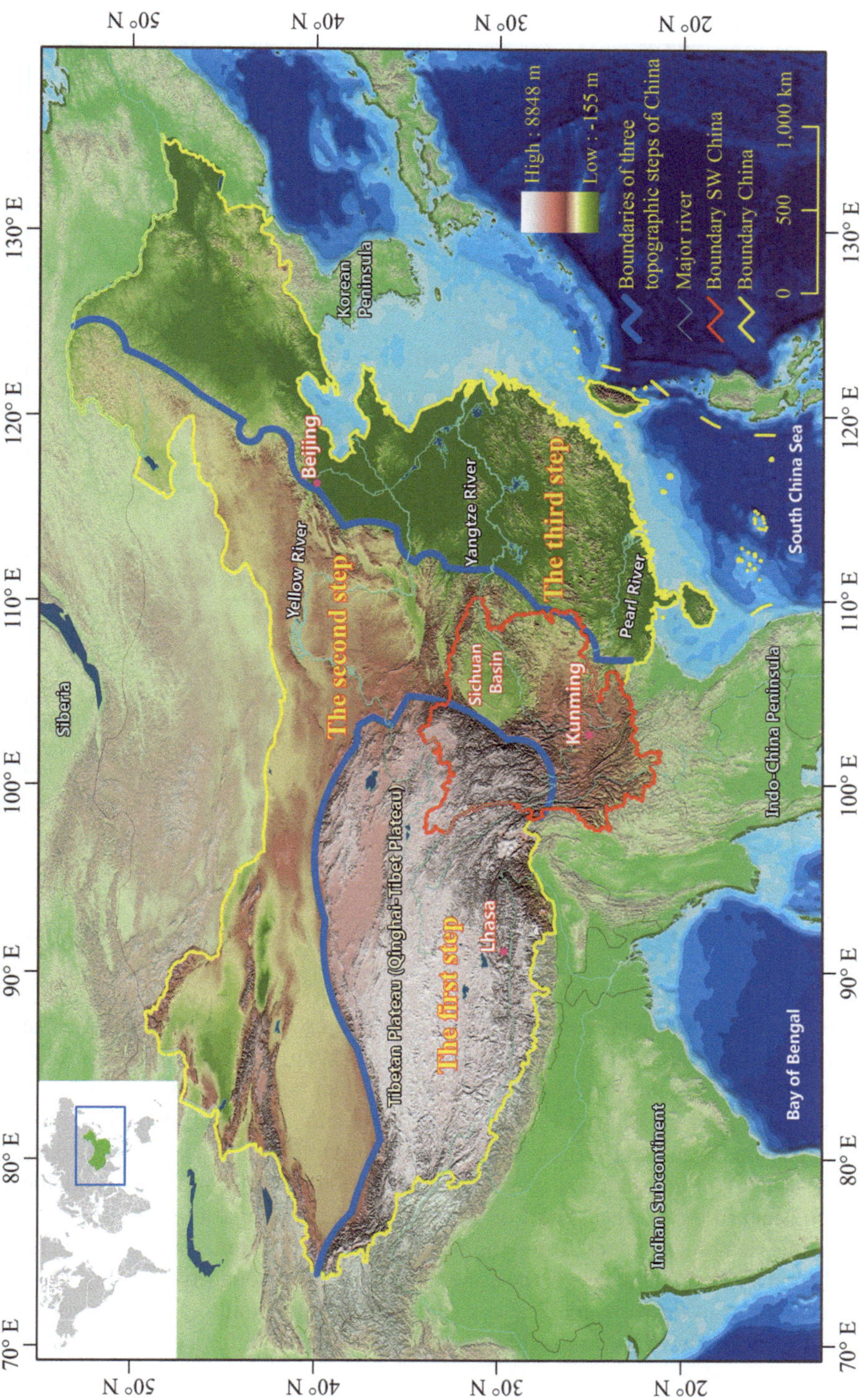

Fig. 1.1 Three topographic steps of China

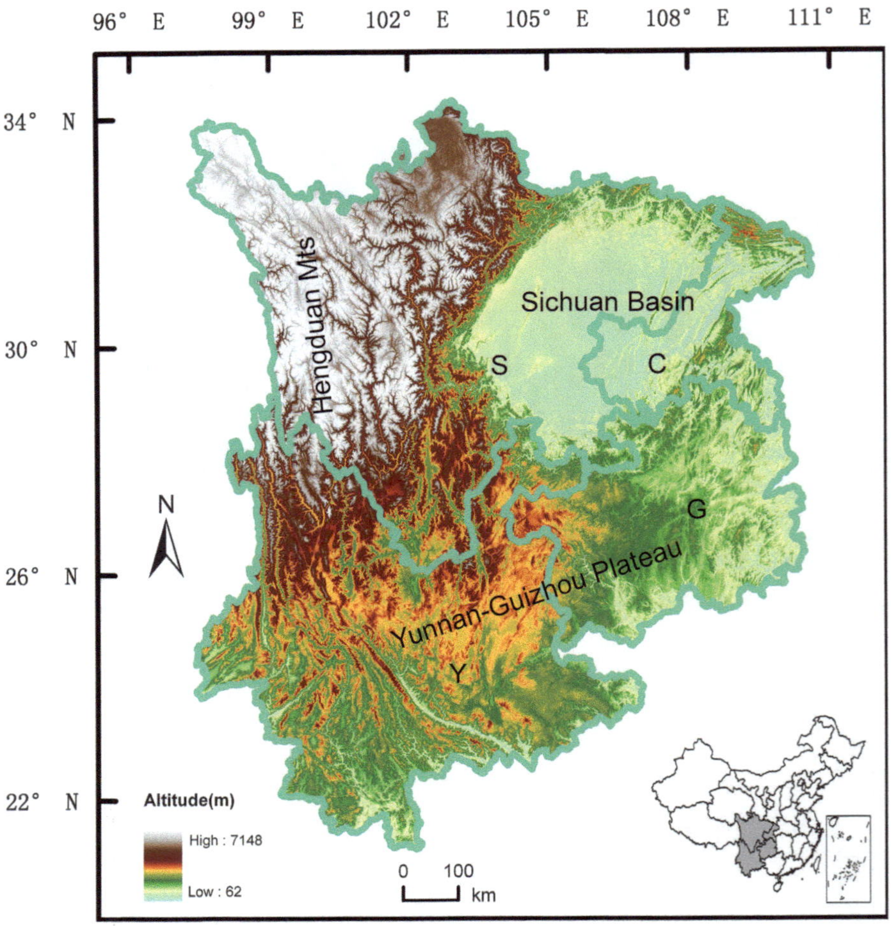

Fig. 1.2 SW China. *Y* Yunnan Province, *G* Guizhou Province, *S* Sichuan Province, *C* Chongqing Municipality

means "traverse cutting mountains". These high ranges are, from west to east, the Gaoligongshan ("shan" means "mountain"), the Nushan, the Yunling ("ling" also means "mountain"), the Daxueshan, the Qionglaishan, and the Minshan, defining the eastern edge of the Tibetan Plateau, western Yunnan, and the western rim of the Sichuan Basin. Branching further out from the Yunling Range are some secondary ranges – the Wuliangshan and the Ailaoshan in the south-central area and the Wumengshan in the northeast part of Yunnan. The Gaoligongshan at the southern end of the Hengduan ranges forms the border between Myanmar and China. The Minshan in western Sichuan is at the eastern end of the Hengduan system (Fig. 1.3). Four rivers run in deep parallel valleys separated by mountains of the Hengduan range (Fig. 1.4). They flow in parallel out of Tibet to the south for a certain distance and then each takes its own way: the Dulongjiang (upper Irrawaddy) and Nujiang (upper Salween) to the Andaman Sea, the Lancangjiang (upper Mekong) to the

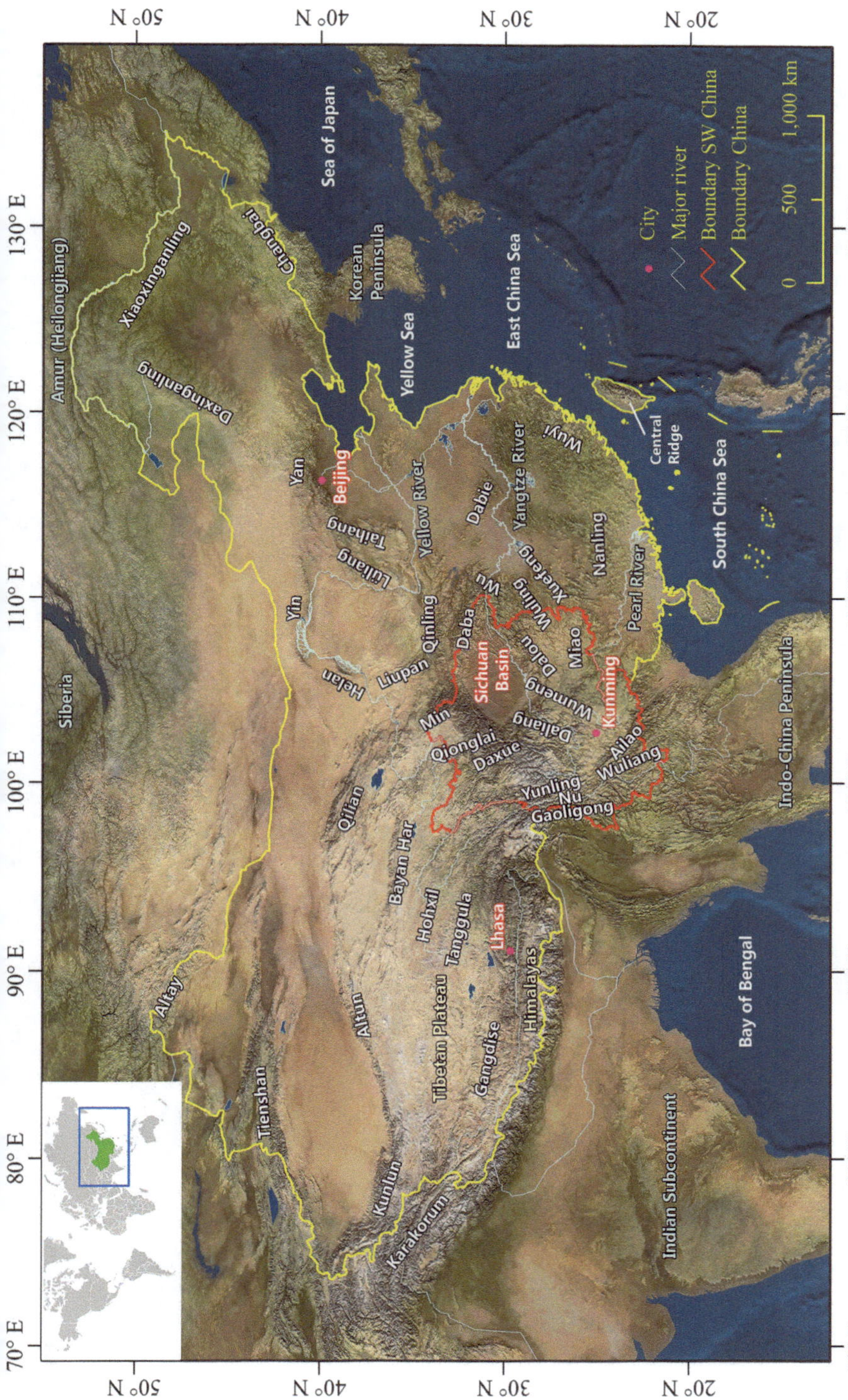

Fig. 1.3 Major mountain ranges of China

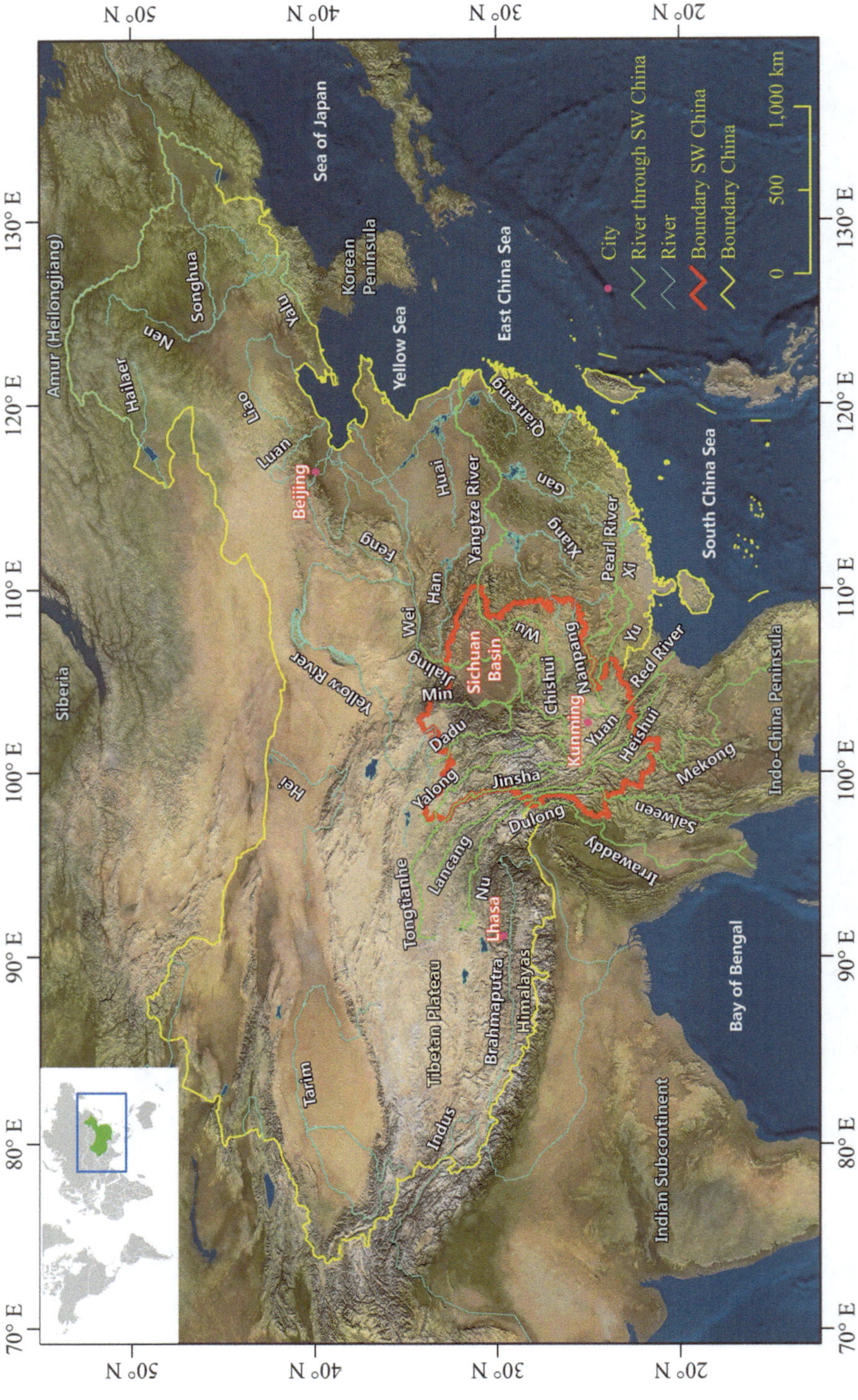

Fig. 1.4 Major rivers of China

South China Sea, the Yangtze River (the Upper Yangtze is called Jinshajiang in Chinese) – flowing for 6,380 km (the longest river in Asia and the third-longest in the world) from glaciers on the Tibetan Plateau eastward across southwestern, central and eastern China before emptying into the East China Sea at Shanghai. The Yangtze, a major transport route, provides water for a large agricultural area where more than half of China's rice and other grains grow.

In the south, southeast and west Yunnan Province borders on Laos, Vietnam, and Myanmar, respectively. The major landforms in Yunnan are mountains (84 %), highlands (10 %), small basins and valleys (6 %). The Yunnan-Guizhou Plateau mainly covers eastern Yunnan Province (the Ailao Mountains serve as a center line dividing the province into the western and eastern parts) and the major part of Guizhou Province. Two distinct areas mark this plateau: one of high plateaus averaging about 2,000 m with mountain peaks as high as 4,344 m (Jiaozishan) and 4,042 m (Yaoshan) in eastern Yunnan, and a second area of rolling hills, deep river-carved gorges, and mountains marked by geological faults in western Guizhou. Easily-eroded limestone underlies the plateau, allowing for spectacular karsts. The highest elevation in Guizhou is 2,900 m at Jiucaiping in Hezhang to the west, while the lowest altitude is 137 m, to the east. The mountainous area, hilly land and river valley account for 61.7, 31, and 7.3 %, respectively of the territory. The Sichuan Basin comprises the central and eastern portions of Sichuan Province as well as Chongqing Municipality. It is an expanse of low hills and plains completely encircled by mountains. Tibetan borderlands lie to the west (e.g. Daxueshan and Daliangshan) and north (e.g. Longmenshan and Mingshan). The Yunnan-Guizhou Plateau extends to the south (e.g. Daloushan). Hills of the Yangtze River valley lie to the east (e.g. Wushan); Dabashan and Qinling to the northeast (Fig. 1.5). The topography in Sichuan is extremely varied and complex. There is a Chinese saying, "The road to Sichuan is harder than the road to heaven". In Sichuan Province, four geomorphic types: mountainous areas, hilly areas, plains and plateaus, account for 77.1, 12.9, 5.3, and 4.7 % respectively of the total area.

In SW China, the altitudinal gradient is one of the longest bioclimatic gradients in the world extending from lower than 500 m asl (e.g. Hekou with 76.4 m in southeastern Yunnan; the central Sichuan Basin area with 250–700 m) to over 6,000 m (e.g. the Meili Mountains with 13 peaks over 6,000 m in northwestern Yunnan; Mt. Gongga or Minya Konka with 7,756 m at the southern section of the Daxue Mountains in western Sichuan. Minya Konka means "the Highest Snowy Mountain" in Tibetan) (Figs. 1.6, 1.7, 1.8, and 1.9). The topographic profiles of Yunnan, Sichuan and Guizhou are shown in Fig. 1.10. Also shown are scenes characteristic of SW China: picturesque limestone karsts and plentiful plateau lakes (Figs. 1.11, 1.12, and 1.13).

Fig. 1.5 The Sichuan Basin encircled by mountains

Fig. 1.6 An Evergreen broad-leaved forest ecosystem at 1,700–2,400 m asl in Bingzhongluo, northwestern Yunnan. Bingzhongluo is located in the northernmost part of Gongshan County; it is one of the largest valleys among the Nujiang gorges. It is also called the garden of the peaches of immortality, where the local people (mainly Nu, Lisu, Tibetan and Dulong) share an appreciation of nature's abundance that seems to have a parallel in their interracial harmony (Photograph courtesy of Huan-Chong Wang)

Fig. 1.7 A cloud forest dominated by evergreen broad-leaved tree species of *Lithocarpus*, *Cyclobalanopsis* and *Castanopsis* along with the Tertiary relict coniferous *Taiwania cryptomerioides* at c. 2,300 m asl in the Gaoligong Mountains, northwestern Yunnan

Fig. 1.8 Yubeng Village at 3,140 m asl in the foot hills of the Meilixue Mountains, northwestern Yunnan. On the mountain slopes, mainly evergreen sclerophyllous *Quercus* and various coniferous trees

Fig. 1.9 The Meilixueshan range at 6,000–6,740 m asl in northwestern Yunnan. A holy site as confirmed by the Tibetan Buddhist Baitas

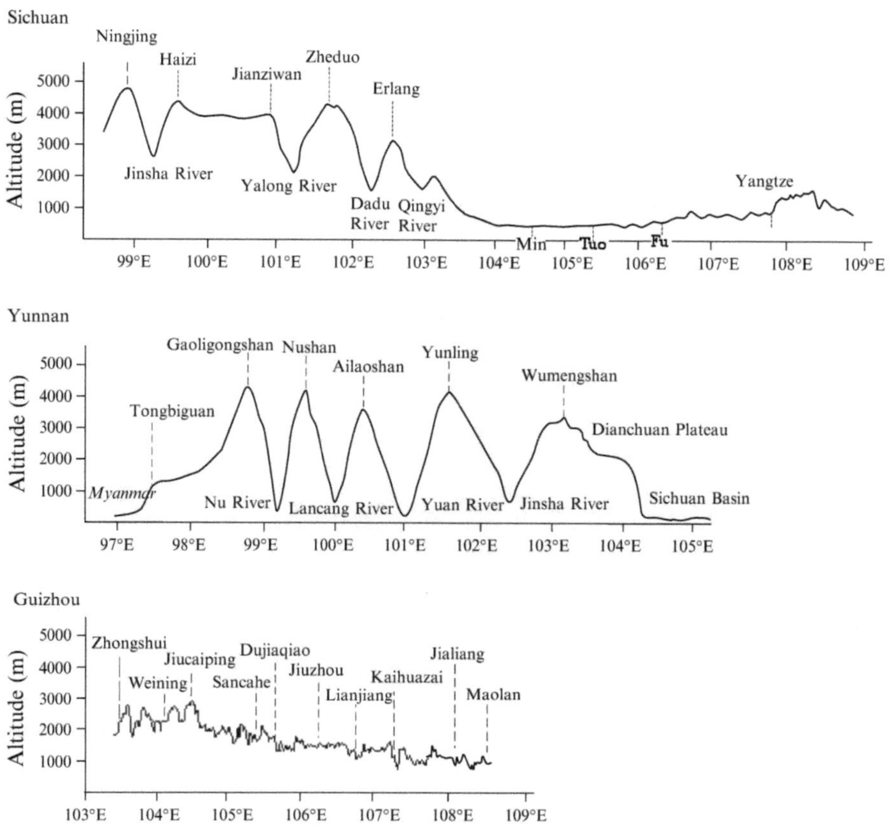

Fig. 1.10 Topographic profiles of SW China

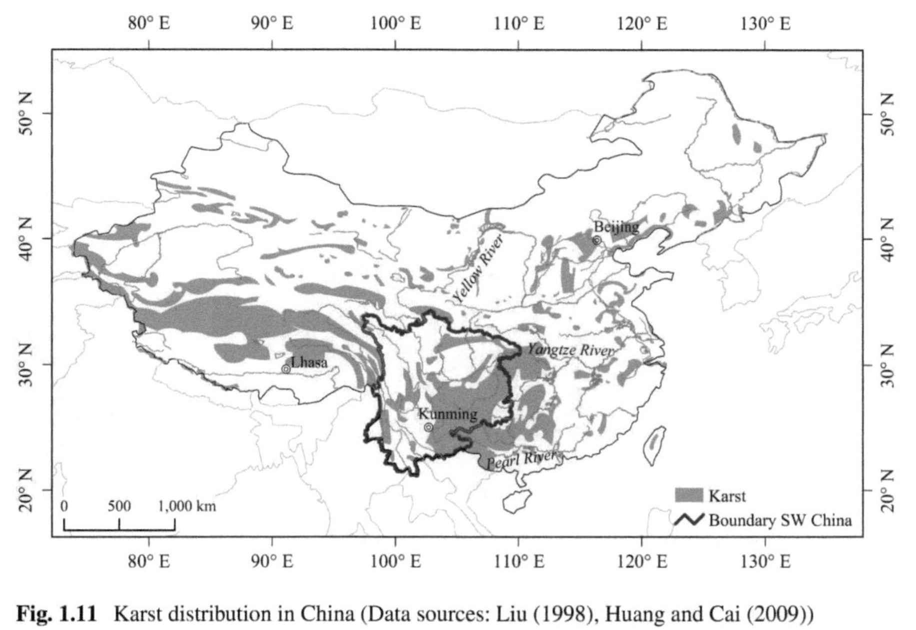

Fig. 1.11 Karst distribution in China (Data sources: Liu (1998), Huang and Cai (2009))

Fig. 1.12 Shilin (Stone Forest) at 1,800 m asl, central Yunnan. One of the world-famous karst landscapes

Fig. 1.13 Luguhu, a plateau lake at 2,685 m, located between Linglong in Yunnan and Yanyuan in Sichuan

1.1.2 Climate

China's subtropical regions are bounded to the north by a line running between the Huai River and the Qinling Mountains (35° N), with a southern limit approaching the Tropic of Cancer. Eastward the subtropics extend to the coastlands and the islands of the East China Sea, the South China Sea and Taiwan; the westward limit is the Chinese national border, extending from the eastern slope of the Tibetan Plateau to southern Yunnan. The subtropical zone thus spans 11–12° from north to south, 28° from east to west, and covers more than 2,400,000 km^2 (Wu 1980). The region is divided into three different zones: (1) the southern subtropics (transitional tropics), extending from latitude 22° to 24° or 25° N; (2) the middle subtropics, which lie between 24° or 25° N and 31° or 32° N; and (3) the northern subtropics (transitional subtropics) which cover from 31° or 32° N to 34° or 35° N (Hou 1983). SW China is located at subtropical latitudes often distinguished as north, middle, and south subtropical (Fig. 1.14). A small part of southernmost Yunnan (e.g. Xishuangbanna) falls into the tropical zone. According to altitude, climates include tropical, subtropical, temperate, alpine and frigid. The climate is dominated by the Asian monsoon system, including the southwestern monsoon (the summer

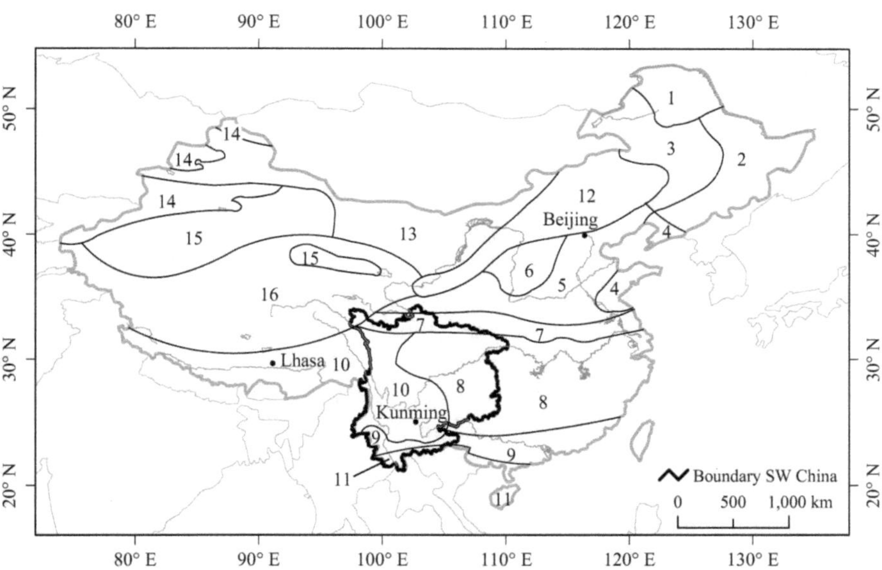

Fig. 1.14 Climate divisions of China. *1* Cold-temperate humid, *2* Middle-temperate humid, *3* Middle-temperate subhumid, *4* South-temperate humid, *5* South-temperate subhumid, *6* South-temperate subdry, *7* North-subtropical humid, *8* Middle subtropical humid, *9* South-subtropical humid, *10* Middle-subtropical subhumid and plateau, *11* Tropical humid, *12* Middle-temperate subdry, *13* Middle-temperate dry, *14* Middle-temperate alpine subdry, *15* South-temperate dry, *16* Plateau (dry and subdry) (Modified from The Normal College of Northwest China (1984), Winkler and Wang (1993))

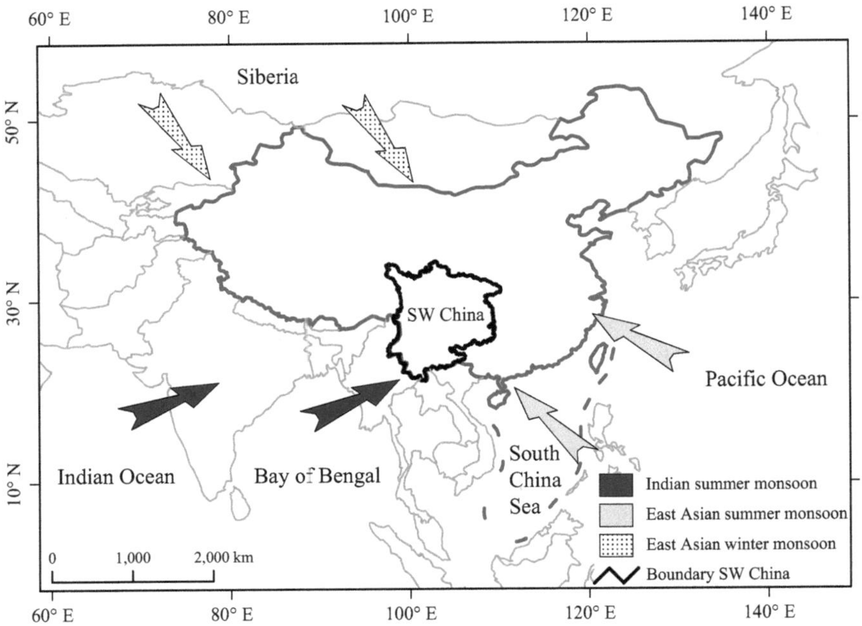

Fig. 1.15 Three monsoon winds in China

monsoon of India), the East Asian summer monsoon, and the East Asian winter monsoon, with dry continental winds in winter and moist oceanic winds in summer (Fig. 1.15). In China, generally the oceanic monsoon season begins earlier and ends later in the south than in the north, giving the south a longer rainy season and milder winter winds (Guo and Werger 2010). In early May, the southwestern monsoon (Indian summer monsoon) coming from the Bay of Bengal and the East Asian summer monsoon coming from the South China Sea advance onto the Yunnan-Guizhou Plateau. These two marine air masses become the source of summer rainfall and create these regions' hot humid summer climate, but they do not penetrate the interior of the Hengduan Mountains. As a result the deep mountain valleys of Yunnan are generally arid. Humidity to the west of the Ailao Mountains is generally higher than to the east, because of the stronger influence of the southwestern wind from the Bay of Bengal (Fig. 1.16). In winter the Tibetan Plateau hinders the invasion of the Siberian cold current into SW China. This is an important reason for the northward deviation of the limit of subtropical vegetation there. The Yunnan-Guizhou Plateau is rarely invaded by cold currents, but because the eastern plain of China lacks the protection of high mountains, the cold current from the Mongolian Cold High can reach to the far south, producing cold waves in the southeastern subtropics (Chang 1983; Wang 2005), for instance in Shanghai.

In Guizhou, the mean annual precipitation is 1,100–1,300 mm. In some areas (e.g. Fanjingshan) of eastern Guizhou, it reaches 2,600 mm. The annual rainfall generally decreases from the east to the west, and from the south to the north (Yang 1983; Wang and Wang 2002).

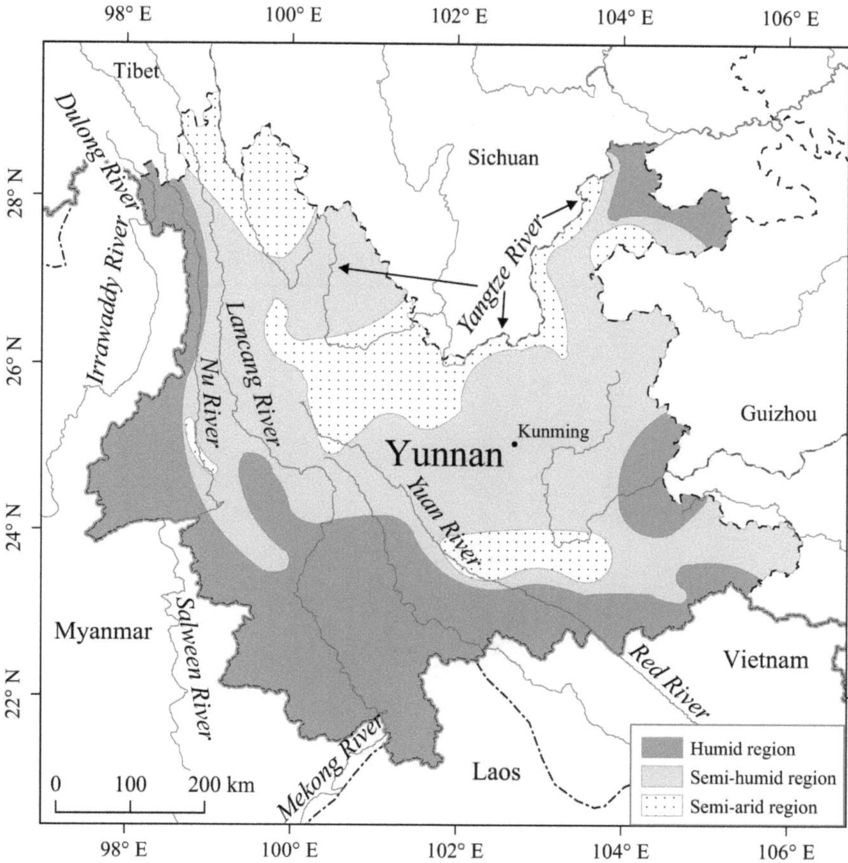

Fig. 1.16 Humidity patterns in Yunnan

The west of Sichuan is more strongly influenced by the southwestern wind from the Bay of Bengal, while the east (the Sichuan Basin) is more strongly affected by the southeastern wind from the South China Sea. The combination of strong seasonal winds and numerous mountain ranges results in many local climatic effects, from seasonally alternating windward and leeward slopes and rain shadows to the unusual, persistently cloudy climate of the Sichuan Basin. The overall characteristics of the climate of Sichuan can be summarized as follows: the warm, rainy monsoon climate is dominant. Regional climate differences are prominent: the east is characterized by cool winters, hot, hazy and rainy summers, humid autumns, almost year-round cloud and fog, and scant sunshine, while the north has moderately cold winters as it is protected from the southward movement of cold air from Central Asia; altitudinal climate variations are significant. The mean annual precipitation is generally 1,000–1,300 mm in the basin, but in the mountainous area at the basin's west rim, it reaches 1,500–1,800 mm. This area is known as the "Rainy Zone of West China". Sunny hours range from 900 to 1,300 per year. A local saying holds that in Sichuan, dogs bark when the sun appears. In contrast, a hot-dry climate predominates in the steeper river valleys, mainly in Yunnan and western Sichuan.

The climatic conditions of SW China in general are presented in Fig. 1.17a–f. Climate variations according to altitude, latitude and longitude in SW China are shown in Fig. 1.18.

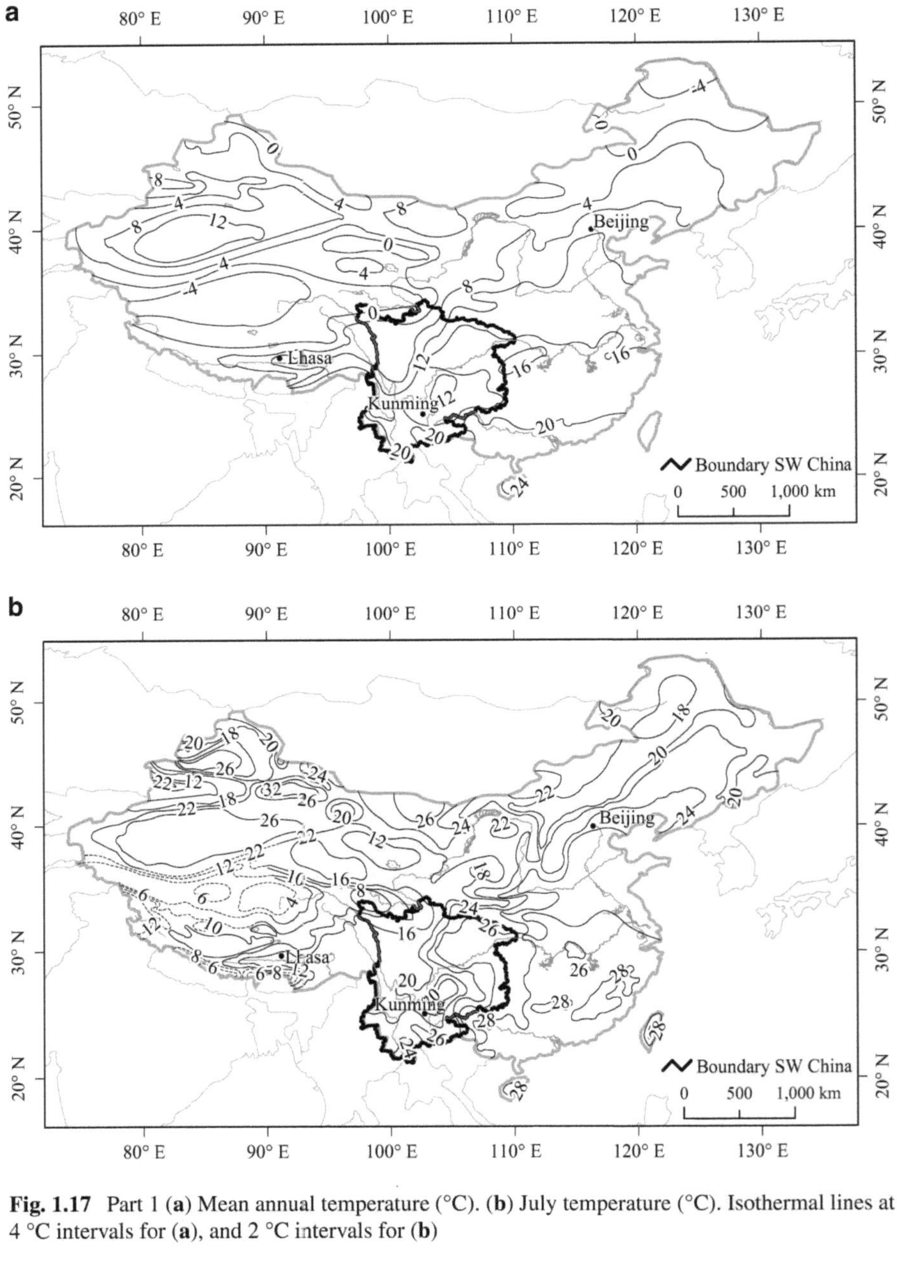

Fig. 1.17 Part 1 (**a**) Mean annual temperature (°C). (**b**) July temperature (°C). Isothermal lines at 4 °C intervals for (**a**), and 2 °C intervals for (**b**)

c

d

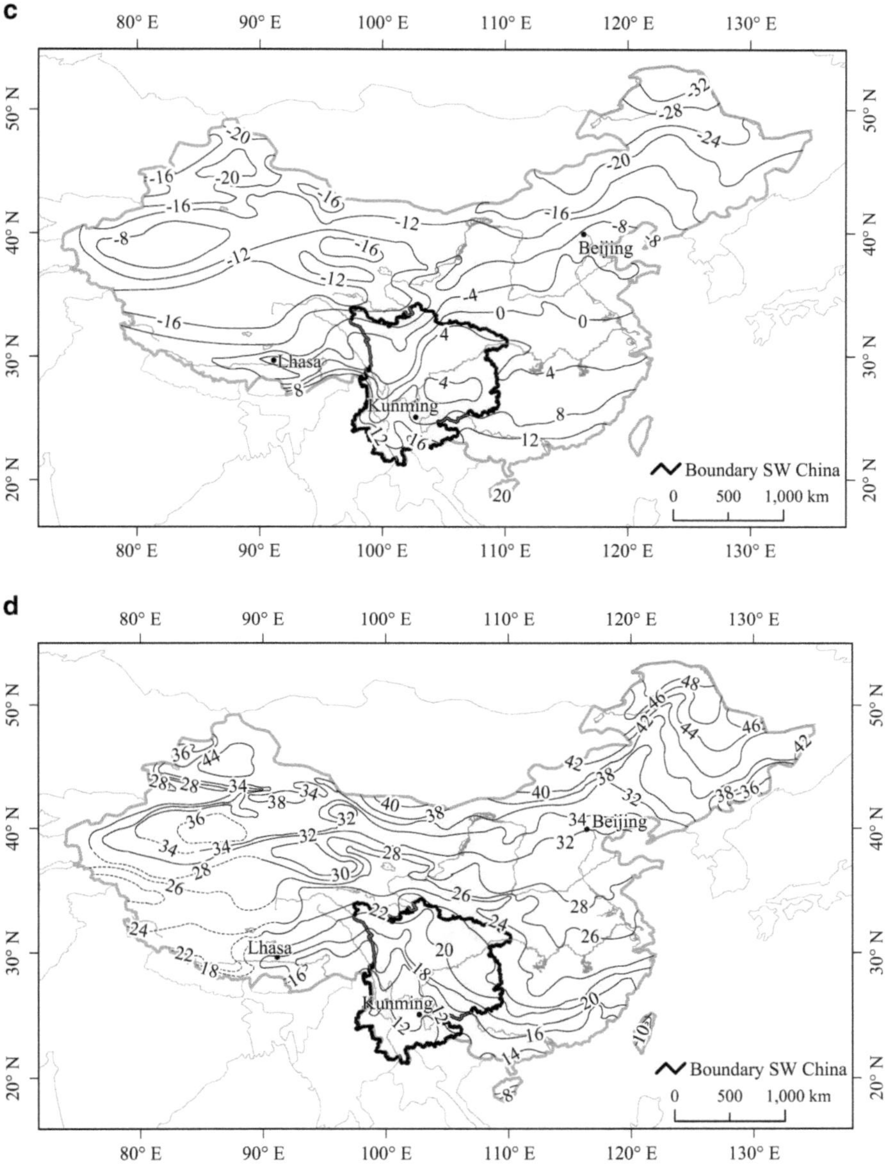

Fig. 1.17 (continued) Part 2 (**c**) January temperature (°C). Isothermal lines at 4 °C intervals. (**d**) Temperature (°C) difference between coldest and warmest months. Isothermal lines at 2 °C intervals

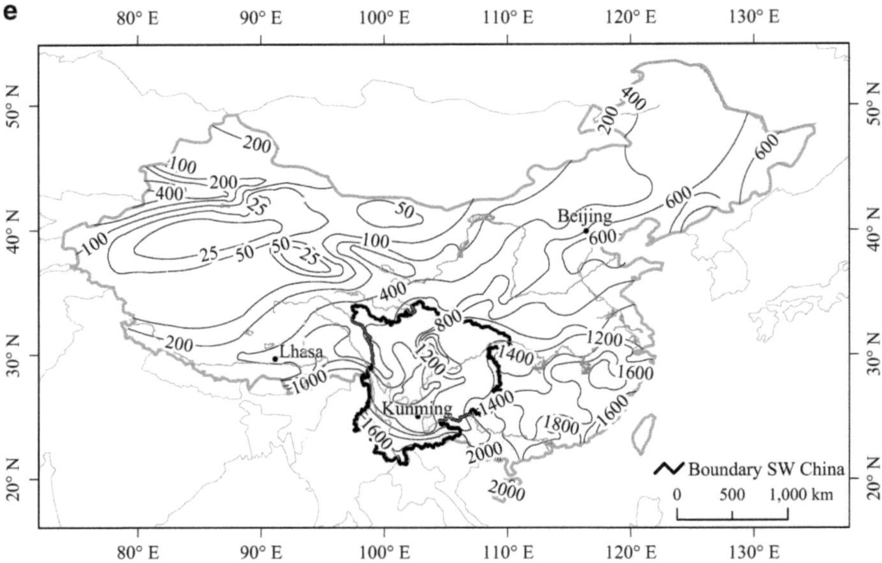

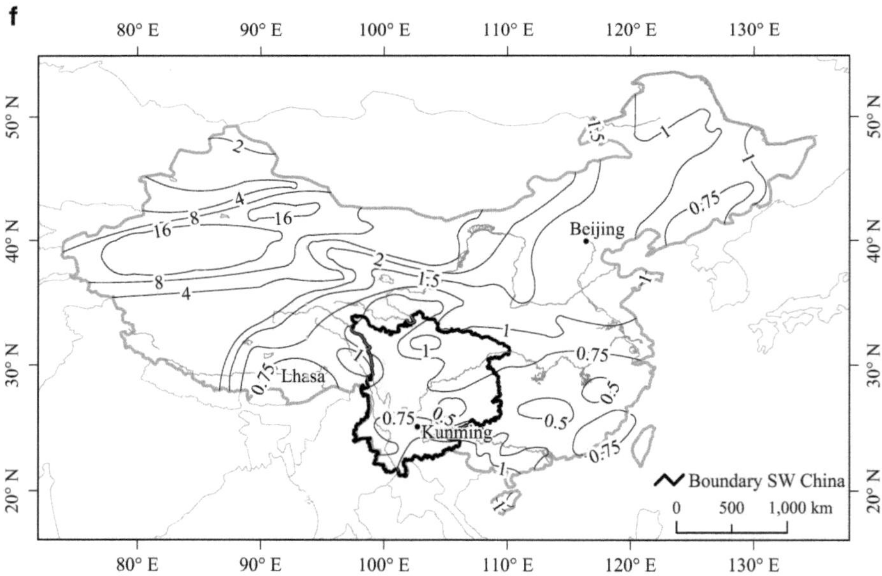

Fig. 1.17 (continued) Part 3 (**e**) Annual precipitation (mm), isohyetal lines at 200 mm intervals for areas that received over 400 mm. (**f**) Aridity index, isolines at 0.25 intervals (**a–f** are Modified from Ren et al. (1985), Winkler and Wang (1993))

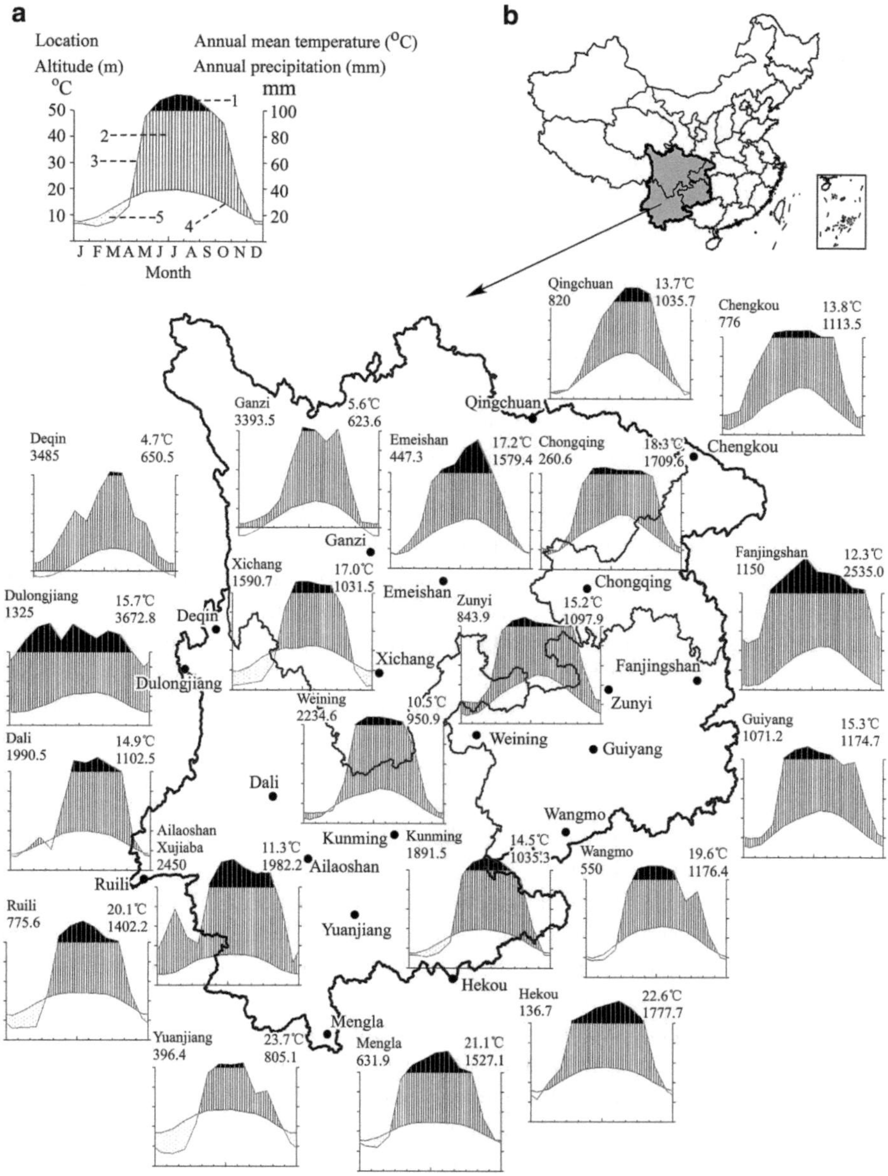

Fig. 1.18 (**a**) Explanatory climatic diagram. *1* Monthly precipitation over 100 mm, *2* humid period, *3* monthly precipitation curve, *4* monthly mean temperature curve, *5* dry period. (**b**) Climatic conditions vary widely across the landscape of SW China

1.1.3 Geology

The complex relationships between the physical and biological components of ecosystems and habitats have challenged ecologists for many decades. Within this complexity it is apparent that geology plays an important role in delineating the ecology of an area, through its contribution to the formation of soils as well as its topographic component. Geology has strong ties with biodiversity and plant life, in that the nature of the substrate, as usually determined by the nature of the underlying rock, is a key factor in the distribution of habitats and plant species (Cottle 2004). Landforms have conditioned the presence and distribution of plant species. For instance, the limestone area of SW China supports high endemism and specific plant species assemblages.

The crust movements of various parts of SW China have undergone tortuous and complicated evolutionary processes over the long development of geological history. Geotectonically Yunnan is situated in the conjoint position between the Tethys-Himalayan and the Marginal-Pacific tectonic domains (YBG 1990). Guizhou is located in the Yangtze massif and the Early Paleozoic South China fold belt (GBG 1984). Sichuan is at the juncture of three tectonic domains, the Tethys-Himalayan, Marginal-Pacific and Paleo-Asiatic (SBG 1982). Chongqing lies across two tectonic units, the Yangtze paraplatform and Qinling geosynclines (Zhu et al. 2014).

A schematic map of tectonics in SW China is shown in Fig. 1.19. The Sichuan-Yunnan region (SYR) is an active transition zone between the Yangtze platform to the east and the Tibetan plateau to the west. This transition zone was formed about 50 Ma when the Indian plate collided with the Eurasian plate. This continental collision created a large-scale tectonic deformation in the Sichuan-Yunnan region that is still ongoing. This is evidenced by the high level of seismicity in this region, which is the most active in China.

The major tectonic units in southwestern China are as follows: I. Bomi-Tengchong Fold System; II. Zuogong-Gengma Fold System; III. Tibet-Yunnan Fold System; IV. Songpang-Ganzi Fold System; V. Yangtze Platform; VI. South China Fold System (Zhan et al. 2001).

The chief geological characteristic of SW China is that the crust movement is strongly linked with the forming of the Hengduan Mountains, complicated in geological structure and active in new tectonic movements. With the Indo-China movement in the Mesozoic Era, these mountains formed, and their northern sector rose above sea level. With the Yanshan Movement in the late Jurassic Period, the whole mountain range became land. The common effects of the north-eastward drift, the subduction of the Paleomediterranean Plate, the north-westward displacement of the Yangtze Plate, and the southward resistance of the North China Plate, forced the geological tectonic line to strike north-south. During the later Mid- and earlier Late-Pleistocene the Hengduan Mountains were swiftly uplifted by neotectonic activity, which has continued throughout much of the Cenozoic, and is associated with Himalayan orogeny. The mountains were disconnected from the common Tibetan Plateau. They dissected down along the north-south tectonic zone, formed the river system, and evolved their

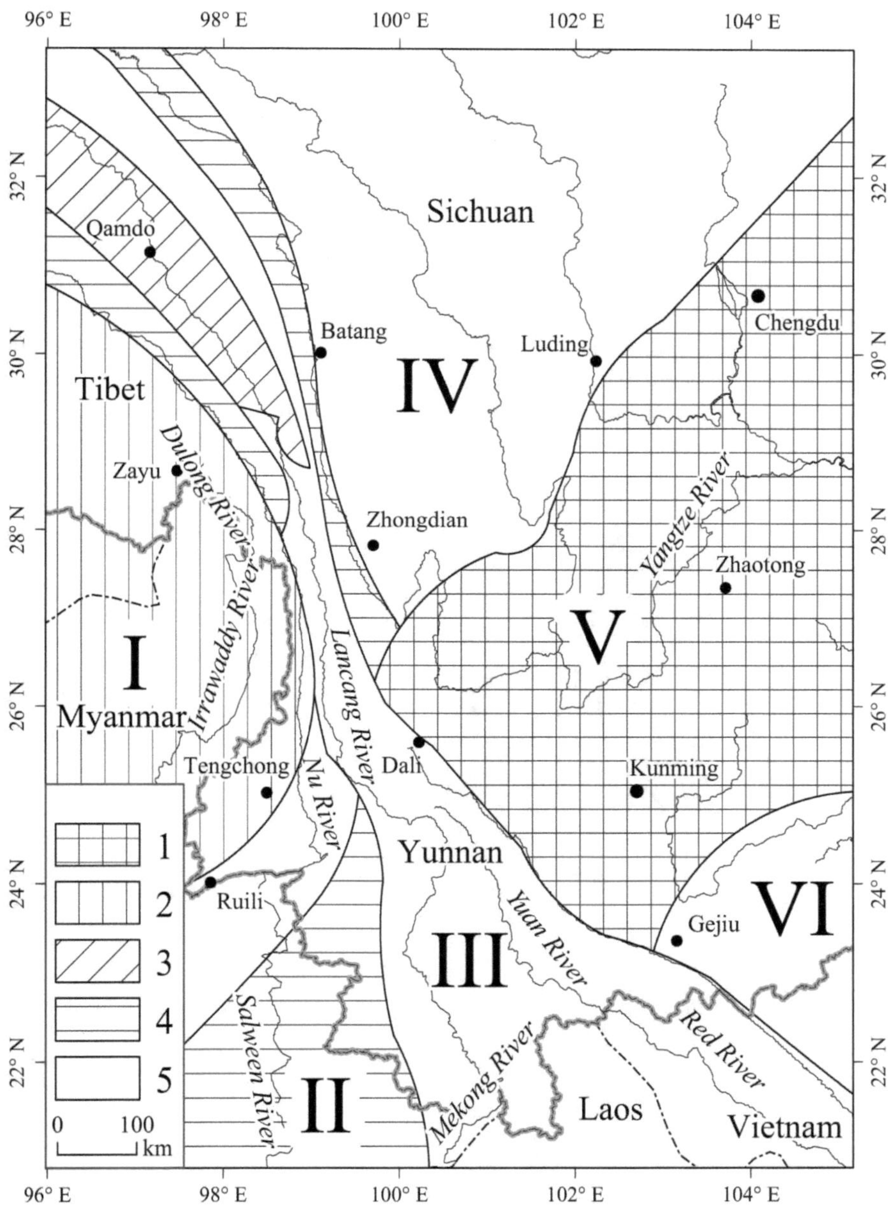

Fig. 1.19 A schematic map of the tectonics of SW China (Adapted from Zhan et al. (2001)). *1* Yangtze Platform, *2* Early Caledonian fold zone, *3* Late Caledonian fold zone, *4* Variscan fold zone, *5* Indo-China fold zone. I–VI are tectonic units named in the text

current characteristics (Li 1986). About 140 Ma, the Sichuan basin was an inland lake. The mountains at the rim of the basin rapidly lifted. At 66 Ma, in the middle and upper reaches of the Yangtze River, the water system opened up and flowed toward the east, which has resulted in today's Sichuan landform (SBG 1982).

The Nujiang (upper Salween)–Lancangjiang (upper Mekong)–Jinshajiang (upper Yangtze) area (also called the three-rivers-area) is an important region in the southeastern Tibetan syntaxis, with strong Cenozoic deformation and structural overprinting. The kinematics and deformational timing of the major shear systems in that syntaxis, such as the Ailaoshan–Red River shear system (ASRRSS), Chongshan shear system (CSSS), Gaoligong shear system (GLSS), Wang Chao shear zone (WCSZ) and Three Pagodas shear zone (TPSZ), have been systematically studied by many researchers (e.g. Leloup et al. 2001; Lee et al. 2003; Wang et al. 2006). As a great example of tectonic structure, Fig. 1.20 shows how the area

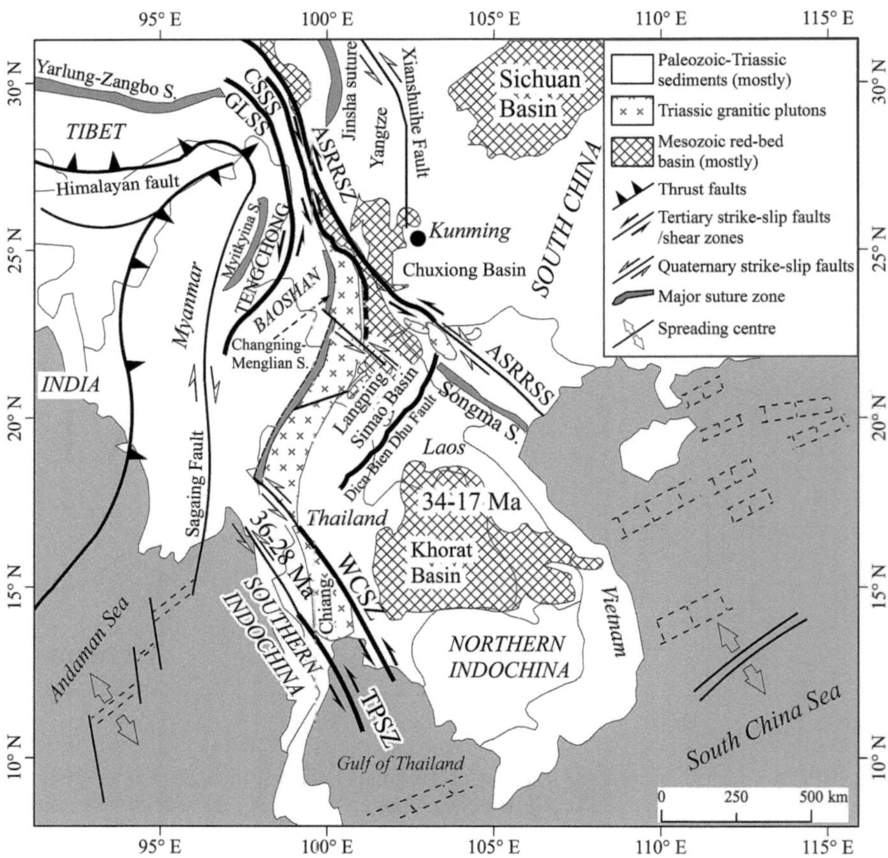

Fig. 1.20 Tectonic sketch map of SE Asia showing major faults/shears (Adapted from Wang et al. (2006)). *ASRRSS* the Ailaoshan–Red River shear system, *GLSS* the Gaoligong shear system, *CSSS* the Chongshan shear system, *WCSZ*: the Wang Chao shear zone, *TPSZ* the Three Pagodas shear zone

is bounded by the ASRRSS to the northeast and the Sagaing fault (the present-day western boundary of the Indochina Block) to the southwest, and has a wedge-shaped geometry with its narrower northwestern end adjoining the Himalaya syntaxis and its broader side facing the southeast. In this region, from west to east, three tectonic units (the Tengchong block, Baoshan block and Lanping–Simao fold belt) are separated by the GLSS and CSSS, respectively (Wang et al. 2006).

The complex geological history has fostered high biodiversity through complex patterns of exchange, isolation, adaptation, extinctions, and speciation. Of particular importance have been the mountain ranges and the powerful rivers that have led to great taxonomic diversity among plants.

1.1.4 Soils

Soils influence vegetation type and distribution. Parent rocks (both in physical properties such as texture, color and weatherability, and in chemical content), climate, and vegetation, have profoundly influenced the development of soils. SW China harbors a variety of soils with pH that varies from acidic types to limestone-derived alkali types. On the Yunnan-Guizhou Plateau surface, red, yellow, yellowish brown, and brown soils predominate in mountain areas. Sichuan has rich soil types. Yellow, yellowish brown and purple soils prevail in the Sichuan Basin. The soils of altitudinal zones, from lower to upper elevations, are: mountain yellow and red earth, mountain yellow-brown earth and mountain gray-brown, subalpine/alpine meadow soils. Figure 1.21 shows the altitudinal soil zones in the subtropical areas, as represented by the Gaoligong Mountains in western Yunnan, Mt. Gongga in northwestern Sichuan, Mt. Emei on the western rim of the Sichuan Basin, Mt. Xilong in southeastern Yunnan, and Mt. Fanjing in eastern Guizhou.

1.2 Vegetation Changes over Geological Time

It is widely accepted that according to fossil records, plant life did not appear on land until about 475 Ma. Figure 1.22 shows the origin and development of plant life and major crust movements during different geological periods. Mosses and ferns evolved from species of algae that lived in the sea and in fresh water. About 420 Ma, during the Silurian Period, ancient plants began to occupy the land. Over the millions of years that followed, these land colonizers developed and adapted to their new habitat. The first forests were dominated by giant horsetails, club mosses, and ferns that stood up to about 15 m tall. The first seed plants or gymnosperms evolved in the Late Paleozoic. They became the dominant vegetation of the Mesozoic. By the end of the Mesozoic, they were largely displaced by the newly evolved angiosperms, the flowering plants. The first flowering plants date from 160 Ma. They diversified enormously during the Lower Cretaceous and became

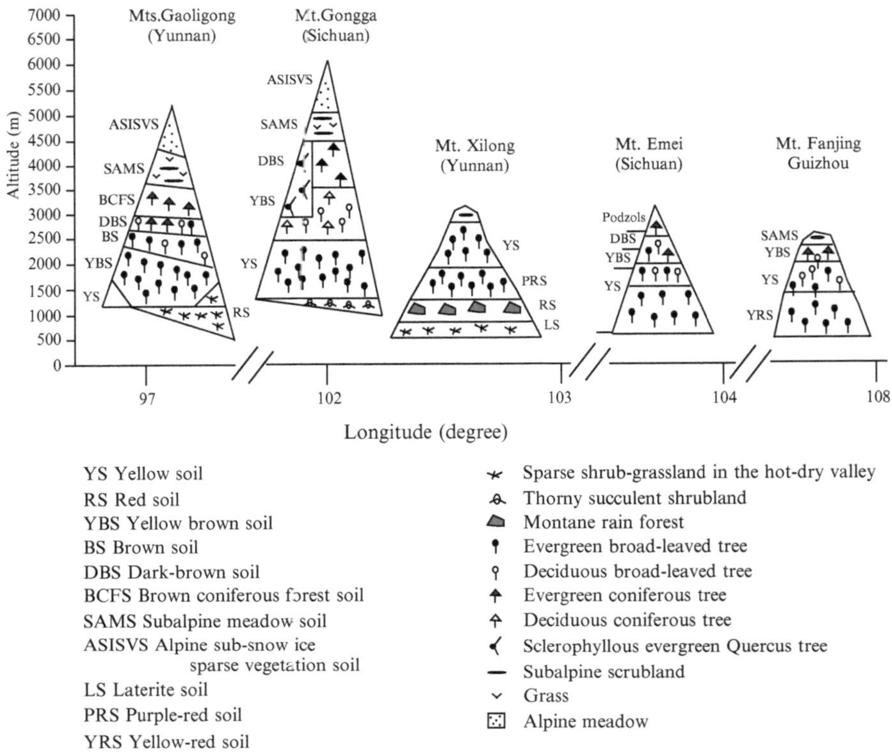

Fig. 1.21 Altitudinal soil zones (Data sources: Zhou (1992), Tang and Ohsawa (1997), Tang (2006) and YFB (1998))

widespread around 120 Ma. Just as gymnosperms forced non-seed plants into the ecological background, the further evolution of angiosperms, around 60–100 Ma, forced the more primitive gymnosperms into restricted habitats. Presently, gymnosperms generally dominate cold and dry ecosystems such as boreal forests and the alpine forests at higher mountain elevations, but in all other habitats, flowering plants rapidly became the dominant plant life.

Changes in land masses over geological time have provided the impetus for the evolution of a great variety of taxonomic groups. As the original continent Pangaea split into Laurasia and Gondwana and as these land masses developed, organisms became separated from each other, and thereafter followed different evolutionary paths. It is generally thought that one of the major mechanisms of the evolution of new species was the stronger and divaricating effect of natural selection following geographic isolation of populations belonging to the same species, which might eventually diverge sufficiently to be considered separate species.

Recent geological, palynological, and archeological studies in China have provided much evidence for changes in environment and vegetation that took place over geological time. Vegetation changes since the Early Tertiary are summarized below.

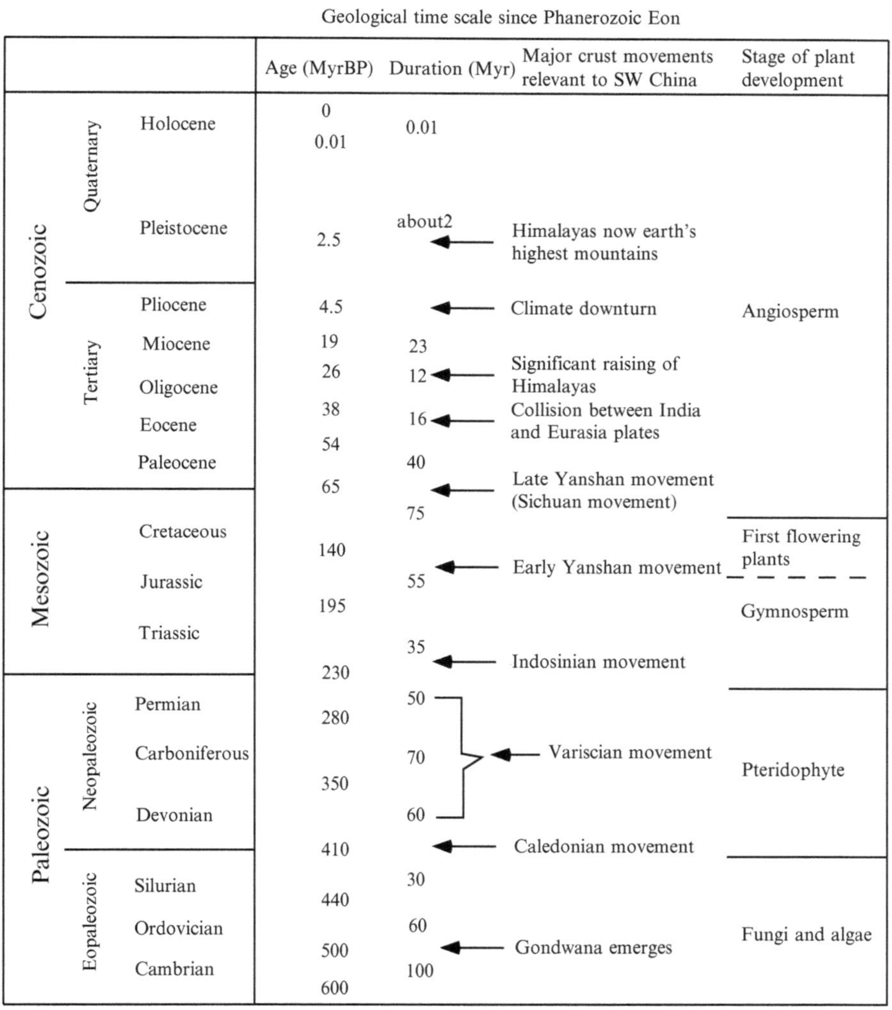

Fig. 1.22 Plant life development over geological time scale (Data sources: SBG (1982), GBG (1984), YBG (1990), ECFC (1997), Chapman and Wang (2010), Wikipedia (2012) and Gondwana (2014))

1.2.1 Vegetation in the Early Tertiary

By the Early Tertiary period of the Cenozoic Era, angiosperms were the dominant plants. The distribution of vegetation types was influenced by a shift from dry to wet climates. The vegetation of China in the Early Tertiary is shown in Fig. 1.23. In eastern Yunnan, Sichuan, Chongqing and Guizhou there were sparse subtropical dry and semi-dry woodlands. In Ganzi, western Sichuan, fossil species of *Eucalyptus,*

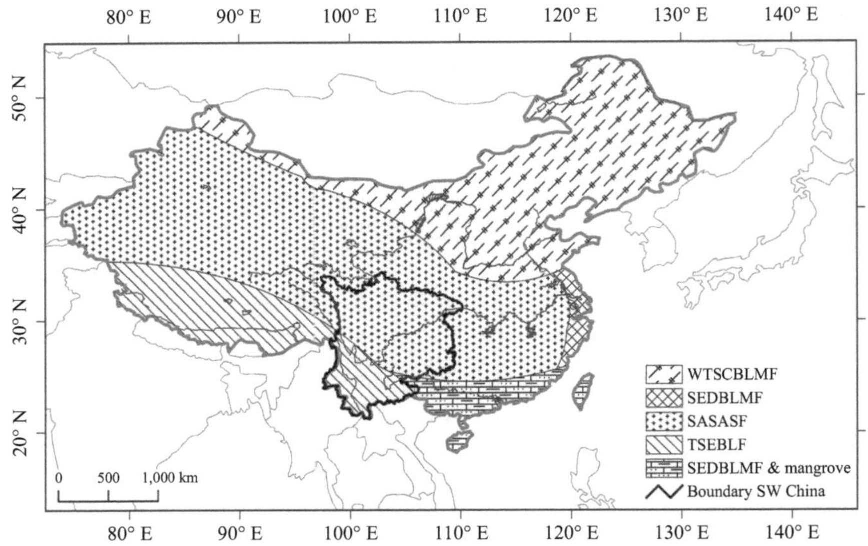

Fig. 1.23 Vegetation of China in the Early Tertiary (ECFC 1997). *WTSCBLMF* warm-temperate and subtropical coniferous and broad-leaved mixed forest, *SEDBLMF* subtropical evergreen and deciduous broad-leaved mixed forest, *SASASF* subtropical arid steppe and arid sparse forest, *TSEBLF* tropical and subtropical evergreen broad-leaved forest, *SEDBLMF* subtropical evergreen and deciduous broad-leaved mixed forest

Lonicera, Pistacia, Albizia, Viburnum, etc. have been found (Chen et al. 1983). The Panxian River of Guizhou has yielded fossil species of *Cinnamomum, Phoebe, Actinodaphne, Sassafras, Lindera, Litsea, Quercus, Castanea, Fagus, Ulmus, Cercidiphyllum, Zelkova, Celtis, Juglans, Tilia, Taxodium, Podocarpus, Cryptomeria, Tsuga, Pinus* and *Cedrus*. Serrated leaves in the smaller sizes of broad-leaved plants reflect a warm and semi-dry climate (Mao 1992).

Most parts of Yunnan and Tibet had warm-rainy winters and hot-dry summers, supporting subtropical evergreen broad-leaved forests. According to fossil pollen analysis for the Lunpola Basin of central Tibet, species of Myrtaceae, Lauraceae, Araliaceae, *Myrica, Engelhardia, Ilex, Camellia, Magnolia, Platanus, Cycas, Ginkgo, Podocarpus, Cyathea, Pteris*, etc. have been identified (Wang et al. 1975).

By the end of the Early Tertiary, in southern Yunnan the climate became warm and dry in winter, but hot and moist in summer, wetter than at the present. Evergreen broad-leaved species occupy 2/3 of the fossil plant assemblage of the Jinggu area. They include species of *Lithocarpus, Cinnamomum, Phoebe, Machilus, Nothaphoebe, Quercus, Rehderodendron, Celtis, Juglans, Rosa, Sorbus, Rhus, Zelkova, Erythrophleum, Celocedrus, Cephalotaxus, Annona*, and *Terminalia*, among others (Xu 1978; Xu and Li 1980).

1.2.2 *Vegetation in the Late Tertiary*

In the Late Tertiary, the vegetation of China became more complex. The deciduous broad-leaved forest thrived, and the coniferous forest spread northward. The grassland spread from the northwest to the east of China. The vegetation of SW China, except for southern Yunnan, became a subtropical evergreen and deciduous broad-leaved mixed forest (Fig. 1.24). Late Tertiary pollen data from the Binghe Coal Basin in Guizhou indicate that the forest was dominated by deciduous *Liquidambar, Alnus, Quercus, Trochodendron*, and some evergreen broad-leaved trees, along with coniferous *Cryptomeria*. On the mountain slopes, *Abies* and *Picea* have also been found (Mao 1992).

On Mt. Emei, on the western fringe of the Sichuan Basin, fossil pollen records of *Ginkgo, Salix, Myrica, Juglans, Alnus, Betula*, and species of the Fagaceae and Magnoliaceae, *Pinus, Picea, Cryptomeria,* Cupressaceae, as well as warm- and wet-adapted fern species of *Osmunda, Alsophila, Hicriopteris* show a resemblance to the flora living at present time (SRSEG 1978; Guan 1990).

In Jianchuan, northwestern Yunnan, component species were of Fagaceae, Lauraceae, *Zelkova, Acer, Pistacia, Quercus, Pinus*, etc. (Xu 1978). Southern

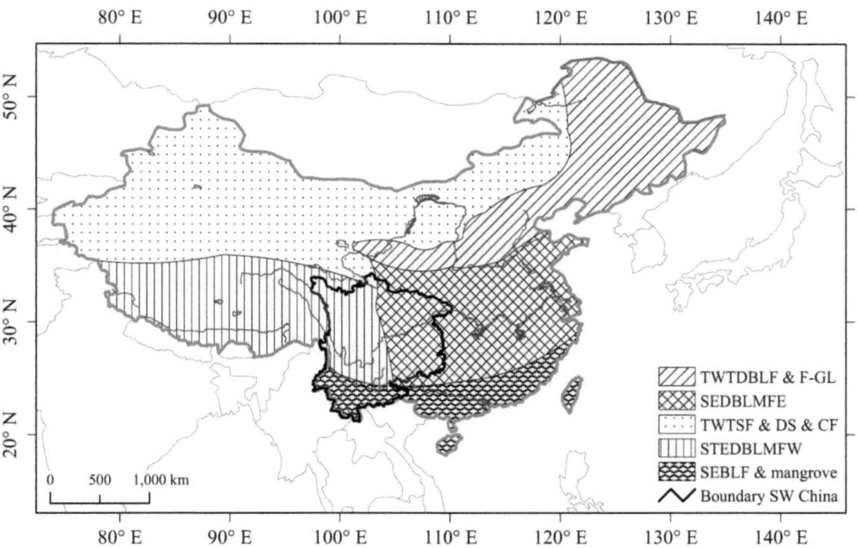

Fig. 1.24 Vegetation of China in the Late Tertiary (ECFC 1997). *TWTDBLF & F-GL* temperate and warm-temperate deciduous broad-leaved forest and forest-grassland, *SEDBLMFE* subtropical evergreen and deciduous broad-leaved mixed forest of eastern China, *TWTSF & DS & CF* temperate and warm-temperate sparse forest and desert steppe and coniferous forest, *STEDBLMFW* subtropical evergreen and deciduous broad-leaved mixed forest of western China, *SEBLF* subtropical evergreen broad-leaved forest

Yunnan was dominated by the subtropical evergreen broad-leaved forest. In Tengchong, abundant fossil plants have been found, e.g. *Cinnamomum glandulifera, Lindera prelatilia, Litsea populifolia, Lithocarpus dealbata, Castanopsis cerata-cantha, Fagus chinensis, Sophora miojaponica, Carpinus subcordata, Meliosma buchanannifolia, Pinus, Keteleeria* (Tao and Du 1982).

Recently, Zhang et al. (2012) have provided a study on vegetation and climatic changes in Yunnan resulting from the uplift of the Tibetan Plateau in the Middle Miocene. Based on fossil pollen records from four localities at different altitudes along the Ailao Mountains, they concluded that (1) the vegetation in those mountains in the Middle Miocene was composed of mixed evergreen and deciduous broad-leaved forests with some conifers, (2) vegetation in the Ailao Mountains suggests a subtropical climate, though that climate was cooler than it is today.

1.2.3 Vegetation in the Quaternary

Fossil pollen records from Lake Er Hai in Yunnan indicate that after 11,750 cal. years BP a warming climate coupled with enhanced summer monsoon precipitation resulted in the expansion of *Tsuga* and evergreen broad-leaved trees (*Cyclobalanopsis, Lithocarpus* and *Castanopsis*). The climatic optimum occurred at c. 8,400 years BP and lasted until at least 6,370 years BP. The early- to mid-Holocene warm and wet climate regime represented a regional climate across the Yunnan Plateau (Shen et al. 2006).

During the 1980–1990s, researches (e.g. Liu et al. 1986; Lin et al. 1986; Sun et al. 1986; Walker 1986) included paleoclimatic and paleovegetational studies on SW China (primarily Yunnan). The results of this work have been summarized by Winkler and Wang (1993), as follows.

The fossil pollen records in the Menghai lakes (1,280 m asl) (Liu et al. 1986) of southern Yunnan, represent high-rainfall vegetation that since 18,000 years BP has included the broad-leaved Fagaceae *Lithocarpus*, *Cyclobalanopsis*, *Castanopsis* and the coniferous Podocarpaceae *Dacrydium* and *Dacrycarpus*.

At Dian Chi (1,886 m asl) near Kunming, central Yunnan, several taxa reached their present northern and altitudinal limits (Sun et al. 1986; Walker 1986). High-altitude forest present before about 9,500 BP included *Tsuga dumosa*, which now grows only above 2,300 m in northwestern and northeastern Yunnan. Precipitation before about 10,000 years BP may have been higher than at present, but the present monsoonal precipitation pattern was in place by about 9,500 years BP. Many vegetation changes were occurring at about 10,000 years BP, and the composition of the mixed evergreen and deciduous forest was more complex before 7,000 years BP. Between about 7,000 and 4,000 years BP evergreen broad-leaved forests expanded in the region around Dian Chi, and the climate was warmer and wetter than it is today.

At Xi Hu (1,980 m asl) near Er Yuan, in the foothills of the mountains of NW Yunnan, the climate was perhaps drier than at present, and colder, until about 10,000 years BP (Lin et al. 1986; Walker 1986). Alpine forests of *Picea*, *Abies* and

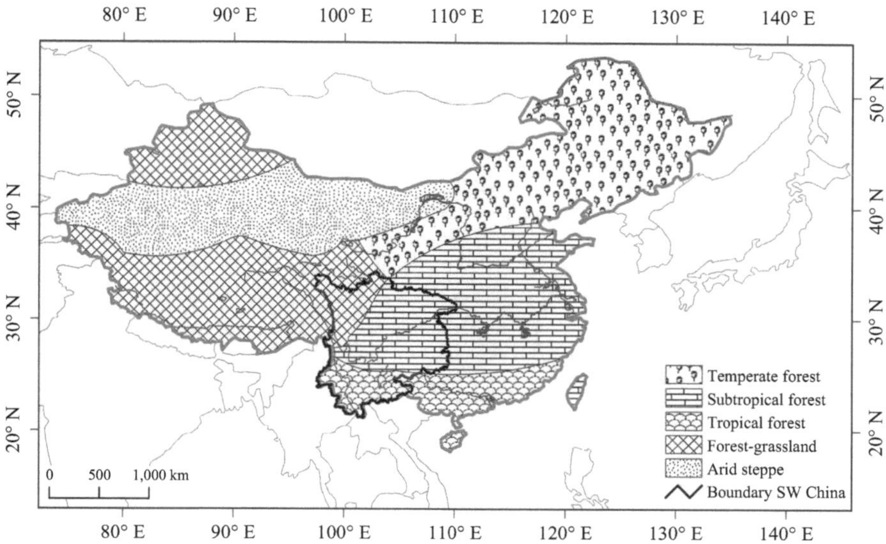

Fig. 1.25 Vegetation of China 6,000 years ago (Data sources: The Normal College of Northwest China (1984), Ren et al. (1985), Winkler and Wang (1993))

Tsuga occupied the slopes at altitudes at least 500 m lower than at present from 17,000 years BP, and an altitudinal limit of 3,000 m for these forests was gradually attained by 10,500 years BP.

It is noteworthy that SW China has never been covered by extensive unified ice-sheets (Shi et al. 1986; Liu 1988). Although the western mountain glaciers contracted after about 10,000 years BP, they are still features of the Asian landscape. In general, since the Holocene, the vegetation types of SW China have changed little (Winkler and Wang 1993; ECFC 1997). The vegetation of China in 6,000 years BP is shown in Fig. 1.25.

1.3 Floristics and Vegetation

The flora of SW China skirts the boundary between the extensive Holarctic and Paleotropical floristic kingdoms of the east Asiatic region (de Laubenfels 1975; Takhtajan 1986). SW China includes the Sikang-Yunnan and central Chinese floras, and is closely related to the southeastern Chinese, Ryukyus, Japanese-Korean, the eastern Himalayan, northern Burmese and Khasi-Manipur floras (Fig. 1.26). According to the more detailed analysis of Wu et al. (2011), SW China is located at the southern limit of the Sino-Himalayan and Sino-Japanese floristic regions, representing the transition between the flora of West and East Asia. This rich and complex flora results from a combination of elements from the Sino-Japanese,

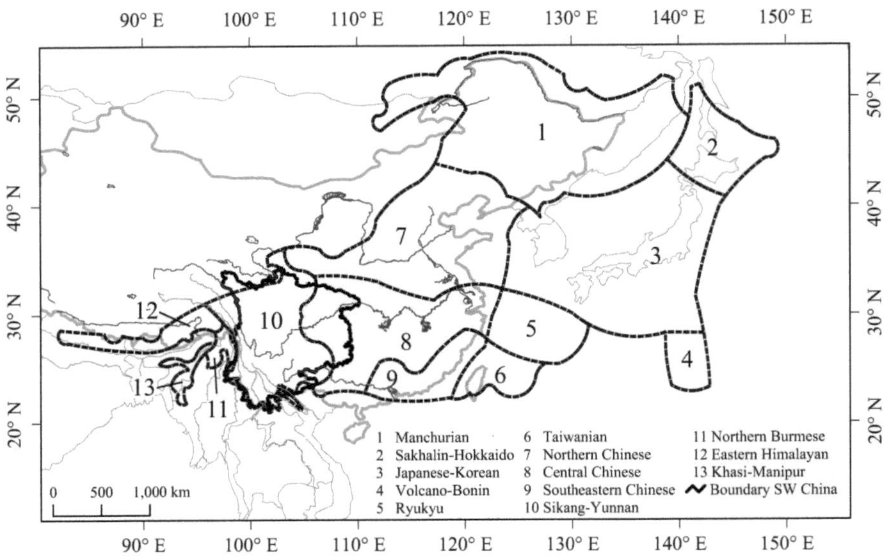

Fig. 1.26 Floristic region of East Asia (Takhtajan 1986)

Sino-Himalayan, African and Mediterranean, and Indo-Malesian floristic regions (Wu et al. 2011). On the map of the world's 25 biodiversity hotspots (Myers et al. 2000), some parts of Yunnan fall into the Indo-Burma biodiversity hotspot, while the Yangtze River valley area on the boundary between Chongqing Municipality, Sichuan, Guizhou, Hubei and Hunan provinces falls into the South-Central China biodiversity hotspot. China is ranked number 3 in the world for plant resources – second only to Malaysia and Brazil. Southwestern China is one of the most botanically diverse terrestrial regions in the nation (Wang 1992). Yunnan is one of the world's most diverse. Yunnan supports a total of 14,822 species of native seed plants (not including subspecific rank) (Wu 1977–2006), amounting to 49 % of the total seed plant species in China (30,270 species) (SEPA 1998; López-Pujol et al. 2006). Sichuan (including Chongqing) has 8,790 species (ECFS 1981–2005; WCPCEC 1991; Li and Zhang 2002) with 29 %, Guizhou has 5,374 (ECFG 1982–2004) with 18 % of the total for China. As a whole, SW China has a disproportionate amount of the nation's overall seed plant species (c. 60 %) with 18,309 native seed plant species, and a high level of endemism (Editorial Committee of Flora Reipublicae Popularis Sinicae 1959–2004; Yang et al. 2013). The Hengduan Mountains (northwestern Yunnan and western Sichuan), southeastern Yunnan and the Yangtze River region have been identified as important centers of endemism in China (Ying et al. 1993; Ying 2001; López-Pujol et al. 2011; Huang et al. 2012). It is geographically interesting that, although most tree species are of subtropical and tropical genera, such as *Castanopsis, Lithocarpus, Cyclobalanopsis, Machilus, Cinnamomum, Lindera, Manglietia, Michelia, Schima,* almost all the existing temperate genera of woody plants, including *Acer*, *Betula*, *Juglans*, *Carpinus*, *Quercus*, *Tsuga*, *Picea* and *Abies*, also grow at various altitudes in this region.

There are an exceptional number of ancient and relict tree species of particular value and interest. Since SW China was not strongly influenced by the last ice sheet, it is an important glacial refuge supporting high diversity, especially as it includes distinctive paleoendemic species, such as *Ginkgo biloba*, originating from the Permian Era of the Paleozoic and dating back some 270 Ma, and *Podocarpus* spp., *Dacrydium pierrei*, *Cephalotaxus* spp., *Taxus wallichiana* var. *chinensis*, *Amentotaxus argoenia* and *Pseudotaxus* spp., that all originate from the Mesozoic to the Early Tertiary. Other Tertiary relict broad-leaved species include deciduous *Liriodendron chinense*, *Tetracentron sinense*, *Cercidiphyllum japonicum*, *Eucommia ulmoides*, *Camptotheca acumimata*, *Bretschneidera sinensis*, *Davidia involucrata*, *Nyssa sinensis*, *Emmenopterys henryi*, *Tapiscia sinensis* and evergreen broad-leaved *Pachylarnax sinica* etc. in the subtropical evergreen broad-leaved forest zone.

According to Hyde et al. (2003), there are six forest regions in China (Fig. 1.27). Tibet, Yunnan, Guizhou, Sichuan and Chongqing together make up the Southwest forest region. However, Tibet is not considered as a part of SW China in this study. The present vegetation of China is shown in Fig. 1.28. Along the altitudinal gradient the biogeographic patterns within SW China are: tropical–subtropical–temperate–subalpine–alpine. It has a wealth of vegetation types (Figs. 1.29, 1.30, 1.31, 1.32, 1.33, 1.34, 1.35, 1.36, and 1.37), surpassing all other regions of China, and includes

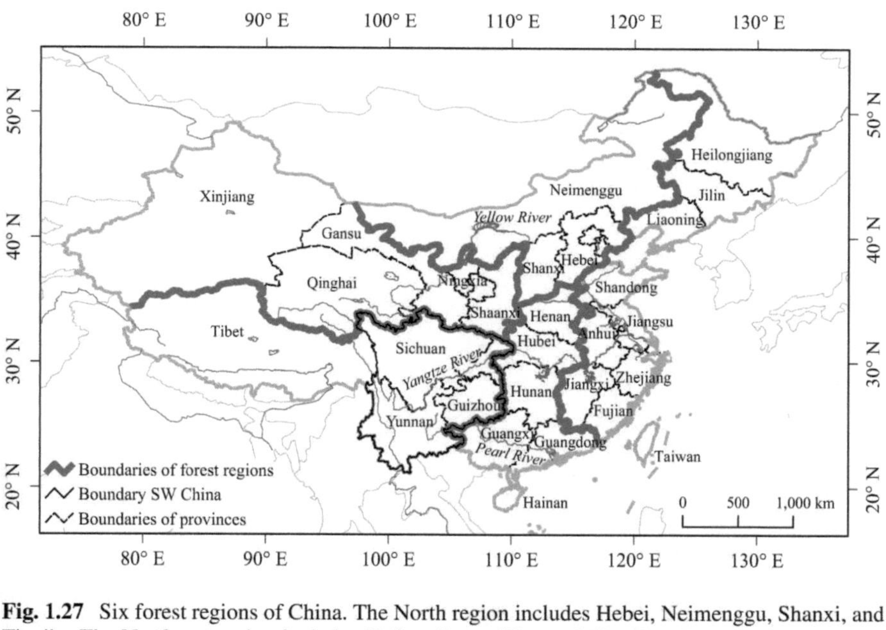

Fig. 1.27 Six forest regions of China. The North region includes Hebei, Neimenggu, Shanxi, and Tianjin. The Northeast region includes Heilongjiang, Jilin, and Liaoning. The Northwest region includes Gansu, Ningxia, Qinghai, Shaanxi, and Xinjiang. The south Central region includes Guangdong, Guangxi, Hainan, Henan, Hubei, and Hunan. The Southeast region includes Anhui, Fujian, Jiangsu, Jiangxi, Shandong, and Zhejiang. The Southwest region includes Guizhou, Sichuan, Yunnan and Xizhang (Data source: Hyde et al. (2003))

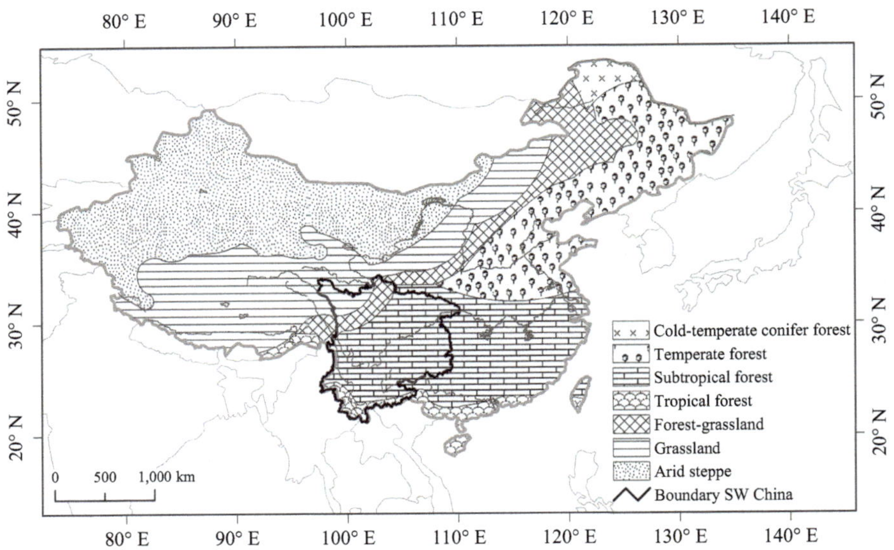

Fig. 1.28 Modern vegetation of China (Data sources: Hou (1979), The Normal College of Northwest China (1984), Ren et al. (1985), Winkler and Wang (1993), ECFC (1997))

Fig. 1.29 Jinsha River First Bend Valley at 1,850 m asl in northwestern Yunnan. A typical hot-dry valley

Fig. 1.30 A temperate-dry valley at 2,800 m asl in Adong, NW Yunnan, with sparse small-leaved scrubs on the mountain slope. The inhabitants of the valley are Tibetans

Fig. 1.31 A tropical seasonal rain forest dominated by *Parashorea chinensis* and *Pometia pinnata* at 450 m asl in the Dawei Mountains in Hekou, southern Yunnan

Fig. 1.32 A mid-montane moist evergreen broad-leaved forest dominated by species of *Lithocarpus* and *Castanopsis* at 2,200–2,800 m asl in Zhelong, eastern Ailao Mountains

Fig. 1.33 A cold-temperate coniferous forest dominated by various species of *Picea* and *Abies* at c. 3,300 m asl in Jiuzhaigou, Sichuan

Fig. 1.34 A *Tsuga dumosa* forest at 2,900 m asl in the Gaoligong Mountains, northwestern Yunnan

Fig. 1.35 An alpine grassland predominated by species of *Roscoea, Juncus, Euphorbia, Potentilla, Carex, Poa, Avena, Swertia* and *Artemisia* at 3,500 m asl in Shiangri-la (Zhongdian), northwestern Yunnan

Fig. 1.36 An alpine scrubland dominated by dwarf *Quercus pannosa* and species of *Rhododendron* at 3,500 m in the Baimaxue Mountains, northwestern Yunnan

Fig. 1.37 The alpine timberline (*Abies georgei*) at c. 4,400 m asl in the Baimaxue Mountains, northwestern Yunnan

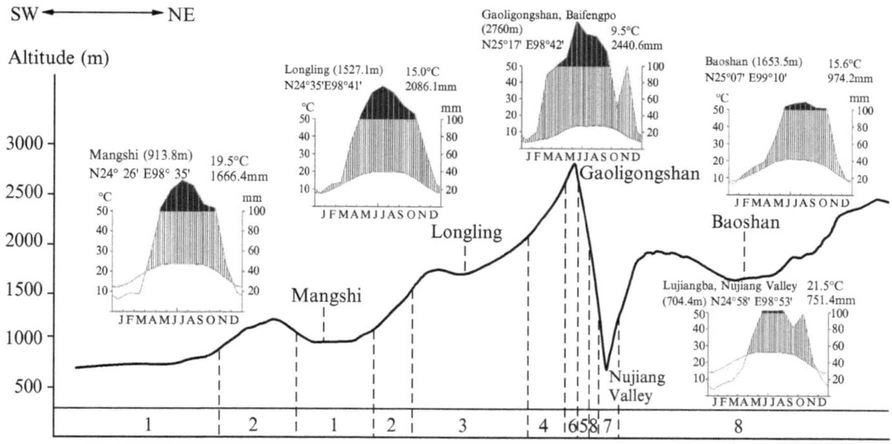

Fig. 1.38 Relationships among vegetation, topography and climate in western Yunnan. *1* Seasonal rain forest, *2* monsoonal evergreen broad-leaved forest, *3* humid evergreen broad-leaved forest, *4* mid-montane moist evergreen broad-leaved forest, *5 Tsuga* forest and coniferous and evergreen broad-leaved mixed forest, 6 mossy dwarf plant community, *7* semi-savanna in the hot-dry valley, *8* semi-humid evergreen broad-leaved forest and *Pinus yunnanensis* forest (Partial data from Wu et al. (1987))

tropical rain forest, subtropical evergreen broad-leaved forest, warm-temperate deciduous broad-leaved forest, temperate coniferous and broad-leaved mixed forest, cold-temperate coniferous forest, alpine scrub, meadow and wetland, as well as small-leaved scrubland, thorny shrubland, sparse shrub-grassland (savanna) in temperate- or hot-dry valleys (Wang 1961; SVEG 1980; Wu et al. 1987; Huang et al. 1988; Jin 1992); however, by far most of the vegetation consists of what is commonly referred to as the subtropical evergreen broad-leaved forests. Northern tropical rain forests exist in the tropical border areas of southeastern, southern and southwestern Yunnan. They have some obvious transitional characteristics from tropical to subtropical (Zhu 1997; Zhu et al. 2006). The various altitudinal vegetation zones can be seen in Figs. 1.21 and 1.38.

In this geographic zone, the subtropical vegetation, including the evergreen broad-leaved forests, several other forest types and minor plant communities, varies with differing environmental conditions and levels of disturbance (natural and human). Subtropical primary evergreen broad-leaved forests dominated by species of *Cyclobalanopsis, Castanopsis, Lithocarpus, Machilus, Phoebe, Cinnamomum, Schima*, etc. are now restricted to less accessible steep slopes, remote mountains, temple grounds and nature reserves. Also within this zone, there are precious Tertiary relict forests in specific habitats such as scree slopes, limestone areas and stream banks. Secondary stands dominated by species such as *Pinus massonia, Pinus yunnanensis, Alnus nepalensis*, *Quercus acutissima* and *Platycarya strobilacea* now prevail over large areas on foothills and lower mountain slopes. In some cases bamboo communities (e.g. *Phyllostachys pubescens*, *Chimonobambusa quadrangularis* and *Pleioblastus amarus*) appear after the original evergreen broad-leaved forests have been destroyed. Dramatic expansions of coniferous species *Pinus, Keteleeria*, and imported broad-leaved *Eucalyptus* occur.

1.4 Human Influences

Human influences over more than 5,000 years have led to large fluctuations in extent and location of forested areas in China. Though it is not possible to document those influences directly from remote ancient times, palynologic data such as pollen records dating from c. 6,370 years BP in Lake Er Hai, northwestern Yunnan, indicate a decrease in the natural altitudinal vegetation gradient occurring in the catchment during the late glacial period. This change is attributed to selective forest clearance as part of an early phase of shifting agriculture (Shen et al. 2006). China has thousands of years of history of crop cultivation (Deng 1981; Lu 1999). The expansion of croplands in the Sichuan Basin from the first century BC, in northwestern Yunnan and western Sichuan from the sixth to the eighth century AD, and in Yunnan-Guizhou Plateau from the fourteenth to the nineteenth century AD, is shown in Fig. 1.39.

Vegetational change since the nineteenth century has been regarded as almost entirely the result of human activity. Population pressure and intensive land use have led to serious land degradation in SW China. Native vegetation was excessively exploited. From 1930 to the early 1980s, the Chinese government officially encouraged extensive removal of primary vegetation; it has left most of the

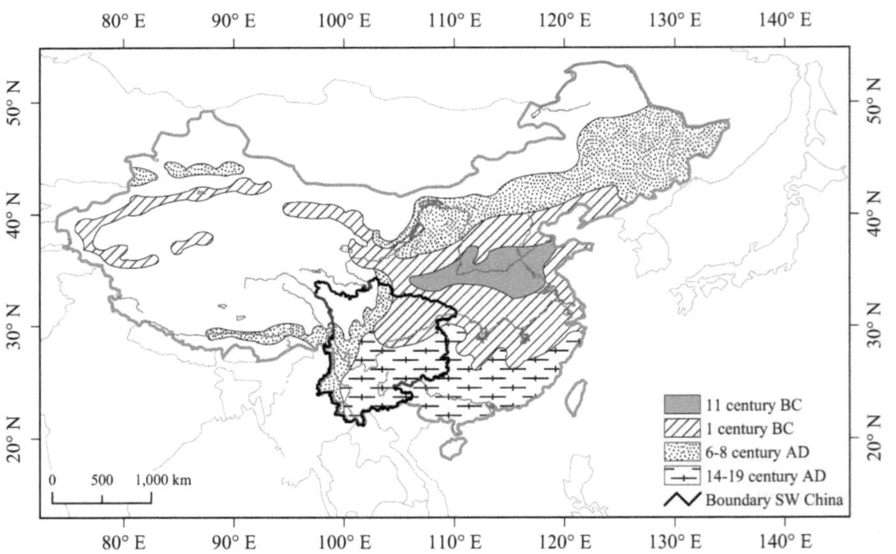

Fig. 1.39 Expansion of cropland in SW China from the eleventh century BC (Deng 1981)

remaining vegetation in patches stretching across the landscape of the region. Especially during the Great Leap Forward (1958–1960), intensive steel-making and tree-cutting on a national scale reduced the forest cover and impoverished the native plant diversity, resulting in great changes to the subtropical ecosystem. Agricultural development in the 1930–1980s in SW China was overwhelmingly the most important such factor. Today, the original evergreen broad-leaved forest ecosystems on hills and low mountains in subtropical/tropical areas have almost disappeared despite the national logging ban. Even in some areas of remote montane subtropical evergreen broad-leaved forests, shifting agricultural techniques have caused the extinction of the natural ecosystem. The last 85 years have seen a continuing development of farming and other land uses, leaving only few areas, difficult of access, where natural vegetation remains. The majority of semi-natural types are located in the lowlands. People continue to cut down the trees for firewood and house construction. Primeval forests have been converted to agricultural fields and to tree plantations for economic gain, generally with *Pinus yunnanensis, Pinus massonia, Cunninghamia lanceolata, Cryptomeria japonica* var. *sinensis* (formerly *Cryptomeria fortunei*) and *Eucalyptus* species. Urban–industrial growth, including urbanization (the fully developed area of a city and its suburbs that includes no rural land), roads, railways, mining and dam building for energy generation on all the main rivers, has directly wiped out large areas of vegetation, leaving degraded habitats. Excess herb medicine collection and recreational pressures (e.g. tourism) in natural forest areas have brought increasing disturbance. Complex mosaics of stands of varying age, composition, and structure intermingle with farmlands or mono-species plantations (Figs. 1.40, 1.41, 1.42, 1.43, 1.44, 1.45, 1.46, and 1.47). Some species have been almost eliminated. These destructive effects continue, with increasing habitat degradation, deforestation, desertification, farmland reduction, soil erosion and other serious consequences.

In recent years, recognizing the serious environmental and ecological issues arising from economic booms, the Chinese government has implemented a series of policies to promote ecological restoration. For example, the Grain to Green Program (GTGP, also known as the Sloping Land Conversion Program and the Farm to Forest Program), launched in 1999, is the largest land retirement program in the developing world, and uses a public payment scheme that directly engages millions of rural households as core agents of project implementation. This is distinct from China's other soil and water conservation and forestry programs, being one of the first, "payment for ecosystem services" programs in China, and is certainly the most ambitious (Bennett 2008; Li 2009). Since the 1990s, conservation measures have included setting up many nature reserves to boost populations of native plant species. In addition, certain traditional forest management policies have limited the disturbing impact on plant diversity. Cultural practices and traditional sustainable forest resource uses (e.g. in village fengshui forests) support some primary forests' regeneration niches and diversity. Holy mountains (some called Dragon mountains by local people) involving popular religions, also play a role in natural forest conservation (Figs. 1.48 and 1.49).

Fig. 1.40 Forestland turned into pasture at 2,550 m asl in Luquan, central Yunnan

Fig. 1.41 The Yuanyang terraces in the southern Ailao Mountains, Yunnan, built by the Hani people. Terrace width is adapted precisely to variations in the natural landscape

Fig. 1.42 Patchy agricultural fields in Nujiang Gorge, western Yunnan

Fig. 1.43 Diverse field types of paddy land-use, including rice paddies, coffee, sugarcane and herb medicine fields, in Baihualing of Baoshan in the southern Gaoligong Mountains, western Yunnan

Fig. 1.44 A tea garden in the Dalou Mountains, Guizhou

Fig. 1.45 Cultivated tobacco and *Cryptomeria japonica* var. *sinensis* are economic crops at c. 1,400 m asl in the Dalou Mountains, Guizhou

Fig. 1.46 Developed suburban karst landscape of Guiyang, Guizhou

Fig. 1.47 Developed Zitong County supports a population of 380,000 in a plain measuring 1,442 km^2 in the Sichuan Basin

Fig. 1.48 An ethnic Dai village in Xishuangbanna, southern Yunnan. In the Dai tradition, humans are a part of nature

Fig. 1.49 The Dai people's Holy Longshan Forest (the dragon forest) where many indigenous plants are found

References

Bennett MT (2008) China's sloping land conversion program: institutional innovation or business as usual? Ecol Econ 65:699–711
Chang DHS (1983) The Tibetan Plateau in relation to the vegetation of China. Ann Mo Bot Gard 70(3):564–570
Chapman GP, Wang Y-Z (2010) The plant life of China: diversity and distribution. Springer, Berlin
Chen MH, Kong SK, Chen H (1983) On the discovery of old Tertiary flora in western plateau Sichuan and its significance. Acta Bot Sin 25(2):195–196
Cottle R (2004) Linking geology and biodiversity. English Nature Research Reports. Number 562. English Nature, Northminster House, Peterborough PE1 1UA
de Laubenfels J (1975) Mapping the world's vegetation: regionalization of formation and flora. Syracuse University Press, Syracuse
Deng JZ (1981) Agrogeography of China. Science Press, Beijing
ECFC (Editorial Committee of the Forest of China) (1997) The forest of China. Chinese Forestry Press, Beijing
ECFG (Editorial Committee of Flora Guizhouensis) (1982–2004) Flora Guizhouensis, vols 1–10. Guizhou People's Press, Guiyang
ECFS (Editorial Committee of Flora Sichuannica) (1981–2005) Flora Sichuanica, vols 1–17. Sichuan Science and Technology Press, Chengdu
Editorial Committee of Flora Reipublicae Popularis Sinicae (1959–2004) Flora Reipublicae Popularis Sinicae. Science Press, Beijing
GBG (Guizhou Bureau of Geology) (1984) Regional geology of Guizhou province. Geological Publishing House, Beijing
Gondwana (2014) http://en.wikipedia.org/wiki/Gondwana#cite_note-SCAssmbly-3. Retrieved 8 Oct 2014
Guan ZT (1990) Forest changes during geological periods in Southwest China. In: Li CB (ed) Ecological studies on forests of Sichuan. Sichuan Science and Technology Press, Chengdu, pp 187–208
Guo K, Werger MJA (2010) Effect of prevailing monsoons on the distribution of beeches in continental East Asia. For Ecol Manage 259:2197–2203
Hou HY (ed) (1979) Vegetation map of China. Institute of Botany, Academia Sinica. Map Publisher of P.R.C., Beijing
Hou XY (1983) Vegetation of China with reference to its geographical distribution. Ann Mo Bot Gard 70:509–548
Huang Q, Cai Y (2009) Mapping karst rock in Southwest China. Mountain Res Dev 29(1):14–20
Huang W-L, Tu Y-L, Yang L (eds) (1988) The vegetation of Guizhou. Guizhong People Press, Guiyang
Huang JH, Chen B, Liu CR, Lai JS, Zhang JL, Ma KP (2012) Identifying hotspots of endemic woody seed plant diversity in China. Divers Distrib 18:673–688
Hyde WF, Belcher B, Xu J (eds) (2003) China's forests. Global lessons from Market reforms. RFF Press, Washington, DC
Jin ZZ (1992) The natural vegetation types of Yunnan, China. A series of geobotanical monographs. Braun-Blanquetia 8:65–76
Lee HY, Chung SL, Wang JR, Wen DJ, Lo CH, Yang TY, Xie YW, Lee TY, Wu GY, Ji JQ (2003) Miocene Jiali faulting and its implications for Tibetan tectonic evolution. Earth Planetary Sci Lett 205:185–194
Leloup PH, Arnaud N, Lacassin R, Kienast JR, Harrison TM, Phan Trong TT, Replumaz A, Tapponnier T (2001) New constraints on the structure, thermochronology, and timing of the Ailao Shan–Red River shear zone, SE Asia. J Geophys Res 106:6683–6732
Li Z (1986) A preliminary study on the formation of Hengduan Mountain. Exploration Nat 5(1):144–146
Li YC (2009) Grain for Green program is a great concrete action towards the ecological civilization in China: summary of a decade program implementation. Forestry Constr 27(5):3–13

Li R-W, Zhang H-T (2002) Analysis is on the components of seed plant flora in Sichuan region. J Wuhan Bot Res 20(5):381–386
Lin S, Qiao Y, Walker D (1986) Late Pleistocene and Holocene vegetation history at Xi Hu, Er Yuan, Yunnan Province, southwest China. J Biogeogr 13:419–440
Liu KB (1988) Quaternary history of the temperate forests of China. Quaternary Sci Rev 7:1–20
Liu M-G (ed) (1998) Atlas of physical geography of China. Atlas Press of China, Beijing
Liu D, Ding M (1984) The characteristics and evolution of the palaeoenvironment of China since the late Tertiary. In: Whyte RO (ed) The evolution of East Asian Environment. Centre of Asian Studies, University of Hong Kong, Hong Kong, pp 11–40
Liu J, Tang L, Qiao Y, Head MJ, Walker D (1986) Late Quaternary vegetation history at Menghai, Yunnan Province, Southwestern China. J Biogeogr 13:399–418
López-Pujol J, Zhang FM, Ge S (2006) Plant biodiversity in China: richly varied, endangered, and in need of conservation. Biodivers Conserv 15:3983–4026
López-Pujol J, Zhang FM, Sun H-Q, Ying T-S, Ge S (2011) Centres of plant endemism in China: places for survival or for speciation? J Biogeogr 38:1267–1280
Lu TLD (1999) The transition from foraging to farming and the origin of agriculture in China. BAR International Series 774. Museum of Yunnan Province, The 1977: Cultural relic in Yunnan, Yunnan Culture Press, Kunming
Mao ZZ (1992) History of vegetation changes in geological periods in Guizhou. In: Zhou ZX (ed) The forest of Guizhou. Guizhou Science and Technology Press, Guiyang, pp 73–77
Myers N, Mittermeier RA, Mittermeier CG, da Fonseca GAB, Kent J (2000) Biodiversity hotspots for conservation priorities. Nature 403:853–858
Ren M, Yang RZ, Bao H-S (1985) An outline of China's physical geography, China knowledge series. Foreign Languages Press, Beijing
SBG (Sichuan Bureau of Geology) (1982) Regional geology of Sichuan province. Geological Publishing House, Beijing
SEPA (State Environmental Protection Administration) (1998) China's biodiversity: a country study. China Environmental Science Press, Beijing
Shen J, Jones RT, Yang X, Dearing JA, Wang S (2006) The Holocene vegetation history of Lake Erhai, Yunnan province southwestern China: the role of climate and human forcings. The Holocene 16(2):265–276
Shi Y, Ren B, Wang J, Derbyshire E (1986) Quaternary glaciation in China. Quaternary Sci Rev 5:503–507
SRSEG (Sichuan Regional Geostratigraphy Editorial Group) (1978) Southwest China regional geostratigraphy (Sichuan volume). Geological Publishing House, Beijing
Sun X, Wu Y, Qiao Y, Walker D (1986) Late Pleistocene and Holocene vegetation history at Kunming, Yunnan Province, southwest China. J Biogeogr 13:441–476
SVEG (Sichuan Vegetation Editorial Group) (1980) The vegetation of Sichuan. Sichuan People's Press, Chengdu
Takhtajan A (1986) Floristic regions of the world. University of California Press, Berkeley
Tang CQ (2006) Forest vegetation as related to climate and soils at varied altitudes on a subtropical mountain, Mt. Emei, Sichuan, China. Ecol Res 21(2):174–180
Tang CQ, Ohsawa M (1997) Zonal transition of evergreen, deciduous and coniferous forests along the altitudinal gradient on a humid subtropical mountain, Mt. Emei, Sichuan, China. Plant Ecol 133:63–78
Tao JR, Du NQ (1982) Neogene flora of Tengchong Basin in Western Yunnan, China. Acta Bot Sin 24(3):273–281
The Normal College of Northwest China (1984) Maps of the natural geography of China. Department of Geography, Normal College of Northwest China. Special Publications Map Office, Beijing
Walker D (1986) Late Pleistocene-early Holocene vegetational and climatic changes in Yunnan Province, southwest China. J Biogeogr 13:477–486
Wang CW (1961) The Forests of China with a survey of grassland and desert vegetation. Maria Moors Cabot Foundation Publication No. 5, Harvard University, Cambridge

Wang H (1992) Floristic geography. Science Press, Beijing
Wang Y (2005) Yunnan mountain climate. Yunnan Science and Technology Press, Kunming
Wang ZC, Wang HY (2002) Introduction of climatic characteristics and vegetation distribution in Guizhou Province. Guizhou Forestry Sci Technol 30(4):46–50
Wang KF, Yang JW, Li Z, Li ZS (1975) The inference on the Tertiary period and palaeogeography of the Lunpola Basin based on pollen records. Geol Sci 4:366–374
Wang Y, Fan W, Zhang Y, Peng T, Chen X, Xu Y (2006) Kinematics and 40Ar/39Ar geochronology of the Gaoligong and Chongshan shear systems, western Yunnan, China: implications for early Oligocene tectonic extrusion of SE Asia. Tectonophysics 418:235–254
WCPCEC (West China Plant Conservation Experimental Center) (1991) An Enumeration of Plants of Dujiangyan (Guanxian) Sichuan China. Forestry Bureau of Dujiangyan City. Press, Dujiangyan
Wikipedia (2012) http://en.wikipedia.org/wiki/Flowering_plant. Retrieved 13 June 2012
Winkler MG, Wang PK (1993) The late-quaternary vegetation and climate of China. In: Wright HE Jr, Kutzbach JE, Webb T III, Ruddiman WF, Street-Perrott FA, Bartlein PJ (eds) Global climates since the last glacial maximum. University of Minnesota Press, Minneapolis, pp 221–261
Wu ZY (ed) (1977–2006) Flora Yunnanica, vols 1–16. Science Press, Beijing
Wu ZY (ed) (1980) Vegetation of China. Science Press, Beijing
Wu ZY, Zhu YC, Jiang HQ (eds) (1987) The vegetation of Yunnan. Science Press, Beijing
Wu ZY, Sun H, Zhou ZK, Li DZ, Peng H (2011) Floristics of seed plants from China. Science Press, Beijing
Xu R (1978) Plants in the Cenozoic Era in China. Science Press, Beijing
Xu R, Li SY (1980) The development and changes of plants in the Cenozoic Era in China. In: Wu ZY (ed) Vegetation of China. Science Press, Beijing
Yang L (1983) The structure and dynamics of *Cyclobalanopsis stewardiana* forest of the Fanjingshan Mountain. Acta Phytoecologica Sin 7(3):204–214
Yang W, Huang J, Ma K (2013) Taxonomic composition and spatial pattern of endemic seed plant species in Southwest China. In: Hobohm C (ed) Endemism in vascular plants, vol 9, Plants and vegetation. Springer, Dordrecht, pp 281–288
YBG (Yunnan Bureau of Geology) (1990) Regional geology of Yunnan Province. Geological Publishing House, Beijing
YFB (Yunnan Forestry Bureau) (1998) Nujiang nature reserve. Yunnan Art Press, Kunming
Ying T-S (2001) Species diversity and distribution pattern of seed plants in China. Biodiver Sci 9(4):393–398
Ying T-S, Zhang Y-L, Boufford DE (1993) The endemic genera of seed plants of China. Science Press, Beijing
Zhan WW, Wang CY, Mooney WD (2001) 3-D crustal structure in southwestern China. 23rd Seismic research review: worldwide monitoring of nuclear explosions–October 2–5. www.geologie.end.fr. Retrieved 28 Apr 2014
Zhang Q-Q, Ferguson DK, Mosbrugger V, Wang Y-F, Li C-S (2012) Vegetation and climatic changes of SW China in response to the uplift of Tibetan Plateau. Palaeogeogr Palaeoclimatol Palaeoecol 363–364:23–36
Zhou ZX (ed) (1992) The forest of Guizhou. Guizhou Science and Technology Press, Guiyang, pp 73–77
Zhu H (1997) Ecological and biogeographical studies on the tropical rain forest of south Yunnan, SW China with a special reference to its relation with rain forests of tropical Asia. J Biogeogr 24:647–662
Zhu H, Cao M, Hu H (2006) Geological history, flora, and vegetation of Xishuangbanna, southern Yunnan, China. Biotropica 38(3):310–317
Zhu Y, Li D, Zhou Z, Chen CC, Yu H, Luo F, Cheng L (2014) 1516910 lower silurian shale gas potential in Chongqing Southwest China: geological settings and characteristics of longmaxi formation. http://www.searchanddiscovery.com/pdfz/abstracts/html/2013/90163ace/abstracts/extended/zhu3.pdf.htm. Retrieved 5 May 2014

Chapter 2
Evergreen Broad-Leaved Forests

Abstract In SW China, the subtropical evergreen broad-leaved forests are represented by four types, (1) the semi-humid forest dominated only by species of *Cyclobalanopsis*, *Castanopsis* and *Lithocarpus* (Fagaceae), (2) the monsoon forest by species of *Castanopsis, Lithocarpus, Cyclobalanopsis* (Fagaceae), *Machilus, Beilschmiedia, Cryptocarya* (Lauraceae), and *Schima* (Theaceae), (3) the mid-montane moist forest by species of *Lithocarpus*, *Castanopsis*, *Cyclobalanopsis* (Fagaceae), *Machilus*, *Cinnamomum* (Lauraceae), *Schima* (Theaceae), *Manglietia, Alcimandra* and *Michelia* (Magnoliaceae) in Yunnan, and (4) the typical humid evergreen broad-leaved forest by *Castanopsis*, *Cyclobalanopsis, Lithocarpus* and species of Lauraceae (e.g. *Machilus*, *Cinnamomum*, *Lindera* and *Phoebe*), Magnoliaceae (*Manglietia, Michelia*), Theaceae (*Schima* and *Gordonia*), and Elaeocarpaceae (*Elaeocarpus*) in the Sichuan Basin and eastern and central Guizhou. The vertical structures of the forests are multilayered (with emergent, canopy, subcanopy, shrub, and understory layers). The canopy height of the primary forest generally ranges from 23 to 34 m, with umbrella-shaped or rounded crowns. Among the forest types, the monsoon evergreen broad-leaved forest, in mesic-humid sites, exhibits the highest species diversity, followed by the mid-montane moist and humid types, then by the semi-humid type.

The leaf-size patterns of evergreen broad-leaved trees along the altitudinal gradient are exemplified on Mt. Emei, Sichuan. The evergreen broad-leaved forest at low altitude is mostly composed of notophyllous evergreen trees of tropical origin. The leaf size decreases toward the upper altitudinal zones. It also varies with forest stratifications. Canopy trees are mostly notophyllous species, while subcanopy and understory trees are mainly microphyllous species.

The regeneration of the primary forests largely depends on the creation of canopy gaps or forest edges. Four types of gap regeneration behaviors in the major evergreen broad-leaved species in Yunnan are generalized as classified by Yamamoto (Botanical Magazine Tokyo 105:29–45, 1992). Type 1: canopy trees regenerate from saplings existing before gap formation (climax species); Type 2: canopy trees of pioneer species regenerate in gaps from saplings recruited after gap formation; Type 3: subcanopy tree and shrub species regenerate in gaps from saplings recruited before gap formation, but rarely reach the canopy layer; Type 4: trees cannot regenerate in gaps formed under the existing gap disturbance regime.

Because of the history of land uses in SW China, the landscapes exhibit dynamic habitats consisting of mosaics of diverse stands at varying successional stages.

C.Q. Tang, *The Subtropical Vegetation of Southwestern China*,
Plant and Vegetation 11, DOI 10.1007/978-94-017-9741-2_2

In general, during the secondary succession, species of coniferous *Pinus, Keteleeria,* and deciduous *Alnus, Populus, Carpinus* and *Platycarya* act as typical light-demanding pioneers. In the early successional plant community, one or several of the pioneer species may dominate. As succession progresses, some intermediate light-demanding shrubs of *Lyonia*, *Rhododendron*, *Myrica* and *Lindera*, and some shade-tolerant species, mainly of Fagaceae, Lauraceae, and Magnoliaceae, mix with the pioneer species. Gradually, as the forest canopy develops more closure, the short-lived light-demanding pioneer species are usually replaced by late successional shade-tolerant evergreen broad-leaved species such as those of *Castanopsis*, *Cyclobalanopsis*, *Lithocarpus, Machilus, Manglietia*, and *Michelia*, although some long-lived pioneer species, such as *Schima argentea*, *Schima wallichii* and *Schima sinensis* are often found in mature forests. The secondary forests at mid-successional stage tend to have the highest diversity; however, differences between the middle and late stages are not significant. As succession proceeds, the species diversity comes to resemble that of the natural mature forests. The secondary forests, as spatially and temporally dynamic patches of vegetation showing different degrees of disturbance along the successional gradient, exemplify the progress of species diversity toward the original natural patterns in the region. In this connection, the role of soil seed banks along the successional gradient, as exemplified in the Ailao Mountains, is also pertinent.

The subtropical areas of SW China are facing serious loss of plant diversity due to habitat fragmentation and degradation. A clear understanding of the ecological features of the EBLFs can lead to more effective conservation and management of these fragile forests and the mountain ecosystems.

2.1 Forest Types

The subtropical vegetation of East Asia is made up predominantly of evergreen broad-leaved forests (lauro-fagaceous evergreen forests), also known as lucidophylls or laurophylls, referring to their often glossy leaves. These forests, dominated by the genera *Castanopsis*, *Lithocarpus*, *Cyclobalanopsis* (Fagaceae), *Machilus* (Lauraceae), *Schima* (Theaceae), *Manglietia* and *Michelia* (Magnoliaceae), and *Distylium* (Hamamelidaceae) are almost exclusively confined to southern, southwestern, and southeastern mainland China, and Taiwan, southwestern Japan, southern Korea, and some mountain regions of Vietnam, Laos, Thailand, Myanmar, India, Bhutan, and Nepal (Kira 1991; Ohsawa 1993a; Tagawa 1995; Song 1995, 2013; Li 1997; Tang and Ohsawa 2009; Tang et al. 2013). Plants from the laurel family (Lauraceae) may or may not be present, depending on the location. Before 1988, the subtropical evergreen broad-leaved forests were not widely recognized outside East Asia as a basic vegetation type, nor had symposia on the forests been held or detailed studies been published (Box et al. 1998). More recently, climatic delimitation of Asian regions has been discussed by investigators (Wolfe 1979; Ohsawa 1990, 1993a; Kira 1991; Box 1995) who have emphasized the seasonal variations in temperature of the monsoon system as the principal determinant of plant distribution patterns. Ohsawa (1995) has provided an explicit template to depict zonal climate-forest patterns along latitudinal and altitudinal gradients in East Asia. As to latitude, SW China falls into the transitional forest zone from tropical mountain to temperate mountain zonation (Fig. 2.1). Song (2013) asserted that the subtropical evergreen broad-leaved forest of China corresponds to the region from 23° to 33° N and from 98° to 123° E. In fact, within Yunnan the forest extends west to c. 97°32′ E (e.g. Dulongjiang), and within Tibet it extends to 95° E (e.g. Chayu and Motuo in Yaluzangbujiang, though in deep gorges at 100–600 m asl, very limited patches of the tropical rain forest dominated by *Dipterocarpus* are found). The most eastern longitude is 122° E in Taiwan. In sum, 95–122° E is the longitudinal

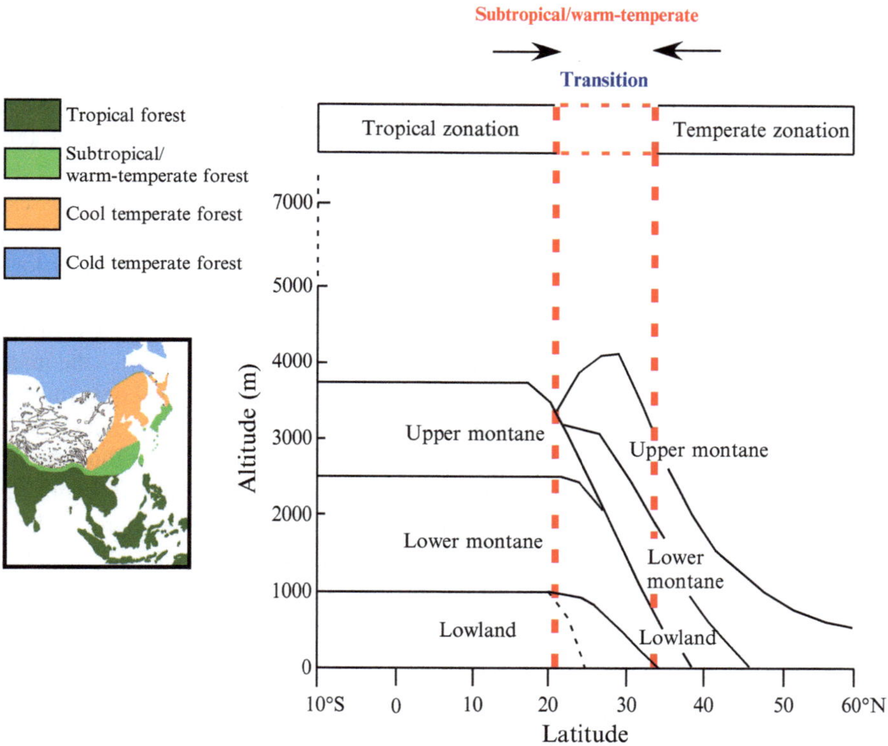

Fig. 2.1 Altitudinal and latitudinal vegetation zonation scheme for East Asia (Ohsawa 1995). The latitudinal patterns of mountain vegetation zonation show a clear boundary ranging from 20° to 30° (33°) N. For the tropical mountains south of 20° N, the altitudinal series includes tropical lowland, tropical lower montane, and tropical upper montane zones. For the temperate mountains north of 30° (33°) N, the series includes temperate lowland, temperate lower montane, and temperate upper montane zones. The mountains located between 20° and 30° (33°) N show a transitional zonation pattern; the lower two zones are comparable to the lower two of the tropical zonation (tropical lowland and tropical lower montane), and the upper two zones are comparable to those of the temperate zonation (temperate lower montane and temperate upper montane). The tropical upper montane zone is not found north of 20°–30° (33°) N, while the tropical lower montane zone reaches down to sea level and so becomes also the temperate lowland zone. Thus the zonation between 20° and 30° (33°) N includes tropical lowland, tropical lower montane/temperate lowland, temperate lower montane, and temperate upper montane zones (Ohsawa 1990, 1995)

range of the subtropical evergreen broad-leaved forest within China. The evergreen broad-leaved forest of southernmost Yunnan (e.g. Xishuangbanna, Xilongshan and Daweishan) is considered to be a tropical montane evergreen broad-leaved forest (Zhu 2007; Zhu and Yan 2009). The subtropical evergreen broad-leaved forests (hereafter EBLFs) are distinguished from the evergreen sclerophyllous forests by the morphology, physiognomy and structure of their leaves (Ovington 1983; Satoo 1983; Hattori 1985; Wu et al. 1987). Primary EBLFs once covered most of the subtropical areas of East Asia, but are now greatly diminished as a result of human activities. A recent study found only 5 % of China's total land area to be covered by natural EBLFs (Chen 1995). In the regions of Yunnan, Sichuan, Guizhou and Chongqing Municipality extensive secondary forests have recently been grown on land used for agriculture or firewood production from the nineteenth to the middle twentieth century, but remnants of the natural EBLFs are still found in remote isolated areas, on steep slopes, in nearly inaccessible or protected areas in nature reserves, and around temples. SW China, especially Yunnan, is home to the most diversified subtropical EBLFs to be found in East Asia (Tang and Ohsawa 2009; Tang 2010; Tang et al. 2013).

Wu et al. (1987) divided the subtropical EBLFs of Yunnan into five subclasses: (1) semi-humid EBLFs, (2) monsoon EBLFs, (3) mid-montane moist EBLFs, (4) montane mossy EBLFs, (5) montane mossy dwarf EBLFs. In view of the physiognomy and ecological phenomena of group 4, including various epiphytes and lianas on tree trunks and branches of the montane mossy EBLFs, are similar to those of the mid-montane moist EBLFs, here I combine the two subclasses into one – the mid-montane moist EBLFs. The montane mossy dwarf EBLFs have formed only in very specific topographic habitats such as near mountain summits. They are not considered as a typical EBLF in this book. In Yunnan, most significantly, the three forest types i.e. semi-humid, monsoon, and mid-montane moist EBLFs, are representative of the major subtropical EBLFs. The subtropical EBLFs exclude sclerophyllous evergreen *Quercus* (subgenus *Quercus*) forests at high altitudes with a dry climate in Yunnan, western Sichuan and western Guizhou, which are similar to the sclerophyllous forests of areas of Mediterranean climate. They also exclude northern tropical rain forests, which are basically tropical in nature but with some obvious transitional characteristics from tropical to subtropical floras, and are found at low altitudes in southernmost Yunnan. However, in this chapter I do include the tropical montane evergreen broad-leaved forest of southern and southeastern Yunnan so as to take into account the altitudinal transition from the tropical rain forest to the subtropical EBLF.

To find the main factors or gradients in large, species-rich, and representative ecological communities, the method of detrended correspondence analysis (DCA) is useful (Hill and Gauch 1980). For SW China, a DCA (Fig. 2.2) for the plots of natural mature EBLFs that have been free of major disturbance during the past 60 years, indicates that along Axis 2 from the low to the upper side with increasing moisture, four forest types are roughly distinguished: the semi-humid EBLF (SH), the monsoon EBLF (MS), the humid EBLF (HE) and the mid-montane moist EBLF (MM). Axis 1 may indicate a gradient of topography or soil properties.

Fig. 2.2 Ordination of plots along the first two detrended correspondence analysis (DCA) axes for SW China. EBLF: evergreen broad-leaved forest. All plots have been free of major disturbance during the past 60 years. The analysis reveals a sequence of 4 EBLF types along Axis 2. Axis 1 may indicate a gradient of topography or soil properties (After Tang (2015). Data sources: Zhou (1992), Zuo (1995), Tang and Ohsawa (1997), Li et al. (2000), Liu and Zhong (2000), Bao and Liu (2002), Peng et al. (2006), Tang et al. (2007a, 2013), Tang and Ohsawa (2009), Shi et al. (2011), Wang et al. (2013))

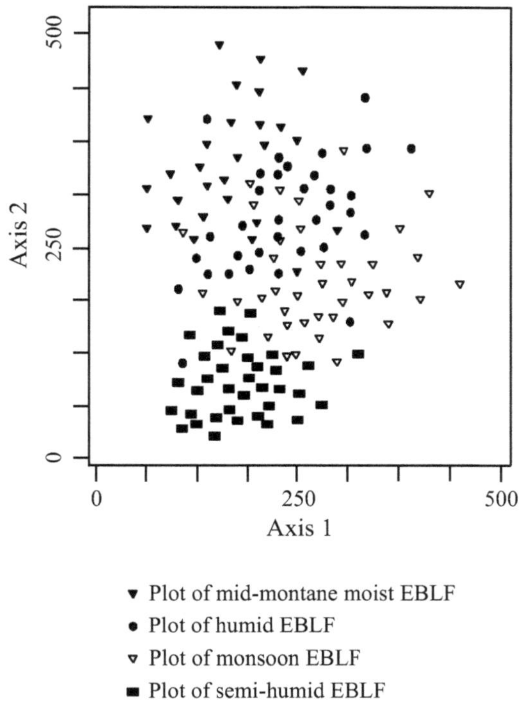

The characteristics and major tree species of the four EBLF types are shown in Table 2.1. The physiognomy of each forest type is shown in Fig. 2.3a–f.

The spatial patterns of distribution of dominant evergreen broad-leaved communities along latitudes, longitudes and altitudes in Yunnan, as represented by 22 mountains, are depicted in Fig. 2.4. The semi-humid EBLFs (SH) dominated by *Castanopsis orthacantha*, *Cyclobalanopsis glaucoides*, *Cyclobalanopsis delavayi*, or *Lithocarpus dealbatus* are found at (1,500) 1,900–2,400 m in central, south-central and eastern Yunnan, including Jizhushan, Huafushan, Shizishan, Qiongzhusi, Wuliangshan, eastern Ailaoshan, Zhujiangyuan, Laojunshan of Wenshan, Wumengshan, and Yaoshan.

The monsoon EBLFs (MS) are mainly distributed at (800)1,000–1,800 m in southern Yunnan (e.g. Tongbiguan, Nangunhe, Beihualin, Yongde Daxueshan, Wuliangshan, Ailaoshan, Xilongshan and Daweishan and Xishuangbanna) and in the province's northwest (e.g. Dulongjiang, Gongshan and Fugong). Most of them are dominated by species of *Castanopsis* and *Lithocarpus* with some *Trigonobalanus* and *Cyclobalanopsis*, as well as Lauraceae and Theaceae, such as *Castanopsis hystrix, Castanopsis fleuryi, Castanopsis calathiformis, Lithocarpus truncatus, Lithocarpus polystachyus, Lithocarpus fenestratus, Trigonobalanus doichangensis, Cyclobalanopsis augustinii, Cyclobalanopsis kerrii, Cryptocarya calcicola, Cryptocarya densiflora, Beilschmiedia yunnanensis, Schima wallichii* and *Anneslea fragrans.*

The mid-montane moist EBLFs (MM) are distributed at (1,600) 1,800–2,500 (2,800) m from the east to the west of Yunnan, and from southern to south-central,

Table 2.1 Characteristics and major tree species (RBA ≥ 1.5 %) of each evergreen broad-leaved forest type in SW China

Forest types	SH	SH	MS	MM	HE	HE
Distribution area	Yunnan	Xichang in W Sichuan	Yunnan	Yunnan	Sichuan	Guizhou
Main altitudinal range (m)	1,800–2,400	1,500–2,400	800–1,800	1,800–2,800	500–1,500	500–1,500
Annual rainfall (mm)	900–1,200	900–1,200	1,100–1,700	1,700–3,700	1,000–1,800	1,000–2,600
Epiphyte and liana	Rare	Rare	Some	Abundant	Not rich	Not rich
Species	RBA (%)	RBA (%)	RBA (%)	RBA (%)	RBA (%)	RBA (%)
Castanopsis orthacantha	28.05					
Cyclobalanopsis glaucoides	23.67	82.36				
Castanopsis delavayi	14.48					
Cyclobalanopsis delavayi	10.15					
Schima argentea	3.48	4.35				4.37
Castanopsis hystrix	3.26		8.06			1.65
Manglietia fordiana	2.29					
Neolitsea homilantha	2.17					
Photinia prionophylla	2.15					
Schima wallichii	2.08		5.18			
Castanopsis calathiformis	1.86		8.67			
Olea yunnanensis	1.75					
Lithocarpus dealbatus	1.54		5.13			
Lithocarpus variolosus	1.51			4.83		
Machilus yunnanensis		1.71				
Cinnamomum wilsonii		1.52				
Actinodaphne emeiensis					3.25	
Machilus gamblei			4.82			
Michelia coriacea			3.64			
Castanopsis rufoto			2.94			
Castanopsis echinocarpa			2.51	3.05		

Castanopsis fabri			2.48	2.32		
Lindera thomsonii			2.32			
Machilus longipedicellata			2.23	1.84		
Sloanea sinensis			2.16			
Cyclobalanopsis chevalieri			1.89			
Lithocarpus truncatus			1.53			
Lithocarpus hancei				9.01	2.37	2.57
Lithocarpus pachyphyllus				5.47		
Cyclobalanopsis lamellosa				5.15		
Castanopsis wattii				5.03		
Lithocarpus xylocarpus				4.90		
Michelia floribunda				3.11		
Manglietia insignis				3.08		
Castanopsis remotidenticulata				2.84		
Machilus viridis				2.27		
Hartia sinensis				2.12		
Castanopsis fleuryi				2.09		
Schima khasiana				2.05		
Cyclobalanopsis myrsinifolia				1.96		
Alcimandra cathcartii				1.73		
Lithocarpus craibianus				1.68		
Phoebe zhennan					19.54	
Castanopsis platyacantha					15.77	
Lithocarpus cleistocarpus					12.37	3.41
Michelia martinii					9.68	1.83
Schima sinense					4.52	
Quercus engleriana					1.99	

(continued)

Table 2.1 (continued)

Forest types	SH	SH	MS	MM	HE	HE
Distribution area	Yunnan	Xichang in W Sichuan	Yunnan	Yunnan	Sichuan	Guizhou
Main altitudinal range (m)	1,800–2,400	1,500–2,400	800–1,800	1,800–2,800	500–1,500	500–1,500
Annual rainfall (mm)	900–1,200	900–1,200	1,100–1,700	1,700–3,700	1,000–1,800	1,000–2,600
Epiphyte and liana	Rare	Rare	Some	Abundant	Not rich	Not rich
Species	RBA (%)	RBA (%)	RBA (%)	RBA (%)	RBA (%)	RBA (%)
Sycopsis triphivenis					1.92	
Cinnamomum szechuanense					1.80	
Turpinia nepalensis					1.62	
Lindera pulcherrima					1.58	
Lithocarpus megalophyllus					1.53	
Illicium henryi					1.51	
Castanopsis carlesii						9.97
Schima superba						8.57
Castanopsis eyrei						7.14
Cyclobalanopsis oxyodon						6.06
Castanopsis fargesii						5.02
Castanopsis carlesii var. *spinulosa*						3.78
Lithocarpus spicatus						3.44
Cyclobalanopsis stewardiana						3.24
Gordonia chrysandra						3.22
Rhododendron sp.						2.74
Liquidambar formosana						2.01
Castanopsis tibetana						1.70
Machilus ichangensis						1.51

Partial data from Tang et al. (2013) for Yunnan, SVEG (1980), Liu (1985), Tang and Ohsawa (1997), Bao and Liu (2002), Tang et al. (2007a) for Sichuan, Huang et al. (1988), Zhou (1992), Zuo (1995) for Guizhou

SH semi-humid EBLF, *MS* monsoon EBLF, *MM* mid-montane moist EBLF, *HE* humid EBLF, *RBA* relative basal area

Fig. 2.3 Part 1 (**a**) The semi-humid evergreen broad-leaved forest dominated by *Cyclobalanopsis glaucoides* and *Castanopsis delavayi* at 2,450 m asl on Mt. Huafu, central Yunnan. (**b**) The monsoon evergreen broad-leaved forest dominated by *Lithocarpus fenestratus*, *Cryptocarya calcicola*, *Castanopsis calathiformis* and *Rehderodendron macrocarpum* around 1,600 m asl on Mt. Dawei, southern Yunnan

Fig. 2.3 (continued) Part 2 (c) The humid evergreen broad-leaved forest dominated by *Castanopsis fargesii, Castanopsis eyrei* and *Castanopsis carlesii* at 1,000–1,100 m asl on Mt. Fanjing, northeastern Guizhou. (d) The mid-montane moist evergreen broad-leaved forest dominated by *Cyclobalanopsis lamellosa, Cinnamomum iners, Machilus longipedicellata* and *Manglietia insignis* at 2,300–2,700 m asl in the Gaoligong Mountains, northwestern Yunnan

e

f

Fig. 2.3 (continued) Part 3 (**e**) The mid-montane moist evergreen broad-leaved forest dominated by *Lithocarpus chintungensis*, *Castanopsis wattii* and *Schima villosa* at 2,300–2,700 m asl in Zhelonghoushan, in the eastern Ailao Mountains, south-central Yunnan. (**f**) The mid-montane moist evergreen broad-leaved forest dominated by *Cyclobalanopsis oxyodon* and *Manglietia gongshanensis* or *Lithocarpus pachyphyllus* at 2,100–2,500 m asl in Dulongjiang, northwestern Yunnan

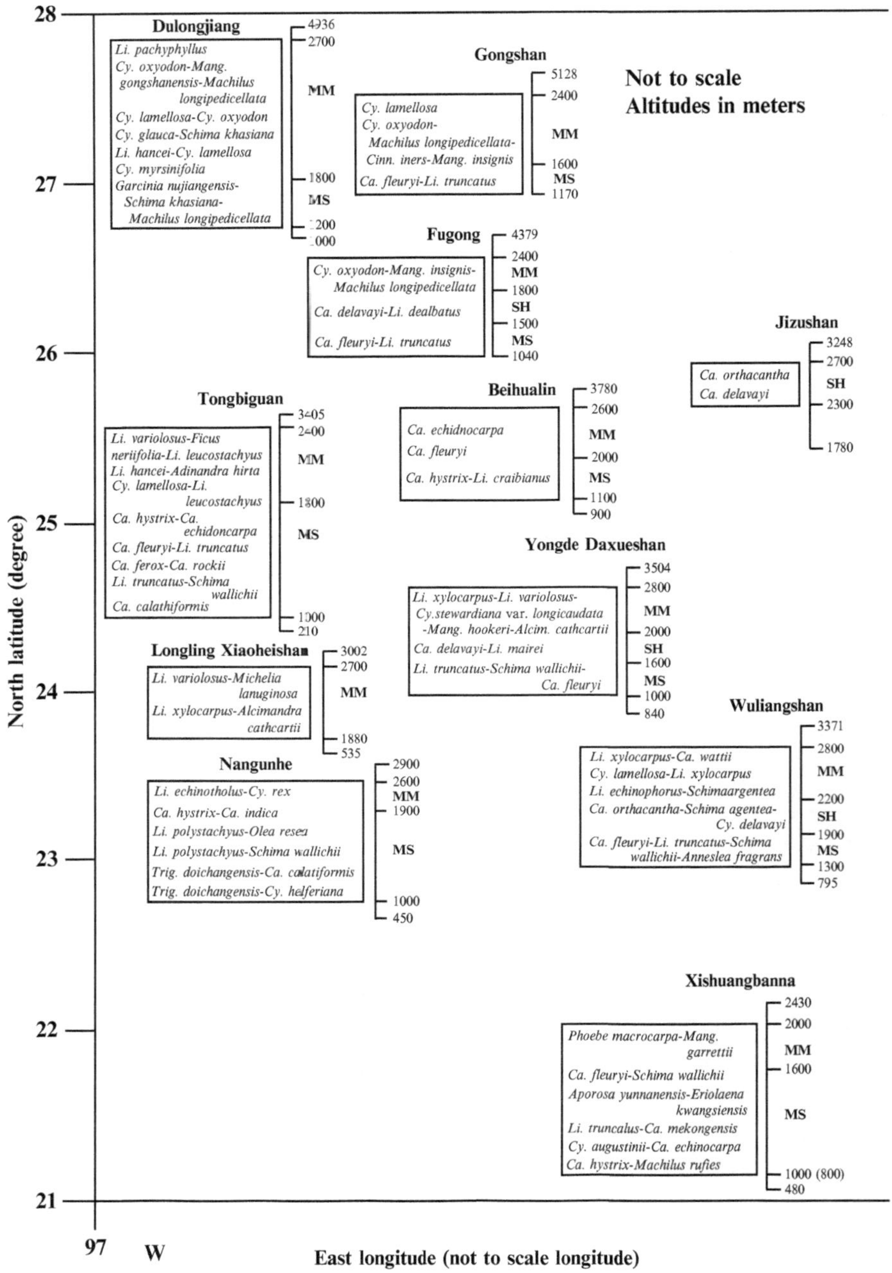

Fig. 2.4 See legend on page 62

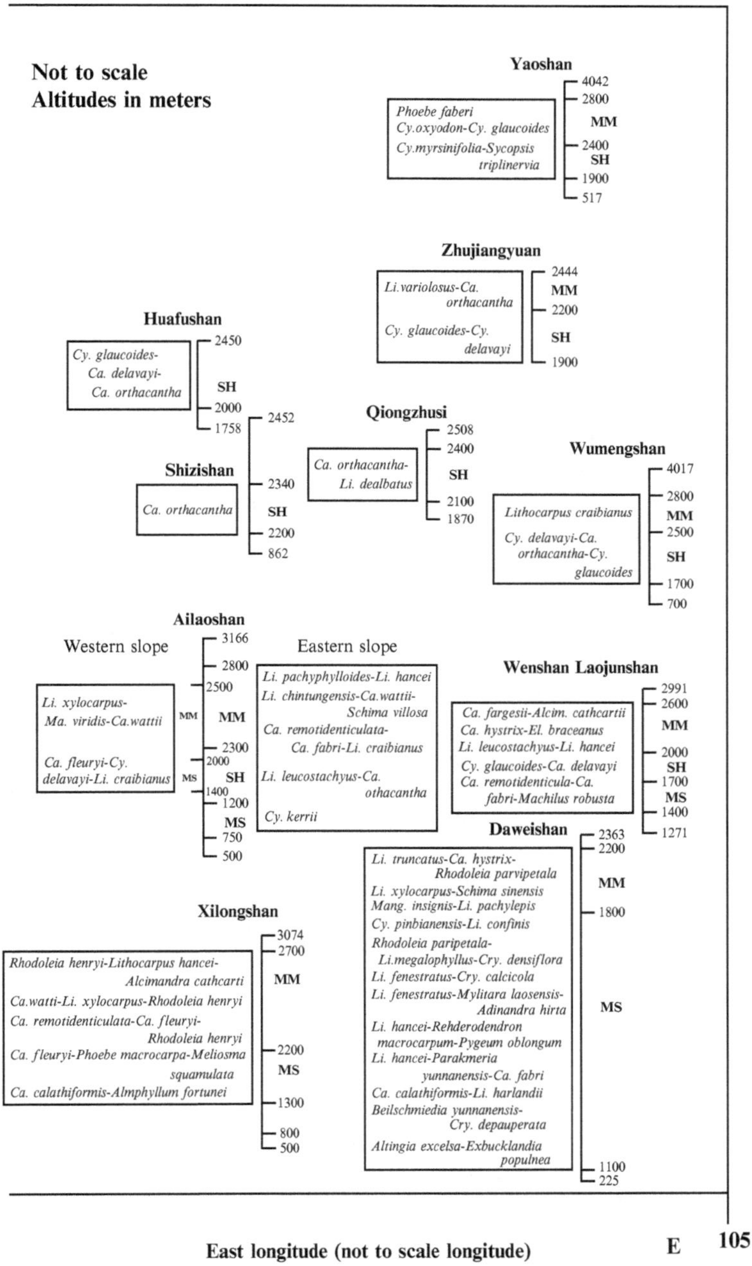

Fig. 2.4 (continued) See legend on page 62

and northern Yunnan, but not in central Yunnan, as on Jizushan, Huafushan, Shizishan and Qiongzhusi, where no such moist EBLFs grow. Compared with the former two forest types, in the mid-montane moist EBLFs are to be found, besides dominant species of Fagaceae (e.g. *Cyclobalanopsis lamellosa, Cyclobalanopsis oxyodon, Cyclobalanopsis myrsinifolia, Lithocarpus variolosus, Lithocarpus hancei, Lithocarpus pachyphyllus, Lithocarpus xylocarpus, Lithocarpus echinotholus, Castanopsis echidnocarpa, Castanopsis wattii,* and *Castanopsis remotidenticulata*), Lauraceae *Machilus longipedicellata, Machilus viridis, Cinnamomum iners, Phoebe faberi*, Theaceae *Schima khasiana, Schima argentea,* and *Schima villosa*, and Magnoliaceae *Manglietia gongshanensis, Manglietia insignis,* and *Alcimandra cathcartii* also appear as co-dominants. The mid-montane moist EBLF is characterized by abundant lianas, ferns and mosses on tree trunks and branches (Fig. 2.5a, b).

In general, the mossy dwarf stands near mountain summits in southern Yunnan are mainly dominated by Ericaceae, with dwarf trees of Fagaceae, Vacciniaceae, Rosaceae, Aceraceae etc. (Wu et al. 1987; Shi and Zhu 2009), which are not shown in Fig. 2.4 since for present purposes they are not considered the subtropical evergreen broad-leaved forest.

In most of the Sichuan Basin and eastern Guizhou, lacking a distinctive dry season throughout the year, the EBLFs are typical humid EBLFs. However, the humid EBLF differs from the mid-montane moist EBLF in lacking rich epiphytes and lianas on tree trunks and branches. Figure 2.6 shows the spatial patterns of distribution of dominant evergreen broad-leaved communities along latitudes, longitudes and altitudes in the areas as represented by 21 mountains. In these areas, natural EBLFs at low altitudes have been even more severely reduced by human activities than in Yunnan. Secondary forests, shrublands and plantations of pines (*Pinus yunnanensis* in the west of Sichuan and Guizhou, *Pinus massoniana* in the east of Sichuan and Guizhou as well as Chongqing) or *Cunninghamia lanceolata* or *Cryptomeria japonica* var. *sinensis* prevail. Some semi-humid EBLFs dominated by *Castanopsis delavayi* or *Cyclobalanopsis glaucoides* are found in western Sichuan (e.g. Jinyangxian, Xiaoxiangling). In these areas, the typical humid EBLF (HE) prevails. The

Fig. 2.4 The spatial patterns of distribution of dominant evergreen broad-leaved communities along latitudes, longitudes and altitudes in Yunnan. *SH* semi-humid EBLF, *MS* monsoon EBLF, *MM* mid-montane moist EBLF, Cy *Cyclobalanopsis*, Ca *Castanopsis*, Li *Lithocarpus*, Ma *Machilus*, Trig *Trigonobalanus*, Mang *Manglietia*, Alcim *Alcimandra*, Cry *Cryptocarya*, El *Elaeocarpus* (After Tang (2015). Partial data sources: Peng and Dang (1998), Hu (1999), Yang and Li (1999), Li et al. (2000), Peng and Wu (2001), Yang and Du (2004), Wang et al. (2005), Li et al. (2006), Peng et al. (2006), Yang et al. (2006), Yang and Du (2006), Liu and Peng (2007), Tang and Zhu (2007), Tang et al. (2007a, b, 2010a, b, 2013), Tang and Ohsawa (2009), Wang et al. (2013), Tang and He (2013). At low altitudes below the EBLF zones on these mountains, secondary forests or shrublands or plantations of conifers (such as *Pinus yunnanensis* and *Pinus armandii*) or eucalyptus, or agricultural fields are all common, while some localities along rivers support hot-dry valley vegetation; also in southern Yunnan there are some patches of tropical seasonal rain forest or tropical rain forest. At higher altitudes above the EBLFs, there are various vegetation types, such as montane mossy dwarf communities, evergreen sclerophyllous Quercus forests, or temperate coniferous (e.g. *Tsuga dumosa*) and broad-leaved mixed forests, or cold temperate forests, alpine scrubland, grassland, meadows)

Fig. 2.5 (**a**) The liana *Rhaphidophora decursiva* (Araceae) (indicated by a *white arrow*) on a tree trunk in a mid-montane moist EBLF at 2,200 m asl in Dulongjiang, NW Yunnan. (**b**) Ferns *Asplenium phyllitidis* (Aspleniaceae) (*yellow arrow*), *Pyrrosia stigmosa* (Polypodiaceae) (*pink arrow*) and *Vittaria linearifolia* (Vittariaceae) (*red arrow*) on a tree branch in a montane moist EBLF at 1,200 m asl in Manhao, Gejiu, SE Yunnan

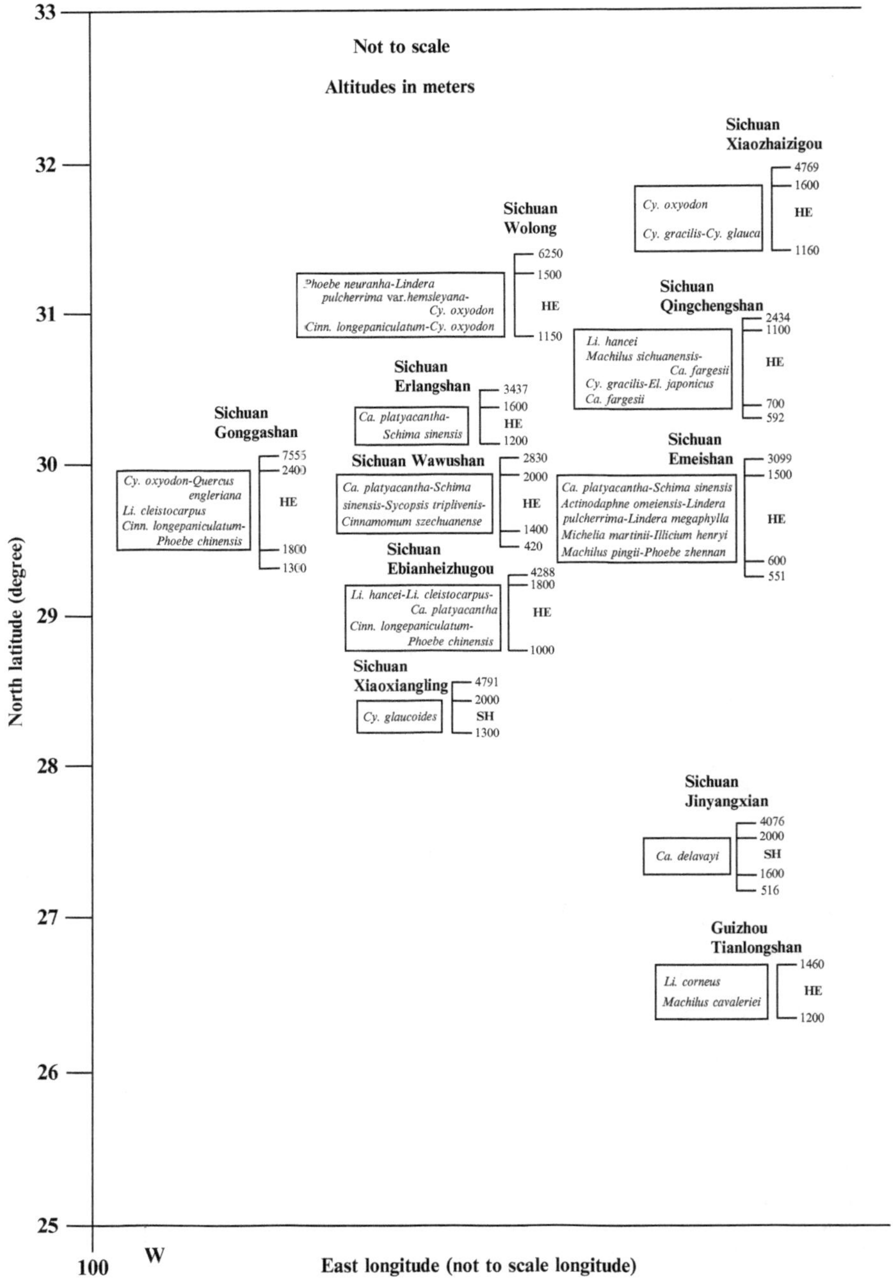

Fig. 2.6 See legend on page 66

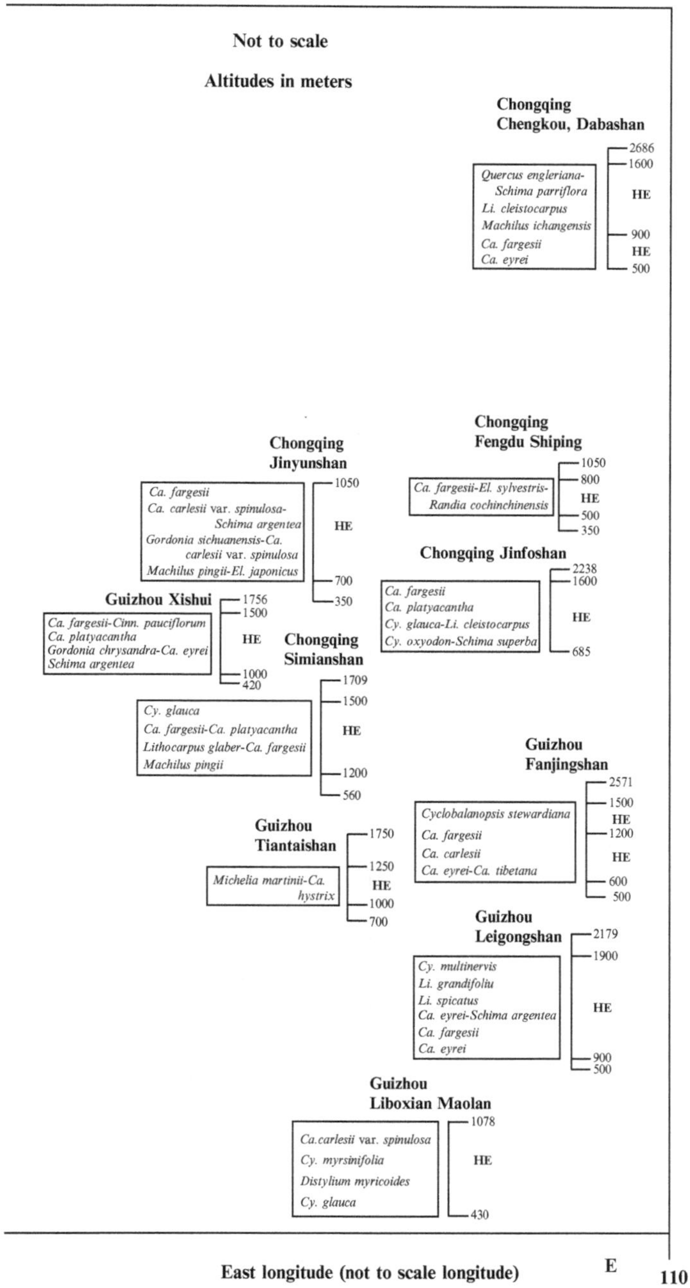

Fig. 2.6 (continued) See legend on page 66

Table 2.2 The vicarious tree species occurring in the evergreen broad-leaved forest zone in the western and eastern areas of SW China

Western area of SW China	Eastern area of SW Sichuan
Yunnan, western Sichuan and western Guizhou	Sichuan Basin and eastern Guizhou
Castanopsis delavayi	*Castanopsis sclerophylla*
Castanopsis orthacantha	*Castanopsis eyrei*
Lithocarpus variolosus	*Lithocarpus cleistocarpus*
Cyclobalanopsis glaucoides	*Cyclobalanopsis glauca*
Cyclobalanopsis delavayi	*Cyclobalanopsis gilva*
Alnus ferdinandi-coburgii	*Alnus cremastogyne*
Albizia mollis	*Albizia julibrissin*
Ehretia corylifolia	*Ehretia dicksonii*
Sapindus delavayi	*Sapindus mukorosii*
Pistacia weinmanniifolia	*Pistacia chinensis*
Pinus yunnanensis	*Pinus massoniana*
Keteleeria evelyniana	*Keteleeria davidiana*

Data source: Li et al. (1990)

dominants are *Castanopsis*, *Lithocarpus* and *Cyclobalanopsis* (e.g. *Castanopsis carlesii, Castanopsis fargesii, Castanopsis eyrei, Castanopsis carlesii* var. *spinulosa, Castanopsis platyacantha, Lithocarpus corneus, Lithocarpus cleistocarpus, Cyclobalanopsis oxyodon, Cyclobalanopsis multinervis*), and many species of Lauraceae (e.g. *Phoebe neurantha, Phoebe zhennan, Lindera pulcherrima* var. *hemsleyana, Lindera megaphylla, Lindera pulcherrima, Cinnamomum longepaniculatum, Machilus pingii, Machilus ichangensis, Actinodaphne omeiensis*), and Theaceae (*Schima superba, Schima argentea, Gordonia sichuanensis, Gordonia chrysandra*), and Magnoliaceae (e.g. *Michelia martinii* and *Manglietia fordiana*) as well as Elaeocarpaceae (e.g. *Elaeocarpus japonica* and *Elaeocarpus sylvestris*).

To summarize: in SW China, the EBLFs are mainly distributed at higher altitudes (1,000–(2,400) 2,800 m) in the western areas of Yunnan and Sichuan, while occurring in lower altitudes (600–1,500 (2,000) m) toward the southern and the eastern areas (Figs. 2.4 and 2.6). Communities dominated by *Castanopsis* are widely distributed over the region of SW China. In Yunnan, many plant communities dominated by *Lithocarpus* and/or *Cyclobalanopsis* are conspicuous compared with those in the Sichuan Basin and eastern Guizhou. The vicarious tree species occurring in the evergreen broad-leaved forest zone in the western and eastern areas of SW China are detailed in Table 2.2.

Fig. 2.6 The spatial patterns of distribution of dominant evergreen broad-leaved communities along latitudes, longitudes and altitudes in Sichuan, Chongqing and Guizhou. *HE* humid EBLF, *SH* semi-humid EBLF, Cy *Cyclobalanopsis*, Ca *Castanopsis*, Li *Lithocarpus*, Ma *Machilus*, Cinn *Cinnamomum*, El *Elaeocarpus* (After Tang (2015). Data sources: SVEG (1980), Yang (1983), Liu (1985), Liu and Qiu (1986), Zhou (1992), Zuo (1995), He et al. (1996), Tang and Ohsawa (1997), Xie and Chen (1998), Liu and Zhong (2000), Tang and Ohsawa (2000), Bao and Liu (2002), Xu et al. (2004), Xu (2005), Tang et al. (2007a), Yang et al. (2008, 2012), Bai et al. (2012), Yang and Xie (2012), Zhang et al. (2013)

2.2 Species Diversity

Measurements of species diversity play a key role in studies of ecology and conservation biology. Aspects most commonly measured are species richness (number of species) and evenness (similarity in abundance among species, sometimes known as equitability). High evenness is conventionally equated with high diversity (Pielou 1975; Magurran 1988). Simpson's Index of Diversity (Simpson's D) ($D = 1 - \Sigma p_i^2$, where p_i is the proportion of the number individuals of the *i*th species to the total number of individuals) is a measure of diversity which takes into account both richness and evenness (Lande 1996). Simpson's D is employed in this section. The EBLFs in SW China are marked by high species diversity.

The forest types of SW China on the x-axis of Fig. 2.7a, b indicate a moisture gradient ranging from semi-humid to mesic-humid to humid, then to mid-montane moist. The semi-humid EBLF has significantly lower values of woody species (greater than 1 cm in DBH in each plot area of 400 m^2) richness (11 on average) and of Simpson's D (0.71 on average) than the monsoon, humid and mid-montane

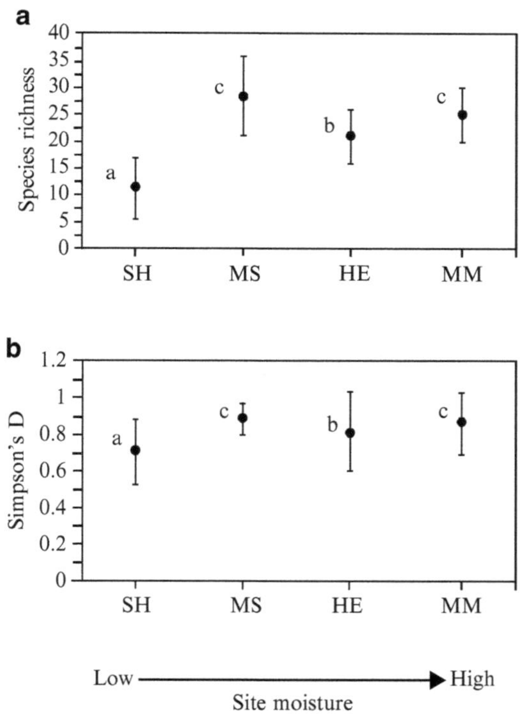

Fig. 2.7 Species diversity for woody species (diameter at breast height ≥1 cm) among the four evergreen broad-leaved forest types in SW China. Significantly different values are indicated by different letters. Forests sharing the same letters do not differ significantly by the non-parametric Kruskal–Wallis all-pairwise comparisons test ($p < 0.05$). *Bar*: Standard deviation. *SH* semi-humid EBLF, *MS* monsoon EBLF, *HE* humid EBLF, *MM* mid-montane moist EBLF. Partial data from Zhou (1992), Zuo (1995), Tang and Ohsawa (1997), Li et al. (2000), Liu and Zhong (2000), Bao and Liu (2002), Peng et al. (2006), Tang et al. (2007a, 2010a, b, 2013), Tang and Ohsawa (2009), Shi et al. (2011), Wang et al. (2013)

moist EBLFs. The monsoon EBLF tends to have the highest values (28 for species richness, 0.88 for Simpson's D), followed by the mid-montane moist EBLF (25 for species richness, 0.86 for Simpson's D), then by the humid EBLF (21 for species richness, 0.81 for Simpson's D), although the differences are not significant between the monsoon and mid-montane moist EBLFs. The monsoon EBLF in the mesic-humid sites tends to have higher species richness and diversity scores. This indicates that site moisture is the best predictor of species composition and diversity among the EBLFs.

The mid-montane moist EBLF can also be called a cloud forest in light of the orographic cloud formation, and it includes abundant ferns, mosses and climbing plants on tree trunks and branches. For example, among 23 woody liana species in 10 plots (20 × 20 m) in the mid-montane moist EBLF (Yuan et al. 2009), the families showing the highest diversity in terms of number of species are Rosaceae and Smilacaceae. The most abundant families in terms of the number of individuals are

Table 2.3 List of lianas of DBH ≥0.5 cm and height ≥2 m as counted in 10 plots measuring 20 × 20 m in the old-growth mid-montane moist evergreen broad-leaved forest of the Ailao Mountains

Family	Species	CT	DT	LF	No. of lianas
Actinidiaceae	*Actinidia callosa*	ST	Z	D	18
Asclepiadaceae	*Marsdenia tsaiana*	ST	AN	E	2
Celastraceae	*Celastrus angulatus*	ST	Z	D	8
	Euonymus vagans	RC	Z	E	13
Elaeagnaceae	*Elaeagnus conferta*	HC	Z	E	4
Hydrangeaceae	*Hydrangea anomala*	RC	AN	D	7
	Schizophragma integrifolium	RC	AN	D	5
Lardizabalaceae	*Holboellia latifolia*	ST	Z	E	6
Loganiaceae	*Gardneria lanceolata*	ST	Z	E	2
Oleaceae	*Jasminum urophyllum*	ST	Z	E	6
Papilionaceae	*Millettia dielsiana*	ST	AU	E	6
Rutaceae	*Zanthoxylum alpinum*	HC	Z	E	6
Rosaceae	*Rosa longicuspis*	HC	Z	E	27
	Rubus paniculatus	HC	Z	E	24
	Rubus pinfaensis	HC	Z	D	1
Sabiaceae	*Sabia yunnanensis*	ST	Z	D	4
Schisandraceae	*Kadsura coccinea*	ST	Z	E	6
	Schisandra henryi	ST	Z	D	1
Smilacaceae	*Heterosmilax yunnanensis*	TC	Z	E	5
	Smilax lebrunii	TC	Z	E	4
	Heterosmilax chinensis	TC	Z	E	6
Vitaceae	*Parthenocissus himalayana*	RC	Z	D	84
	Tetrastigma obtectum	TC	Z	D	8
Total					253

Data source: Yuan et al. (2009)

CT climbing type, *ST* stem twiners, *TC* tendril climbers, *HC* hook climbers, *RC* root climbers, *DT* dispersal type, *Z* zoochory, *AN* anemochory, *AU* autochory, *LF* life form, *E* evergreen species, *D* deciduous species

Vitaceae (36.4 % of the total), followed by Rosaceae (20.6 %). The lianas include four climbing types: stem twiners, tendril climbers, hook climbers, and root climbers (Table 2.3).

2.3 Leaf Size Patterns of Evergreen Broad-Leaved Trees along the Altitudinal Gradient

Leaf size is one of the most plastic traits of a tree and it is common to see great variability within genera, species, individuals within a species, and even between the same individual at different stages of development (e.g. sapling, canopy tree) or on different parts of the same tree (e.g. sun versus shade). Despite this very large variation it is often possible to detect significant associations between a range of environmental variables and leaf sizes (Givnish 1987; Malhado et al. 2009). From tropical to temperate montane forests, the life form of canopy tree species shifts from evergreen to deciduous broad-leaved, and then to needle-leaved (Whitmore 1975; Richards 1996). Ohsawa (1990) discussed how temperature conditions (i.e., temperature sum and the coldest monthly mean temperature) control macro-scale diversification of various leaf types, including evergreen broad-leaved trees, in humid East Asia.

To better understand leaf-size patterns in relation to the distribution of individual evergreen broad-leaved species along environmental gradients, the example of Mt. Emei (Emeishan, 3,099 m asl, 29°34.50′ N, 103°21.50′ E), Sichuan, will serve as a case study. The sacred Mt. Emei is perhaps one of the best places to see the remnants of the once lush Sichuan Basin subtropical forests. In the words of Lancaster (1989), "If there is one mountain in China calculated to set a plantsman's pulse racing, it is Emeishan." Tang and Ohsawa (1997) and Tang (2006) demonstrated that the forest zonation as manifested in different tree leaf-types on Mt. Emei is primarily due to temperature differences along the altitudinal gradient. As for the upper limits of EBLFs, and the limits for the individual trees, the two temperature conditions, i.e., the temperature sum (warmth-index (WI)) of 85 °C·months and the coldest monthly mean temperature (CMT) of −1 °C, are decisive, respectively (Ohsawa 1990). A CMT of −1 °C occurs at 1,600 m while the WI of 85 °C·months corresponds to 1,160 m on Mt. Emei (Fig. 2.8). Thus the CMT of −1 °C is more effective as an upper limit of EBLFs than is the WI of 85 °C·months, on this subtropical mountain, as toward higher altitudes the minimum temperature in winter limits the distribution of most evergreen broad-leaved trees. The climatic forest zone shifts from the low-altitudinal evergreen broad-leaved forest (600–1,500 m) dominated by subtropical evergreen *Machilus*, *Phoebe*, *Actinodaphne* and *Lindera*, to the transitional mixed forest of evergreen/deciduous broad-leaved and coniferous trees at mid-altitude (1,500–2,500 m) co-dominated by evergreen *Lithocarpus*, deciduous *Acer*, coniferous *Tsuga*, *Taxus* and *Abies*, and finally rises to the uppermost temperate coniferous forest near the mountain top (2,500–3,099 m), dominated by *Abies*, with some shrubby evergreen and deciduous broad-leaved species (Fig. 2.8).

While temperature, as a function of altitude, controls the mountain's overall pattern of tree distribution, topography also exercises an important influence on the coexistence mechanisms, by enhancing habitat segregation among these evergreen,

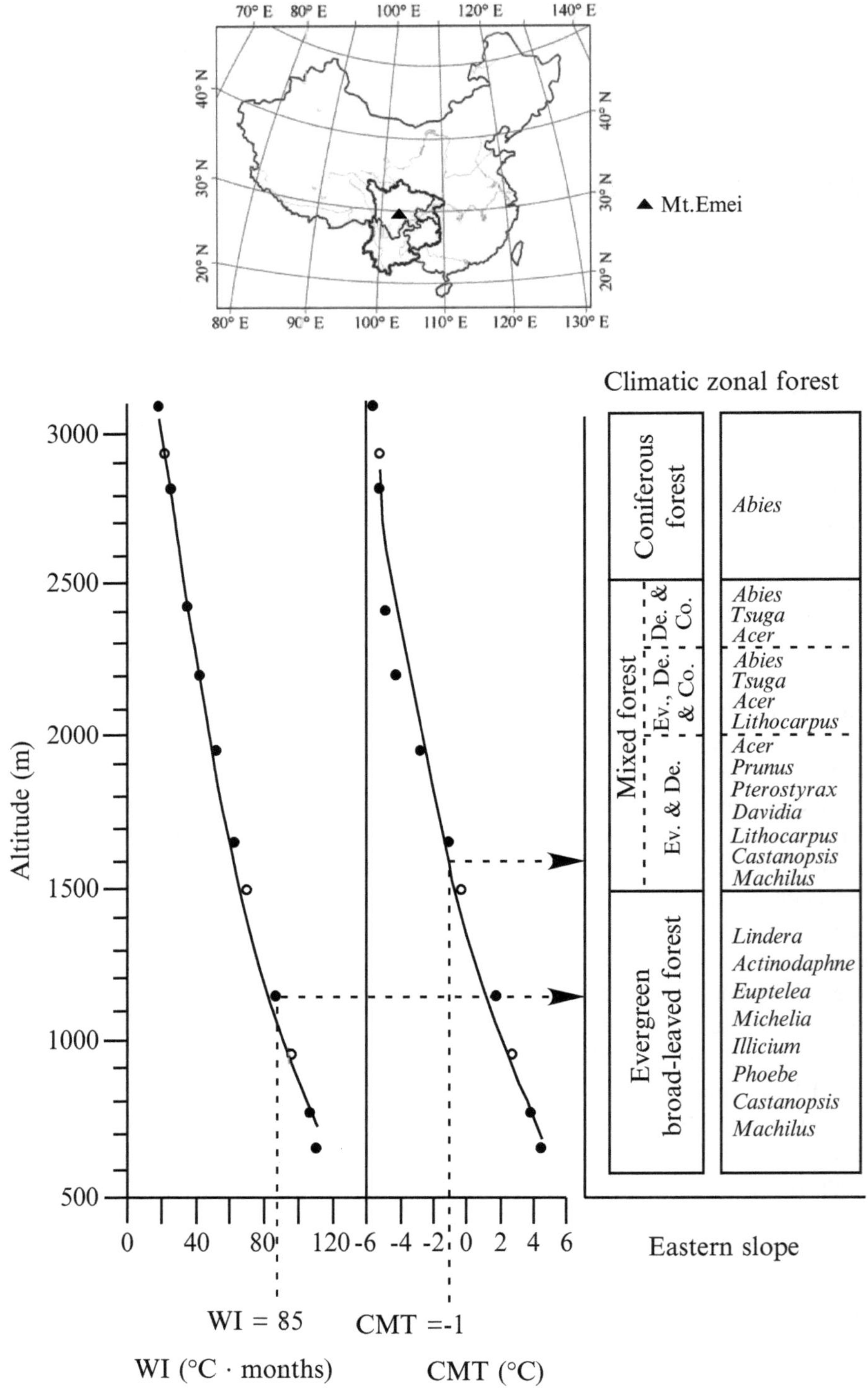
70° E 80° E 100° E 120° E 140° E
40° N
30° N
20° N
80° E 90° E 100° E 110° E 120° E 130° E
▲ Mt.Emei
Climatic zonal forest
Coniferous forest
Abies
Mixed forest
De. & Co.
Abies
Tsuga
Acer
Ev., De. & Co.
Abies
Tsuga
Acer
Lithocarpus
Ev. & De.
Acer
Prunus
Pterostyrax
Davidia
Lithocarpus
Castanopsis
Machilus
Evergreen broad-leaved forest
Lindera
Actinodaphne
Euptelea
Michelia
Illicium
Phoebe
Castanopsis
Machilus
Altitude (m)
3000
2500
2000
1500
1000
500
0 40 80 120
-6 -4 -2 0 2 4 6
WI = 85
CMT =-1
WI (°C · months)
CMT (°C)
Eastern slope

deciduous and coniferous trees in the mid-montane mixed forest (Tang and Ohsawa 2002a), and the Tertiary relict deciduous forest on the scree slope (1,450–1,900 m) (Tang and Ohsawa 2002b). At altitudes below 600 m, where there are direct human influences, secondary forests contain evergreen broad-leaved trees such as *Castanopsis*, *Phoebe* and *Machilus*, and some seral or introduced conifers including *Pinus massoniana*, *Cunninghamia lanceolata* and *Cryptomeria japonica* var. *sinensis*, as human presence causes deviation from natural or mesic conditions.

2.3.1 Floristic Changes in Evergreen Broad-Leaved Trees by Altitudes

In total, evergreen broad-leaved trees of 88 species in 39 genera and 23 families along the eastern slope are recorded (Tang and Ohsawa 1999). The greatest number, 35 evergreen broad-leaved species, occur in the 600–800 m segment; the numbers then decrease toward higher altitudes with 24 spp. at 1,600–1,800 m and 10 spp. at 2,200–2,400 m. Lauraceae species are by far the most frequent at all the lower altitudes (600–1,800 m). The forest at 1,600–1,800 m has the relatively larger number of 24 evergreen broad-leaved species, the dominant families being of Lauraceae (5 spp.), Theaceae (5), Fagaceae (4) and Ericaceae (4), plus some additional species of Symplocaceae, Aquifoliaceae, Oleaceae, Caprifoliaceae, Araliaceae and Elaeagnaceae. At 2,200–2,400 m, the dominant families are Symplocaceae and Ericaceae. Above 2,400 m, the number of evergreen broad-leaved species decreases drastically to only 2 spp. of Ericaceae near the summit. The species richness of both genera and families fluctuates greatly, with the highest species/genus ratio at 1.7 (at 2,200–2,400 m), and the highest species/family ratio at 2.5 (at 600–800 m) (Fig. 2.9a–e).

2.3.2 Leaf Size Changes along the Altitudinal Gradient

Leaf size distribution was studied by Tang and Ohsawa (1999), using the following leaf-size classes: leptophyll (<25 mm^2), nanophyll (25–225 mm^2), microphyll (225–2,025 mm^2), notophyll (2,025–4,500 mm^2), mesophyll (4,500–18,225 mm^2),

Fig. 2.8 Characteristic vegetation with the climatic parameters WI and CMT on the eastern slope of Mt. Emei (the vegetation on the northwestern slope is heavily disturbed by human activity and thus not suitable for analysis of the climatic forest zones). The coldest monthly mean temperature is indicated by CMT. The temperature sum is indicated as the warmth-index (WI) (Kira 1948). As for the upper limits of EBLFs, and the limits for the individual tree species, the two temperature conditions, namely the temperature sum of 85 °C·months, and the coldest monthly mean temperature of −1 °C, are decisive, respectively (Ohsawa 1990). A CMT of −1 °C occurs at 1,600 m while the WI of 85 °C·months corresponds to 1,160 m on Mt. Emei. Thus the CMT of −1 °C is more effective as an upper limit of EBLFs than is the WI of 85 °C·months on this subtropical mountain, as toward higher altitudes the minimum temperature in winter limits the distribution of most evergreen broad-leaved trees. *Ev* evergreen broad-leaved, *De* deciduous broad-leaved, *Co* coniferous (Data sources: Tang and Ohsawa (1997), Tang (2006))

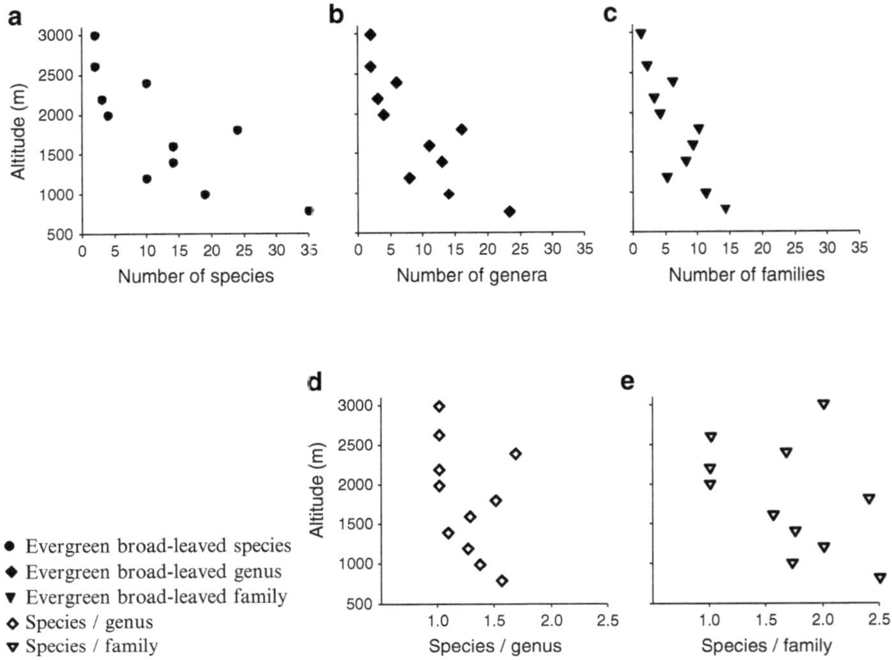

Fig. 2.9 Floristic richness for evergreen broad-leaved species by altitude and taxonomic levels (**a**–**c**) and ratios (**d**, **e**) on Mt. Emei (Data source: Tang and Ohsawa (1999))

macrophyll (18 225–164 025 mm^2) and megaphyll (>164 025 mm^2) (Raunkiaer 1916; Webb 1959; Shimwell 1971). On Mt. Emei, the evergreen broad-leaved trees had a simple leaf shape except that two species (*Brassaiopsis glomerulata* and *Nothopanax davidii*) have compound leaves. For all the evergreen broad-leaved species, the leaf sizes range from 4.80 cm^2 (*Rhododendron monanthum*) to 135.74 cm^2 (*Brassaiopsis glomerulata*). Of the 88 species, 42 (48 %) are notophylls, 32 (36 %) are microphylls, and 14 (16 %) are mesophylls. The number of notophyllous species decreases from 19 at 600–800 m to 3 at 2,200–2,400 m. The number of microphyllous species decreases from 11 to 5 species, and mesophyllous species decrease slightly from 4 to 2 species with increasing altitude (Fig. 2.10). In spite of this trend, the species richness of the trees of the three leaf-size classes reaches a small peak in the 1,600–1,800 m segment. The upper limits for both notophyllous and mesophyllous species occur at c. 2,400 m. Above this altitude, only two ericaceous, shrubby species with microphyllous evergreen leaves are found.

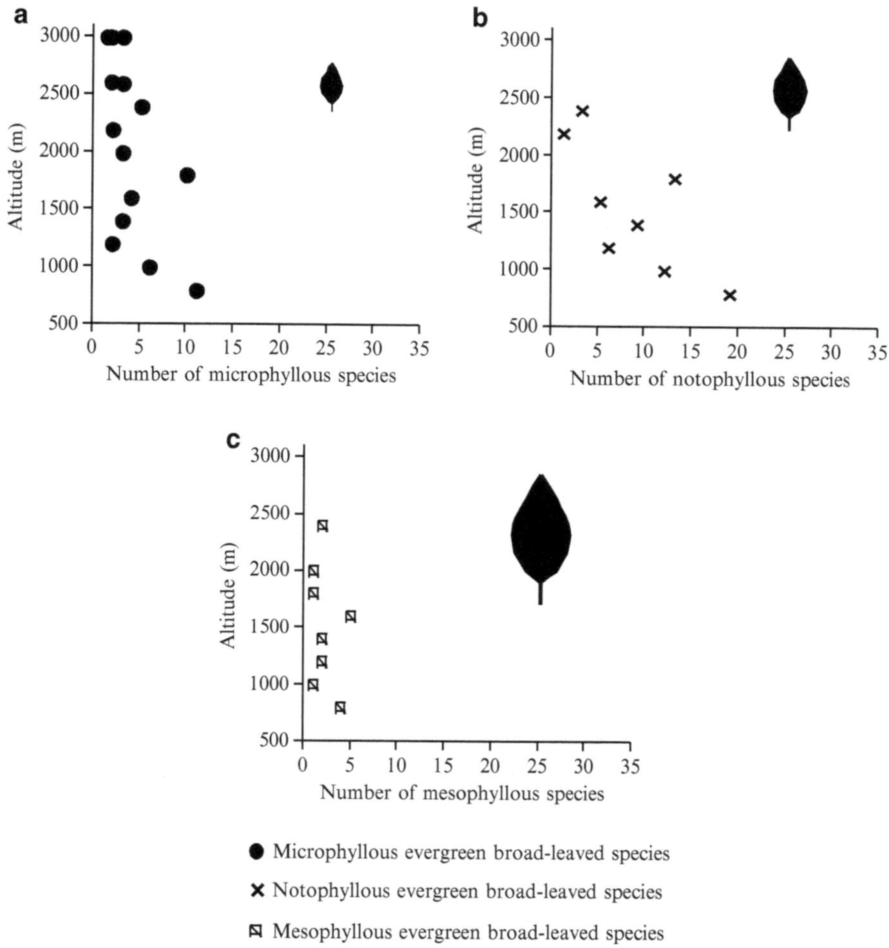

Fig. 2.10 Change in number of evergreen broad-leaved species (**a-c**) according to different leaf-size classes by altitude on Mt. Emei (Data source: Tang and Ohsawa (1999))

The gross patterns of tree life form distribution, in terms of relative dominance (relative basal area, i.e., % RBA), take the following form: the low-altitude evergreen broad-leaved forest zone changes to a mid-altitude mixed forest zone, then to the high-altitude coniferous zone (Fig. 2.11a). Leaf-size spectra are examined for 58 evergreen broad-leaved tree species in the evergreen broad-leaved forest zone, 40 species in the mixed forest zone and 2 species in the coniferous forest zone (Fig. 2.11b).

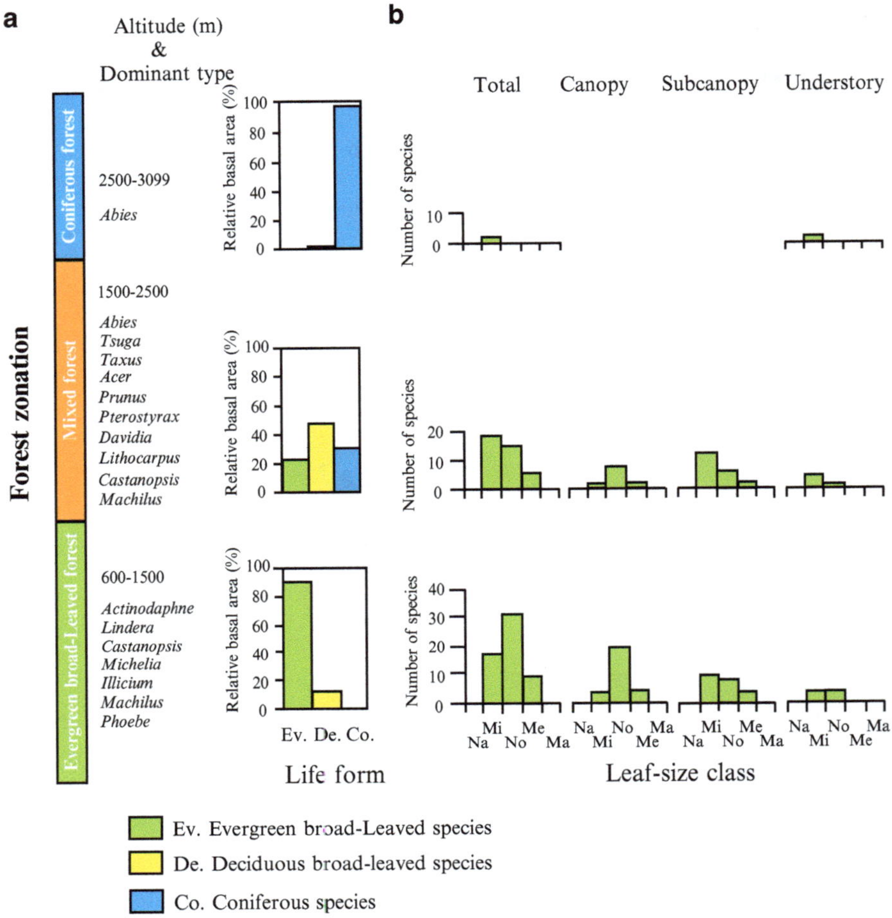

Fig. 2.11 Gross patterns of tree life-form (**a**) and leaf-size (**b**) according to altitude on the eastern slope of Mt. Emei. *Na* nanophyll, *Mi* microphyll, *No* notophyll, *Me* mesophyll, *Ma* macrophyll (Data source: Tang and Ohsawa (1999))

The evergreen broad-leaved forest zone (600–1,500 m) is mainly dominated by evergreen broad-leaved trees of Lauraceae (*Machilus, Phoebe, Actinodaphne* and *Lindera*), as well as by other evergreen species, such as *Castanopsis* of Fagaceae, *Michelia* of Magnoliaceae and *Illicium* of Illiciaceae (89.3 % RBA). Of a total of 58 evergreen broad-leaved species, 19 (65.5 %) of the 29 canopy species are notophylls, 4 (13.8 %) are microphylls, and 6 (20.7 %) are mesophylls. Among the 21 subcanopy species, 9 (42.9 %) are microphylls, 8 (38.1 %) are notophylls, and 4 (19 %) are mesophylls. The eight understory species are equally divided between microphylls and notophylls. The mixed forest zone (1,500–2,500 m) is co-dominated by three life forms, i.e. evergreen *Machilus* of Lauraceae and *Lithocarpus* of Fagaceae (22.9 % RBA), deciduous *Pterostyrax* of Styracaceae, *Acer* of Aceraceae,

Prunus of Rosaceae and *Davidia* of Davidiaceae (47 % RBA), and coniferous *Abies* and *Tsuga* of Pinaceae and *Taxus* of Taxaceae (30.1 % RBA). Among 40 evergreen broad-leaved species, 13 are canopy species. Among them, 8 (61.5 %) are notophylls, 2 (15.4 %) are microphylls and 3 (23.1 %) are mesophylls. Of the 21 subcanopy species, 12 (57.1 %) are microphylls, 6 (28.6 %) are notophylls and 3 (14.3 %) are mesophylls. Among the 6 understory species, 5 (83.3 %) are microphylls and 1 (16.7 %) is notophyll. The coniferous forest zone (2,500–3,099 m) is exclusively dominated by *Abies* (97.7 % RBA). The two evergreen ericaceous, shrubby species are microphylls. The leaf size of the evergreen broad-leaved trees decreases in an overall trend toward upper altitudinal zones, and varies with forest stratifications. Canopy trees are mostly notophyllous species, while subcanopy and understory trees are mainly microphyllous species throughout the evergreen broad-leaved and mixed forests.

2.3.3 Leaf Traits in the Differentiation of Forest Types

Richards (1996) observed that tropical evergreen broad-leaved forest trees often have varying leaf forms, including abundant compound leaves, and many leaves and leaflets have long, distinct drip tips. However, on Mt. Emei three leaf types, the evergreen broad-leaved, deciduous broad-leaved, and coniferous, vary with altitude, and most of the evergreen broad-leaved trees (86 species) have simple leaves of ovoid to oblanceolate shape while only two species had compound leaves. Though there are strongly acuminate leaves among the 88 species, the tips are not as pronounced as the drip-tips of tropical forest. On Mt. Emei, EBLFs at low altitude are mostly composed of notophyllous evergreen trees of tropical origin. Then in the transitional mixed forests at middle altitude, the evergreen broad-leaved trees are mainly microphyllous and notophyllous species. Finally, in the coniferous forests at high altitude, only some shrubby microphyllous evergreen broad-leaved species occur. Differentiation of forest types by leaf traits also occurs along micro-scale environmental gradients (e.g., topographic location of each forest community), and as a result of mechanical stress (e.g., wind-swept sites), and shortage of water or nutrients. Grubb (1977) suggested that in tropical montane forests, leaf life-span, leaf size, leaf form and leaf thickness in the forest may be affected by climatic variables as well as mineral supply on slopes and in gullies. These results suggest that leaf traits are essential factors in the differentiation of forest formations. In general, in subtropical/warm-temperate forests, favorable sites are dominated by noto/mesophyllous evergreen trees, while stressful habitats are dominated by smaller nano/microphyllous evergreen trees (Ohsawa 1993b, 1995; Ohsawa and Nitta 1997). In terms of vertical stratification, on Mt. Emei the canopy has a high proportion of notophyllous species, while both subcanopy and understory layers have a high proportion of microphyllous species. This pattern shows a strong consistency along altitudinal gradients. Within tropical evergreen forests, the most abundant leaf-size class changes from larger to smaller with increasing altitude (Whitmore 1975; Richards 1996).

Turner (2001) proposed that trees with mesophyllous leaf sizes normally constitute approximately three-quarters of all species in lowland tropical rainforests across the globe. The mesophyllous leaf size class dominates in neotropical rainforests (Grubb et al. 1963; Thompson et al. 1992; Kelly et al. 1994; Malhado et al. 2009). Northern tropical lowland forests in southern Yunnan (Zhu 1997), in Hainan, China (Jiang and Lu 1991), in tropical lower montane forests on Mt. Kerinci, Sumatra (Ohsawa and Ozaki 1992), and Mt. Maquiling, Philippines (Brown 1919), are dominated by mesophyllous and notophyllous species. Tropical mid-montane forest on Mt. Maquiling is dominated by mesophylls and notophylls, while tropical upper-montane forests on Mt. Kerinci, Mt. Maquiling, and Mt. Jianfengling are mostly dominated by microphylls and notophylls. Subtropical/warm-temperate lowland forests in Japan are dominated by microphyllous species. The evergreen broad-leaved trees on Mt. Emei include notophyllous (48 %) accompanied by microphyllous (36 %) and mesophyllous (16 %) species (Table 2.4). The evergreen forest zone at 600–1,500 m has the most notophyllous species (54 %), corresponding to the northern tropical lower and upper montane forests on Mt. Jianfengling, Hainan Island in southern China. On Mt. Emei, the mixed forest zone at 1,500–2,500 m mostly has microphyllous species (48 %), corresponding to the tropical upper montane forest on Mt. Kerinci, Mt. Maquiling, and subtropical/warm-temperate forests in Japan. Accordingly, the leaf-size spectrum found on Mt. Emei as a whole represents an intermediate pattern between tropical and temperate montane forests, and the evergreen and mixed forest zones there exhibit affinities with lower- and upper-montane zones of tropical mountains, and also with both subtropical and warm temperate rain forests in East Asia.

2.4 Structure and Dynamics of Primary EBLFs

We use the term primary EBLFs to mean naturally regenerated EBLFs of native species, where there are no clearly visible indications of human activities and the ecological processes are not significantly disturbed. The vertical structures of EBLFs are multilayered. The canopy height of the primary forest generally ranges from 23 to 34 m, with umbrella-shaped or rounded crowns. From above the forest appears as an unending gently waving sea of green, broken only by occasional, taller "emergent" trees, which may reach over 38 m in height. The monsoon, humid and mid-montane EBLFs usually contain emergent, canopy, subcanopy, shrub, and understory layers, while the semi-humid EBLFs are lacking in subcanopy layer.

Figure 2.12 shows the DBH-class frequency distribution of the representative old-growth forests (with trees greater than 1 m DBH and taller than 25 m) of the mid-montane moist, monsoon, and semi-humid EBLFs in Yunnan. Most of the dominant canopy tree species of these forests have the sporadic type of size distribution frequency. The sporadic type, with several peaks for the intermediate size classes, indicates gap regeneration with a poor seedling/sapling bank, as in the dominant species in the mid-montane moist EBLF in Xujiaba in the Ailao Mountains, and the monsoon EBLF in Fadou, Xichou in southeastern Yunnan. In the semi-humid EBLF on Mt. Huafu of central Yunnan, *Cyclobalanopsis glaucoides*, *Castanopsis*

Table 2.4 Latitudinal comparison of leaf size spectra

Forest type	Leaf-size class								Locality	Source
	Lepto	Nano	Micro	Noto	Meso	Macro	Mega	Nsp		
Subtropical/warm-temperate	–	8	<u>**45**</u>	40	8	–	–	40	Mt. Kiyosumi, 35°12′ N, Japan	(2)
Subtropical/warm-temperate	–	9	<u>**45**</u>	35	9	2	–	194	Mt. Takakuma, 31°30′ N, Japan	(2)
Subtropical/warm-temperate	–	–	<u>**62**</u>	30	8	–	–	63	Mt. Yuwan, 28°20′ N, Japan	(2)
Mixed forest zone	–	–	<u>**48**</u>	37	15	–	–	40	Mt. Emei, 29°34.5′ N, China	(5)
Evergreen forest zone	–	–	29	<u>**54**</u>	17	–	–	58	Mt. Emei, 29°34.5′ N, China	(5)
Subtropical montane	–	–	36	<u>**48**</u>	16	–	–	88	Mt. Emei, 29°34.5′ N, China	(5)
Tropical upper montane	–	–	7	<u>**93**</u>	–	–	–	14	Mt. Jianfengling, 18°36.5′ N, China	(4)
Tropical upper montane	–	–	<u>**50**</u>		<u>**50***</u>	–	–	16	Mt. Maquiling, 14°08′ N, Philippines	(3)
Tropical upper montane	–	–	<u>**61**</u>	17	22	–	–	18	Mt. Kerinci, 1°43′ S, Sumatra	(2)
Tropical middle montane	–	–	6		<u>**87***</u>	7	–	70	Mt. Maquiling, 14°08′ N, Philippines	(3)
Tropical lower montane	–	–	2	<u>**80**</u>	18	–	–	40	Mt. Jianfengling, 18°36.5′ N, China	(4)
Tropical lower montane	–	–	4	23	<u>**86***</u>	10	–	92	Mt. Maquiling, 14°08′ N, Philippines	(3)
Tropical lower montane	–	2	20		<u>**49**</u>	7	–	61	Mt. Kerinci, 1°43′ S, Sumatra	(2)
Tropical lowland	–	–	16		<u>**74***</u>	9	1	145	South Yunnan, 21°33′ N, China	(1)

Data sources: (1) Zhu (1997), (2) Ohsawa and Ozaki (1992), (3) Brown (1919), (4) Jiang and Lu (1991), (5) Tang and Ohsawa (1999)

Underlined value indicates the most abundant class. Mesophyll values marked by an asterisk comprise the size class of 20.25–182.25 cm^2 (thus including notophyll)

Lepto leptophyll, *Nano* nanophyll, *Micro* microphyll, *Meso* mesophyll, *Macro* macrophyll, *Mega* megaphyll, *Nsp* number of species in the leaf-size class for evergreen broad-leaved trees

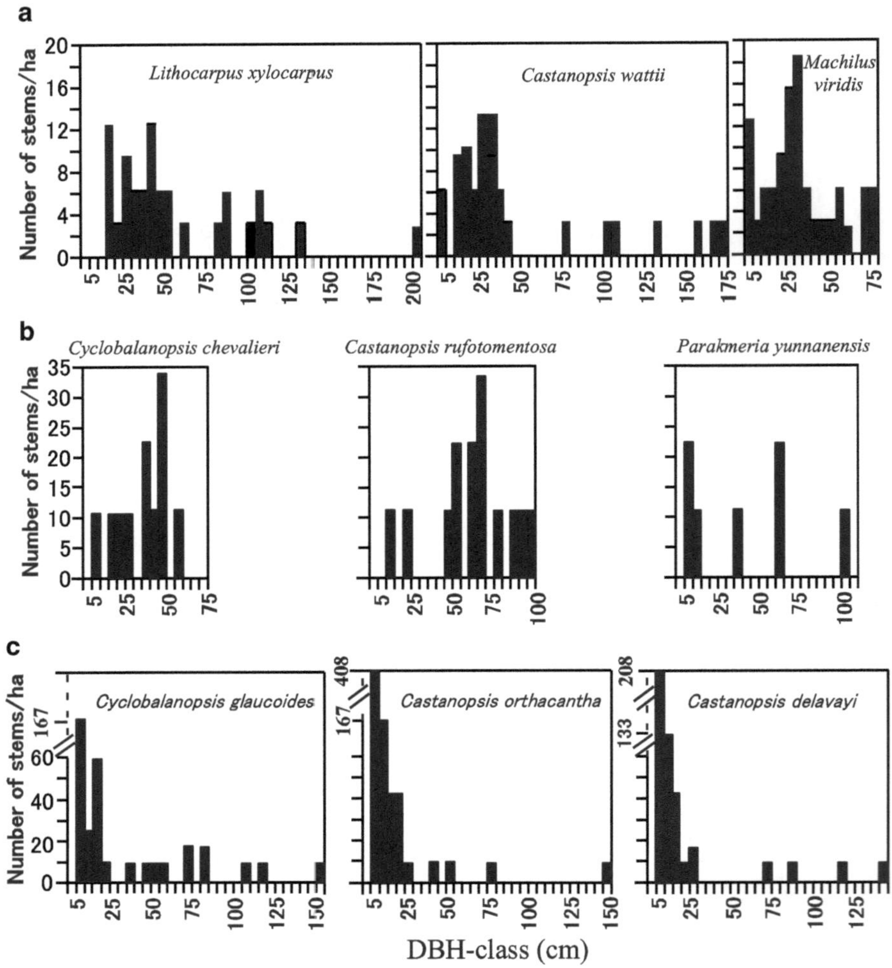

Fig. 2.12 DBH-class frequency distribution for some dominant canopy species of the representative old-growth forests in Yunnan (Data source: Tang and Ohsawa (2009)). (**a**) Subtropical mid-montane moist EBLF in Xujiaba, Mts. Ailao, Yunnan. (**b**) Subtropical monsoon EBLF in Fadou, Xichou, Wenshan, Yunnan. (**c**) Subtropical semi-humid EBLF on Mt. Huafu, Chuxiong, Yunnan

orthacantha and *Castanopsis delavayi* have a great number of stems below 20 cm DBH, most of them being sprouts from the base of trunks.

Forests are dynamic. Seedlings germinate, grow, and compete with each other and with larger trees. Some will survive for hundreds of years. Change is ever ongoing. Which species will be predominant in the future forest depends not only on climate and soils, but also on disturbances (e.g. creation of canopy gaps and forest edges). Creation of canopy gaps causes a series of dynamic events related to primary forest regeneration. Regeneration in canopy gaps is an important mecha-

Table 2.5 Gap regeneration behavior of some tree species in the EBLFs of Yunnan

Regeneration type	Forest type of subtropical/warm-temperate evergreen broad-leaved forests		
	Yunnan		
	MM	MS	SH
1	*Castanopsis wattii*	*Castanopsis fabri*	*Cyclobalanopsis glaucoides*
	Castanopsis echidnocarpa	*Castanopsis calathiformis*	*Castanopsis orthacantha*
	Castanopsis fleuryi	*Castanopsis hystrix*	*Lithocarpus dealbatus*
	Cyclobalanopsis lamellosa	*Machilus gamblei*	
	Lithocarpus hancei	*Machilus longipedicellata*	
	Lithocarpus pachyphylloides	*Lithocarpus fenestratus*	
	Lithocarpus pachyphyllus		
	Alcimandra cathcartii		
2	*Schima villosa*	*Acer davidii*	*Pistacia chinensis*
	Schima argentea	*Bretschneidera sinensis*	*Quercus aliena*
	Acer flabellatum	*Carpinus rupestris*	*Schima argentea*
	Tetracentron sinense	*Cornus wilsoniana*	*Celtis tetrandra*
		Fraxinus trifoliolata	*Alnus nepalensis*
		Schima wallichii	
3	*Hartia sinensis*	*Aucuba chlorascens*	*Ternstroemia gymnanthera*
	Skimmia arborescens	*Ilex longecaudata*	*Photinia glabra*
	Camellia yunnanensis	*Itea macrophylla*	*Pieris formosa*
	Symplocos dryophila	*Luculia intermedia*	*Vaccinium bracteatum*
	Eurya jintungensis		*Viburnum foetidum*
	Eurya nitida		
	Ilex fragilis		
4	*Lithocarpus xylocarpus*	*Castanopsis rufotomentosa*	*Castanopsis delavayi*
	Castanopsis remotidenticulata		*Cyclobalanopsis delavayi*
	Cyclobalanopsis myrsinifolia		

Partial data from Tang and Ohsawa (2009)

MM mid-montane moist EBLF, *MS* monsoon EBLF, *SH* semi-humid EBLF. Types of regeneration are: Type 1: canopy trees regenerate from saplings existing before gap formation (climax species); Type 2: canopy trees of pioneer species regenerate in gaps from saplings recruited after gap formation; Type 3: subcanopy tree and shrub species regenerate in gaps from saplings recruited before gap formation, but rarely reach the canopy layer; Type 4: trees cannot regenerate in gaps formed under the existing gap disturbance regime

nism maintaining the composition and structure of mature stands. Tang and Ohsawa (2009) applied Yamamoto's (1992) four types of gap regeneration behaviors for the Yunnan area (Table 2.5). Type 1: canopy trees regenerate from saplings existing before gap formation (climax species); Type 2: canopy trees of pioneer species regenerate in gaps from saplings recruited after gap formation; Type 3: subcanopy tree and shrub species regenerate in gaps from saplings recruited before gap formation, but rarely reach the canopy layer; Type 4: trees cannot regenerate in gaps formed under the existing gap disturbance regime. In the three types of subtropical EBLF of Yunnan, some tree species of *Castanopsis*, *Lithocarpus*, *Cyclobalanopsis*, *Machilus* and *Alcimandra* follow the Type 1 gap regeneration pattern, regenerating from saplings existing before canopy gap formation. The poor seedling/sapling bank in the mid-montane moist EBLF is apparently due to the presence of dense bamboo stands in the understory. However, *Lithocarpus xylocarpus, Castanopsis remotidenticulata* and *Cyclobalanopsis myrsinifolia* in the mid-montane moist EBLF, *Castanopsis fabri* and *Castanopsis rufotomentosa* in monsoon EBLF, in addition to *Castanopsis delavayi* and *Cyclobalanopsis delavayi* in semi-humid EBLF, showed Type 4 gap regeneration behavior i.e., very little or no regeneration. The regeneration mechanisms for these species have not been clarified. Though they lack productive seedlings/saplings, all the species of Fagaceae in the Yunnan study sites can sprout vigorously from injured trunks. Evergreen tree species of *Schima* and deciduous tree species of *Acer*, *Carpinus*, *Pistacia* etc. are pioneer species showing Type 2 gap regeneration behavior. Species of subcanopy trees and shrubs such as *Hartia*, *Camellia*, *Eurya*, *Symplocos*, *Lyonia*, and *Ternstroemia* exhibit Type 3 gap regeneration behavior.

Gaps may promote not only regeneration of some tree species but also growth of understory vegetation such as dwarf bamboos. The presence of dwarf bamboos (e.g. *Sinarundinaria nitida* and *Bambusa stenostachya*) in the understory is a characteristic of the mid-montane moist EBLFs and some humid EBLFs at middle altitudes in Yunnan and Sichuan, as in the Ailao Mountains and on Mt. Emei. The bamboos in these forests have increased to such a degree that the regeneration of canopy trees is impaired. As an overall trend, with increasing coverage of bamboos in the understory, the number of seedlings of woody species decreased ($r^2=0.086$–0.448, $n=96$, $P<0.01$) (Fig. 2.13) (Tang et al. 2007a). This suppression of tree regeneration would continue until the time of flowering and subsequent mass death of the bamboos, which occurs at intervals of approximately 55 years in *Sinarundinaria nitida* and 60 years in *Bambusa stenostachya*. Tree regeneration can be improved by the death of bamboos, which reduces understory competition and releases resources, especially available light (Nakashizuka 1987; Taylor et al. 1995; Holz and Veblen 2006). As an analogous instance, in Japanese temperate forests, the regeneration of *Fagus crenata*, a canopy dominant, seems to occur only after the bamboo *Sasa palmata* flowers and dies in the understory (Nakashizuka and Numata 1982). González et al. (2002) reported that both reorganization of advance regeneration and new seedling recruitment were observed in gaps after the synchronous mass flowering and death of the bamboo *Chusquea quila* in *Nothofagus*-dominated forest in south-central Chile. Taylor and Qin (1993) have reported that for the bamboo *Bashania fangiana* recovery of patches typically requires 15–20 years, with an initial dormancy of 0–5 years. In old-growth *Abies-Betula* forests of Sichuan, the period of potential high recruitment of the bamboo is probably less than 10 years after the bamboo died. Bamboo height and culm density in 10-year-old stands of *Fargesia scabrida* in the Min Mountains, Sichuan, seems great enough to reduce

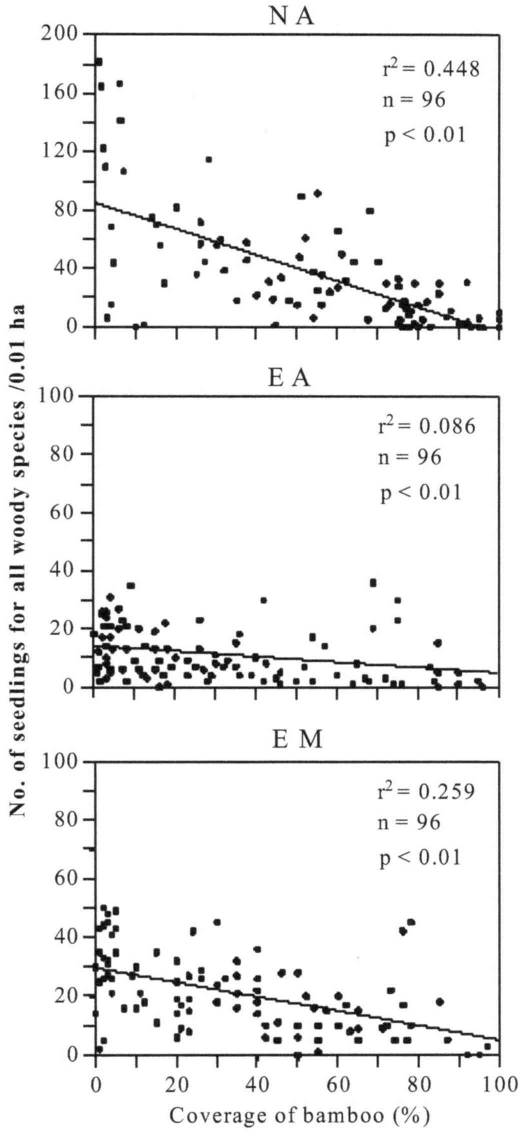

Fig. 2.13 Variations in numbers of seedlings of woody species as a whole according to coverage of bamboo in the understory of the old-growth forest on the Ailao Mountains, Yunnan and Mt. Emei, Sichuan. *NA* northern crest of the Ailao Mountains, *EA* eastern slope of the Ailao Mountains, *EM* Mt. Emei (Data source: Tang et al. (2007a))

tree seedling establishment (Taylor and Qin 1988). Thus if the forests are not likely to be disturbed by catastrophic events and human impacts, the tree regeneration (including the response of formerly suppressed seedlings and saplings as well as sprouting, and the response of new seedling recruitment) would be closely related to the life-cycle of the bamboos (Fig. 2.14).

In recent decades, the primary forests have been fragmented and forest edges have been created by human activities (e.g. shifting cultivation, logging and fuel-wood and wood charcoal production). Distinctive changes in vegetation structure and

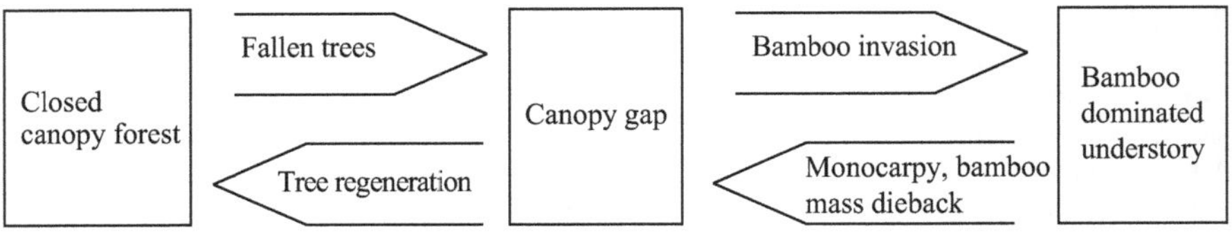

Fig. 2.14 Schematic diagram of tree species regeneration as related to the life cycle of bamboos in the forest understory

species composition along the forest-field margins can be observed. In the tropics, natural disturbances and human activities have been documented to increase liana abundance at forest edges, in tree-fall gaps, forest fragments and secondary forests (Schnitzer et al. 2000; Laurance et al. 2001; Letcher and Chazdon 2009). Lianas are often therefore regarded as light-demanding growth forms (Roeder et al. 2012). Li et al. (2010a) studied regeneration patterns of primary forest species across forest-field gradients in the subtropical Ailao Mountains. In the mid-montane moist EBLF area, the stem density of primary species is higher at distances below 50 m from the forest edge into the forest exterior, and dominant canopy trees *Lithocarpus xylocarpus*, *L. jingdongensis*, and *Castanopsis wattii* reproduce only at distances of less than 50 m. This may be because the edges have better radiation, moisture, and temperature conditions for the establishment and survival of seedlings (Jose et al. 1996; Cadenasso and Pickett 2001). The buffer zone supporting an abundance of thorny lianas, thorns, spines, and prickles provides physical defensive protection against browsing herbivores, thereby facilitating forest regeneration (Fig. 2.15) (Li et al. 2010a).

2.5 Secondary Succession

2.5.1 Successional Trends

Secondary succession involves a process of species turnover and a replacement of one community by another after a major disturbance, involving changes in floristic composition, species diversity, biomass, community structure and function, and soil properties. Secondary succession is a process easily observed, but difficult to quantify (Blatt et al. 2005). Certainly there are relationships among forest characteristics (in terms of ages, structure and species composition), disturbances and previous land uses. To understand the subtropical EBLF ecosystem, it is essential to account for responses to disturbances and the mechanisms of succession.

The vegetation of SW China has been under continuous change because of anthropogenic and natural disturbances. During the twentieth century, most primary EBLFs at lower altitudes were converted into agricultural fields or pastures. The habitats were severely degraded. Since 1985, economic and social development have resulted in the migration of a large proportion of the rural population to the

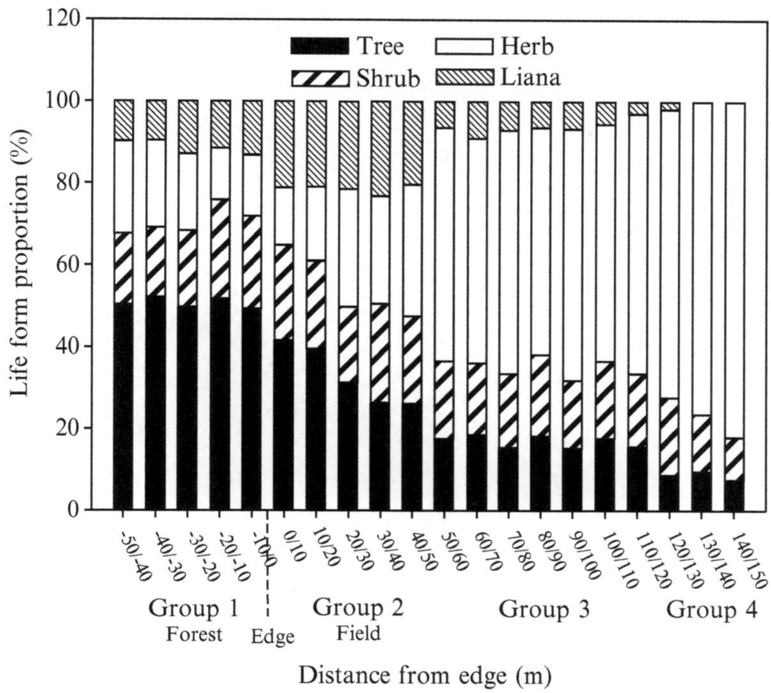

Fig. 2.15 Changes in life form proportion across the forest-field gradient. The numbers along the X-axis indicate the distance running from the forest interior (−50 m) through the edge (0 m) to the exterior (i.e. field: 150 m), perpendicular to that edge. Major canopy trees in Group 1 (−50 to 0 m) are evergreen broad-leaved species *Lithocarpus xylocarpus*, *Castanopsis wattii*, and *Lithocarpus jingdongensis*. Group 2 (0–50 m) is dominated by *Alnus nepalensis* and *Lyonia ovalifolia*. The dominant species in Group 3 (50–120 m) are *Lyonia ovalifolia*, *Rhamnus heterophylla*, and *Pinus armandii*. Finally, Group 4 (120–150 m) is dominated by *Lyonia ovalifolia*, *Hypericum uralum*, and the herb *Eupatorium adenophorum* (Data source: Li et al. (2010a))

cities. In addition, the Chinese government has implemented the Natural Forest Conservation Program (NFCP) and the Grain to Green Program (GTGP) between the middle 1980s and late 1990s. Most of the degraded agricultural fields and pastures were no longer productive and neither grazed nor cultivated. Once abandoned, they were quickly invaded and taken over by light-demanding, fast-growing species. For example, *Pinus yunnanensis* establishes itself as a pioneer species in the abandoned areas of Yunnan, western Sichuan (e.g. Xichang), and western Guizhou, while *Pinus massoniana* appears as a pioneer in the Sichuan Basin and eastern Guizhou. Secondary EBLFs ("secondary forests" here referring to naturally regenerated forests) are abundant after large-scale destruction of primary forests in the subtropical area (Fig. 2.16a–c).

A great example of this process can be seen in the succession in the Kunming area of central Yunnan, where because of the history of land uses, the landscapes are marked by dynamic habitats composed of mosaics of diverse stands at varying successional stages. Near Kunming, the natural zonal vegetation consists of the subtropical semi-humid EBLF dominated either by *Cyclobalanopsis glaucoides*,

Fig. 2.16 (**a**) *Alnus nepalensis* invaded the abandoned farmland in Luquan, central Yunnan. (**b**) A secondary forest utilized as firewood supply 10 years ago in Luquan, central Yunnan. (**c**) A secondary forest after clear-logging for timber 15 years ago in Tuanjiexiang, central Yunnan

Cyclobalanopsis delavayi, *Castanopsis delavayi*, or *Castanopsis orthacantha*. Over time, the original vegetation was destroyed in most areas by logging and agricultural development. Since 1985, originally forested mountainous areas have been allowed to return to their natural state. The resulting secondary forests have been in the natural process of plant succession potentially in the direction of semi-humid EBLF. Mature natural forests exist now only in patches, almost exclusively in sites surrounding temples.

Space-for-time substitution can yield an understanding of the directions and pathways of the succession taking place. A CCA ordination (canonical correspondence analysis) shows the relationship of the sampling plots and the environmental variables (soil chemical properties and degrees of naturalness DN) for Luquan and Qiongzhusi in the Kunming area (Fig. 2.17a, b) (Tang et al. 2010a). In Luquan, the Monte-Carlo tests for both the eigenvalues for each axis ($p=0.01$ for axis 1 and $p=0.03$ for axis 2) and the species–environment correlations ($p=0.02$ for axis 1 and $p=0.01$ for axis 2) are highly significant (Fig. 2.17a). The second axis is closely correlated with available N (r=0.62), total P (r=0.54) and DN (r=−0.51). The suc-

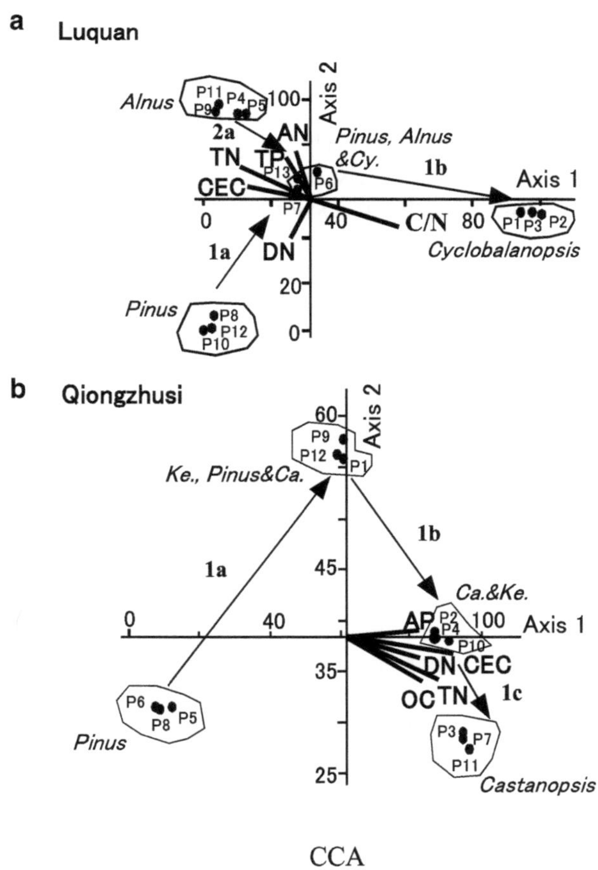

Fig. 2.17 CCA ordination for Luquan (**a**) and Qiongzhusi (**b**) in central Yunnan. *P* plot, *Cy. Cyclobalanopsis*, *Ke. Keteleeria*, *Ca. Castanopsis*, *DN* degree of naturalness, *AN* available N, *TP* total P, *TN* total N, *CEC* cation exchange capacity, *C/N* organic C/total N, *AP* available P, *OC* organic C (Data source: Tang et al. (2010a))

cessional pathway has been from the pioneer *Pinus yunnanensis* community and the deciduous broad-leaved *Alnus nepalensis* community to the seral mixed community of *Alnus nepalensis* and *Pinus yunnanensis*, evergreen broad-leaved *Cyclobalanopsis delavayi* and *Lithocarpus confinis*, and finally to the late-successional evergreen broad-leaved *Cyclobalanopsis delavayi* community (shown by arrows: from 1a to 1b; from 2a to 1b).

In Qiongzhusi, the Monte-Carlo tests for the eigenvalue are significant for axis 1 ($p=0.01$), but not for axis 2 ($p=0.34$) (Fig. 2.17b). The tests for the species–environment correlations for each axis are significant ($p=0.02$ for axis 1 and $p=0.01$ for axis 2). The first ordination axis is strongly correlated with CEC (r=0.68), total N (r=0.66), organic C (r=0.63), DN (r=0.52), available P (r=0.47). In Fig. 2.17b, axis 1 of the CCA shows that the pathway is from the coniferous *Pinus yunnanensis* community to the early seral mixed coniferous and evergreen broad-leaved community, then to a late seral mixed evergreen broad-leaved and coniferous community, finally to the evergreen broad-leaved *Castanopsis orthacantha* community (shown by arrows: from 1a to 1b, then to 1c).

Figure 2.18a, b shows the proportion of basal area at breast height of different tree life forms in each successional phase. In the seral phase in Luquan and Qiongzhusi the increasing relative basal area of evergreen broad-leaved trees shows their replacement of pioneer species to be in progress. The evergreen broad-leaved tree (EBLT) absolutely dominates (>92 % in RBA) in the pre-mature phase, while the coniferous tree (CT) and deciduous broad-leaved tree (DBLT) have nearly disappeared (<7 %). The successional pathway in Luquan can be clearly seen in the photograph in Fig. 2.19. The species composition of each successional phase in Luquan and Qiongzhusi is shown in Tables 2.6 and 2.7.

Most early colonists begin reproduction at a young age and have a fast growth (Grime 2001). By contrast, late-successional species can arrive early or late, and have a slow growth rate, are more shade-tolerant, and live longer than the colonists (Bazzaz 1979). Successful forest regeneration depends upon a range of site and species attributes, including the traits of plants and the site's proximity to seed sources (Guariguata et al. 1995; Zimmerman et al. 2000) or the resprouting ability of the species (Bond and Midgley 2001, 2003; Vesk and Westoby 2004). Stand development has proceeded through the pioneer phase for at least 55 years in Luquan and Qiongzhusi. This phase is characterized in the two study sites by large numbers of pioneer trees (*Pinus*, *Alnus* and *Keteleeria*) and many snags as weaker ones die off. The invasion of *Alnus nepalensis* in abandoned farmland in Luquan is similar to what happens in southeastern Tibet (Sun et al. 1996) and the Himalayas (Singh and Sing 1987). Seral stands are characterized by shade-intolerant pioneer trees and shade-tolerant late-successional tree species as well as shade-tolerant shrubs. *Keteleeria evelyniana* in Qiongzhusi has unique traits in that it is shade-intolerant, has a longer life span and slower growth rate than *Pinus*. Similar

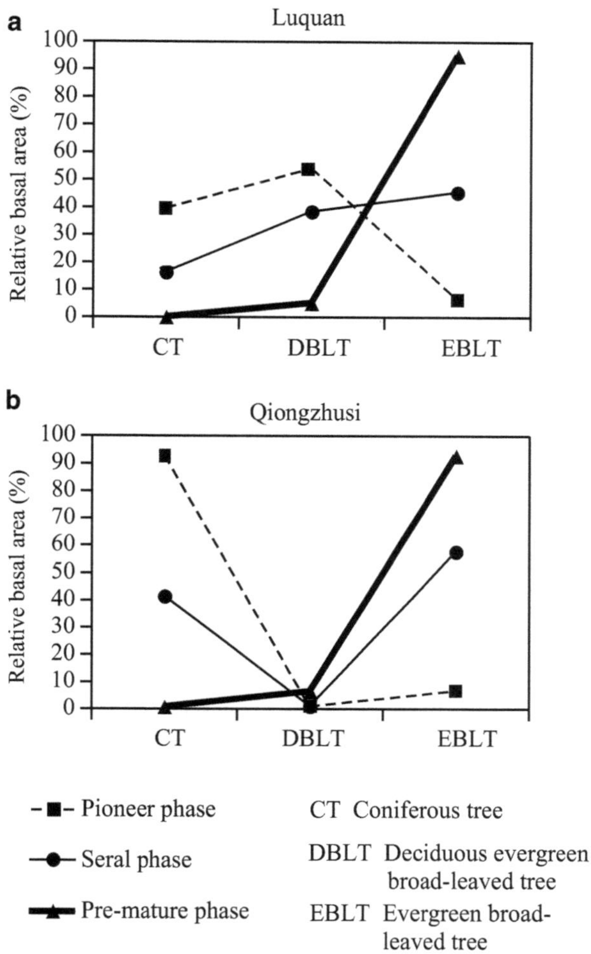

Fig. 2.18 The proportion of tree life forms in each successional phase in Luquan (**a**) and Qiongzhusi (**b**). *CT* coniferous trees, *DBLT* deciduous broad-leaved trees, *EBLT* evergreen broadleaved trees (Data source: Tang et al. (2010a))

traits characterize *Keteleeria davidiana* var. *formosana* in Taiwan, which is a shade-intolerant species that is usually found in open places in association with some evergreen broad-leaved trees such as species of *Machilus* and *Castanopsis* (Lin et al. 2003). In Yunnan, late-successional tree species of *Cyclobalanopsis* and *Castanopsis* have slow growth rates and are able to germinate and establish in some degree of shade. At some point later in the successional process they will become the dominant overstory tree species. The ordination (Fig. 2.17a, b) suggests that the initial successional mechanism after human disturbance in Qiongzhusi and Luquan, in the Kunming area, is the invasion of *Pinus*, *Alnus* and

Fig. 2.19 Secondary succession pathway in Luquan, central Yunnan

Table 2.6 Woody species composition of each community in Luquan

Altitude (m)	2,370–2,460			2,340–2,401				2,350			2,540–2,580		
Community types	Co.			De.				Co. De. & Ev.			Ev.		
Plot	P 10	P 12	P 8	P 11	P 9	P 4	P 5	P 6	P 7	P 13	P 2	P 3	P 1
Relative basal area (%)	RBA	RBA	RBA	RBA	RBA	RBA	RBA	RBA	RBA	RBA	RBA	RBA	RBA
Pinus yunnanensis	**80.5**	**78.43**	**73.6**	2.39	3.75			**19.3**	**16.04**	**13.2**			0.09
Alnus nepalensis	15	14.45	17.1	**90.77**	**85**	**86.9**	**88**	**32.5**	**36.99**	**42.51**	0.35		0.57
Cyclobalanopsis delavayi	2.43	4.48	4			3.19	7.2	**26.1**	**21.54**	**22.36**	**76.55**	**88.14**	**97.4**
Lyonia ovalifolia	1.18	1	0.29			5.29		0.55	2.48	0.32	0.69	0.21	0.79
Berberis ferdinandi	0.46	1.64		0.87	2.8					1.72			
Rosa longicuspis	0.45			1.82	3.11								
Lithocarpus confinis			5			4.58	4.2	**20.7**	**22.39**	**19**			
Cotoneaster pannosa				4.16	5.38								
Rhododendron delavayi						0.05		0.07	0.52	0.05		0.26	
Ternstroemia gymnanthera							0.2	0.36	0.048	0.47			
Lithocarpus dealbatus								0.43		0.37	9.69		0.01
Quercus gilliana											12.49		
Rhododendron decorum											0.23	1.85	1.11
Castanopsis orthacantha												9.53	
Elaeagnus pungens												0.02	

Data source: Tang et al. (2010a)
The dominants are indicated by boldface
Co. coniferous, *De.* deciduous broad-leaved, *Ev.* evergreen broad-leaved, *P* plot

Table 2.7 Woody species composition of each community in Qiongzhusi

Altitude (m)	2,170–2,200			2,230–2,244			2,155–2,290			2,070–2,100		
Community type	Co.			Co. & Ev.			Ev. & Co.			Ev.		
Plot	P 5	P 6	P 8	P 1	P 9	P 12	P 4	P 2	P 10	P 3	P 7	P 11
Relative basal area (%)	RBA	RBA	RBA	RBA	RBA	RBA	RBA	RBA	RBA	RBA	RBA	RBA
Pinus yunnanensis	**89.4**	**95.84**	**93**	**21.9**	**20.2**	**21.5**				2.2		
Castanopsis orthacantha	6.14	1.67	3.51	**26.78**	**25.87**	**25.76**	**80.63**	**79.8**	**79**	**89.4**	**82.82**	**86.8**
Lithocarpus dealbatus	3.56	1.59	1.59	4.31	3.1	5			0.77	1.43	6.92	2.94
Ternstroemia gymnanthera	0.47		0.48			0.02	0.008	0.01	0.01	0.02	0.01	0.004
Lyonia ovalifolia	0.24	0.64	0.2	1.37	2.08	1.68				0.41		
Rhododendron spiciferum	0.16							0.09	0.07			
Myrica nana	0.02			0.01	0.03	0.1			6.6			
Michelia yunnanensis		0.27	0.23	0.05	0.01	0.03					0.002	0.005
Keteleeria evelyniana				**45.42**	**48.02**	**45.43**	16.82	17.8	11.2			
Cyclobalanopsis glaucoides				0.33	0.53	0.32						
Cupressus duclouxiana				0.12	0.18	0.02						
Prunus pseudocerasus				0.1	0.02	0.1						
Ilex polyneura				0.006		0.006	2.06	2.06	2.12		0.01	0.01
Eurya nitida				0.002	0.01	0.002	0.18	0.07	0.07		0.06	0.04
Vaccinium sprengelii							0.14	0.11	0.09	0.15		
Dichotomanthus tristaniaecarpa							0.1				3.4	3.24

Rhododendron ciliicalyx							0.06	0.08	0.07			
Camellia yunnanensis								0.02	0.03		0.01	0.03
Alnus nepalensis										3.72		
Schima argentea										2.2	0.012	0.01
Vaccinium bracteatum										0.39		0.002
Cornus officinalis										0.08		
Quercus aliena											4.6	3.8
Lindera communis											2.11	2
Elaeagnus pungens											0.12	0.23
Photinia glabra											0.1	0.9
Craibiodendron henryi											0.01	0.01
Myrsine africana												0.01
Pieris formosa												0.002

Data source: Tang et al. (2010a)
The dominants are indicated by boldface
Co. coniferous, *Ev.* evergreen broad-leaved, *P* plot

Keteleeria, which possess higher vitality and grow faster than associated evergreen broad-leaved species. Stands of *Pinus*, *Alnus*, and *Keteleeria*, with simple structure and low coverage, provide better conditions for invasion by *Cyclobalanopsis* and *Castanopsis* than does abandoned bare land, because the pioneer trees modify the light and the edaphic conditions of the sites, leading to an environment more favorable to the mesophytic late-successional species. As succession advances, the mixed coniferous and broad-leaved forests are formed. Then, as shade-intolerant pioneer *Pinus*, *Alnus* and *Keteleeria* disappear, and as shade-tolerant tree species of *Cyclobalanopsis* and *Castanopsis* increase their dominance, finally the communities tend toward the late-successional stage.

Regeneration and growth of vegetation are affected by a variety of conditions of the soil. During succession, organic matter and microbial biomass in the soil undergo change as the forest ages (Jia et al. 2005). In Jinyunshan in Chongqing, the soil fertility is improved with succession, and that can vary from place to place. Shifts in soil fertility improvement can affect the adaptability of species to the environment and the competition among species (Song et al. 2005). In the Kunming area, the result of the CCA ordination confirms that diverse plant communities are distinguished by soils containing different nutrient levels at various successional stages. Pine litter contains a large quantity of recalcitrant compounds that resist decomposition. As a result, soil nutrients are low under the pine forests in the study sites. The study shows that the nutrients tend to increase as succession proceeds from the *Pinus* community in Qiongzhusi and Luquan, except for the pioneer *Alnus* community of Luquan. Certain fungi living on the roots of *Alnus nepalensis* have the function of nitrification, which can enrich soils (Sharma and Ambasht 1984). The contents of total and available nitrogen are higher in the pioneer *Alnus* community than in the other forests of Luquan. Symbiotic N-fixers can directly enhance growth of associated species in succession (del Moreal and Wood 1993; Binkley 1992). Pioneer *Alnus* improves soil conditions, supporting the regeneration of late-successional species.

Here I integrate the data on Luquan and Qiongzhusi with results from two other study sites, i.e. Xiashan from Tang et al. (2010a) and Shilin from Tang et al. (2014), to provide a general model representing the pathways of secondary succession in the semi-humid EBLF of central Yunnan (Fig. 2.20). Both the severity (magnitude) and the intensity (frequency) of disturbance influence the process of succession in that region. In the pioneer and seral (middle) phases, cutting for fuel produces many disturbances, leading the broad-leaved species to develop many sprouts and so producing coppices. Sometimes the degree of intensity of disturbance is high, but with many stumps remaining which are then being allowed to resprout. The late successional phase in various plant communities is characterized by different late-successional species, *Cyclobalanopsis glaucoides, Cyclobalanopsis delavayi, Castanopsis delavayi*, and *Castanopsis orthacantha*. The result depends on the availability of seed sources, land use, and forest-management history. In the

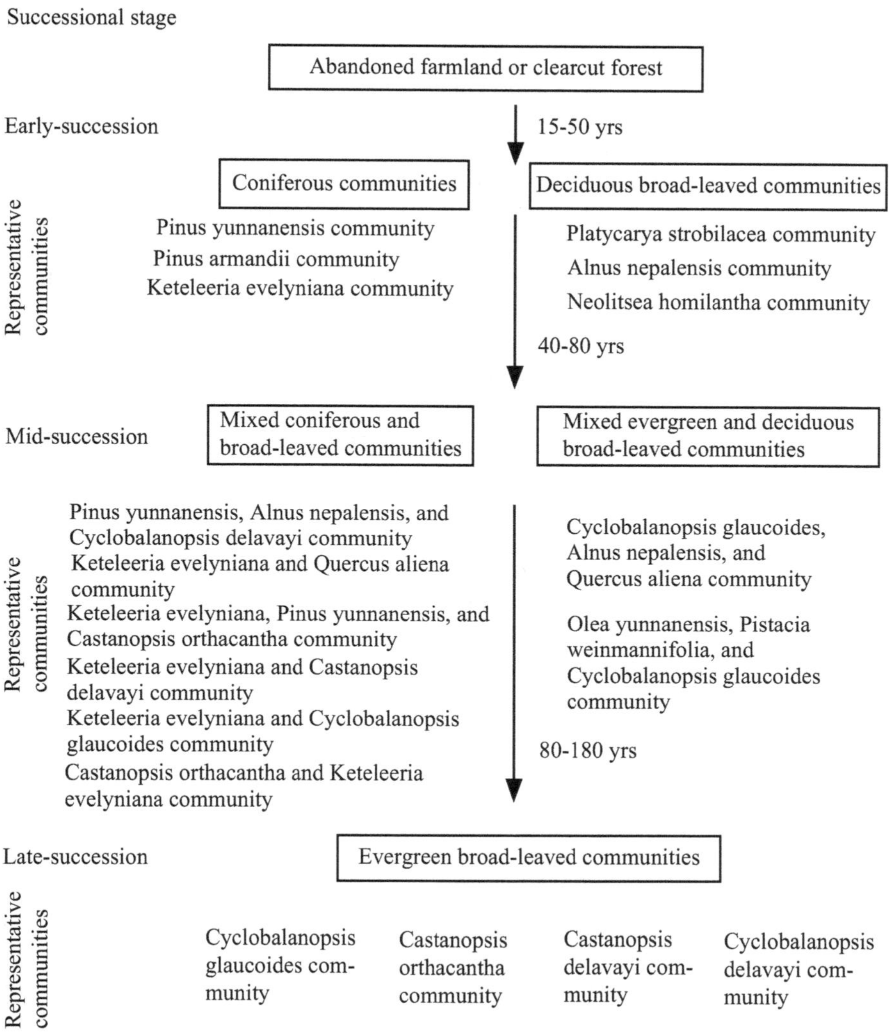

Fig. 2.20 A succession model of the semi-humid evergreen broad-leaved forest in central Yunnan (Data sources: Tang et al. (2010a, 2014)

Kunming area late-successional EBLFs are found only surrounding temples, where no distinctive human disturbance has occurred within the past 40 years.

At the low altitudes (550–650 m) of Mt. Emei, Sichuan, *Pinus massoniana* invades, often as a pioneer, in sites severely disturbed by human activities. As succession progresses, a mixed *Pinus massoniana*, *Mallotus tenuifolus* and *Castanopsis hystrix* forest forms, and then is succeeded by the EBLF dominated by *Castanopsis hystrix*.

In Jinyunshan in Chongqing, the secondary succession is from the *Pinus massoniana* stand to the coniferous and broad-leaved stage (e.g., *Pinus massoniana*,

Schima argentea, and *Castanopsis fargesii*), then to the EBLF dominated by *Castanopsis fargesii* (Guo et al. 1999).

In northwestern Guizhou, a common process of the secondary succession leads from the pioneer species *Pinus yunnanensis* and *Pinus massoniana* or *Populus adenopoda* to a mixed coniferous and broad-leaved forest (e.g. *Pinus yunnanensis, Pinus massoniana, Quercus variabilis, Schima wallichii, Castanopsis delavayi, Cyclobalanopsis glaucoides*), then to the semi-humid EBLF dominated by *Castanopsis delavayi* and/or *Cyclobalanopsis glaucoides* (Zhou 1992).

In SW China, after the EBLFs have been destroyed, sometimes other light-demanding deciduous broad-leaved trees (e.g. *Populus*, *Betula*, *Alnus* and *Quercus*), or bamboos (e.g. *Phyllostachys pubescens*, *Chimonobambusa quadrangularis* and *Pleioblastus amarus*) also occur as pioneer communities.

The general trend of the secondary succession in plant communities in the areas of SW China is comparable to those of Tiantongshan in Zhejiang, eastern China (Ding and Song 1997; Wang et al. 2007), and Dinghushan in Guangdong, southern China (Peng and Fang 1994; Zhou et al. 2005). The process advances from the pioneer species (e.g., *Pinus massoniana* in Tiantongshan and Dinghushan) to the coniferous and broad-leaved forest (e.g., *Pinus massoniana*, *Schima argentea*, and *Castanopsis fargesii* in Tiantongshan, *Pinus massoniana*, *Schima superba*, and *Castanopsis chinensis* in Dinghushan), then to the evergreen broad-leaved forest (e.g., dominant *Castanopsis fargesii* in Tiantongshan; *Castanopsis chinensis* and *Cryptocarya chinensis* in Dinghushan).

2.5.2 Species Diversity along the Successional Gradient

For forest ecosystems, various temporal trends in diversity have been observed during succession, including increases, decreases and one or more peaks with passing time (e.g., Monk 1967; Habeck 1968; Auclair and Goff 1971; Peet 1978; Halpern and Spies 1995; Turner et al. 1997). Bruelheide et al. (2013) reported that species spread through a continuous random immigration over time during secondary forest succession in subtropical eastern China. Relationships between diversity and succession are complex, and should be seen as system-specific. In Yunnan as a whole, *Pinus yunnanensis, Keteleeria evelyniana, Alnus nepalensis, Populus bonatti, Sapium chihsinianum, Carpinus mobeigiana, Platycarya strobilacea* and *Neolitsea homilantha* appear as typical light-demanding pioneer species. At the early successional stage, one of the pioneer species dominates (or several dominate) the plant communities. As succession progresses, some intermediate light-demanding plants, such as *Lyonia ovalifolia*, *Rhododendron delavayi* and *Lindera elongata*, and some shade-tolerant species, mainly of Fagaceae, Lauraceae, and Magnoliaceae, mix with the pioneer species. Gradually, as the forest canopy develops more closure, the short-lived light-demanding pioneer species are replaced by late successional shade-tolerant evergreen broad-leaved species such as species of *Castanopsis*, *Cyclobalanopsis*, *Lithocarpus*, *Machilus*, *Manglietia* and *Michelia*, although some long-lived pioneer species, such as *Schima argentea* and *Schima wallichii*, are often found in mature EBLFs.

Species diversity along the successional gradient is summarized in Table 2.8, based on 98 sampled plots (20×20 m each), partial data from Tang et al. (2013), secondary forests at the early (younger secondary forests, 34 plots), middle (older secondary forests, 34 plots), and late (natural mature forests, 30 plots) successional stages of the EBLFs of Yunnan. As shown by the species richness value (R) and the Simpson's D, species diversity is significantly higher at the middle stage than earlier. In the middle stage, the mixed stands of tree species, characterized by varying attributes including fast growth, slow growth, shade-intolerance and shade-tolerance, contribute to the high species diversity. As the forest further develops, species diversity in the late stage tends to decrease except in the mid-montane moist EBLF in Yunnan; however, differences between the middle and late stages are not significant. As succession proceeds, the species diversity comes to resemble that of the natural mature forests. The secondary forests, with spatially and temporally dynamic patches of vegetation showing different degrees of disturbance along the successional gradient, reveal the progress of species diversity toward the original natural patterns. Such are the effects of disturbance and succession on the plant species diversity patterns in Yunnan.

2.5.3 Soil Seed Banks

The soil seed bank is defined as the total number or density of viable seeds stored in the soil in a given period. It connects the past, present and future above-ground plant populations and the community dynamics of differing life histories and varying seed attributes. As a case study on seed banks in the Ailao Mountains, we have elucidated changes in the distribution of life forms according to differing ecological traits such as light requirements at various successional stages (Li et al. 2010b). In the overstory of the shrubland, the early pioneer stage of the secondary succession is dominated by light-demanding *Lyonia ovalifolia* and *Hypericum uralum*, with participation of *Populus bonatii*, *Ilex corallina*, *Ternstroemia gymnanthera*, and *Vaccinium duclouxii*. The *Populus bonatii* secondary forest, representing the late pioneer stage of the secondary succession, is dominated by light-demanding *Populus bonatii*, intermediate *Vaccinium duclouxii*, and shade-tolerant *Lithocarpus chintungensis* along with 23 other species. The *Lithocarpus* regrowth forest, representing the middle stage of the secondary succession, has 26 species. This regrowth forest is dominated by shade-tolerant *Lithocarpus chintungensis*. In the primary forest, the late stage of the succession, 28 species are found. It is dominated by the shade-tolerant trees *Lithocarpus chintungensis*, *Lithocarpus xylocarpus*, *Machilus bombycina*, and *Manglietia insignis*. In primary forest, more shade-tolerant (note: "shade-tolerant" is meant to clarify "intermediate" in Li et al. 2010b) species are found, including *Illicium macranthum* and *Machilus yunnanensis* among others. In short, the relative basal areas (RBA) of shade-tolerant species increase while the

Table 2.8 Changes in levels of diversity (mean ± SD) of woody species (DBH ≥ 1 cm) at different successional stages in Yunnan

Region	Forest types/zones	Area	Diversity	Forests during succession		
				Early stage	Middle stage	Late stage
Yunnan	Semi-humid EBLF	Luquan, Qiongzhusi, Xishan, Chuxiong	R	8.01 ± 3.59^{a} (18)	13.16 ± 2.14^{b} (18)	11.39 ± 1.94^{b} (18)
			D	0.58 ± 0.04^{a} (18)	0.82 ± 0.06^{b} (18)	0.68 ± 0.03^{a} (18)
	Monsoon EBLF	Xiaoqiaogou in Wenshan	R	11.63 ± 1.27^{a} (9)	29.10 ± 5.99^{b} (7)	27.80 ± 3.70^{b} (6)
			D	0.84 ± 0.04^{a} (9)	0.96 ± 0.11^{b} (7)	0.88 ± 0.06^{a,b} (6)
	Mid-montane moist EBLF	Eastern Slope of Mts. Ailao	R	10.27 ± 3.69^{a} (7)	21.16 ± 6.73^{b} (9)	24.61 ± 5.73^{b} (6)
			D	0.70 ± 0.03^{a} (7)	0.88 ± 0.05^{b} (9)	0.91 ± 0.02^{b} (6)

Partial data from Tang et al. (2013)
Forests at different successional stages bearing different letters differ significantly ($p < 0.05$)
R species richness, *D* Simpson's D

light-demanding species decrease in the overstory as succession progresses from the shrubland to the *Populus bonatii* secondary forest, then to the *Lithocarpus* regrowth forest, and finally to the primary forest (Fig. 2.21a).

In the understory, the shrubland is dominated by light-demanding species, whereas intermediate and shade-tolerant species are rare. As succession progresses, the number of shade-tolerant seedlings of shrubs and trees increases, and finally no seedlings of light-demanding plants are found in primary forest (Fig. 2.21b). The number of germinable tree seeds increases throughout succession, and the number of shrubs decreases with the increasing age of the plant communities (Fig. 2.21c). Light-demanding tree species contribute the highest proportion of germinable seeds in the primary forest. Light-demanding shrub species dominated the germinable seed bank in the shrubland, the *Populus bonatii* secondary forest, and the *Lithocarpus* regrowth forest, but not the primary forest (Fig. 2.22).

Higher seed density and species richness are recorded in the Ailao Mountains as compared to similar studies conducted in East Asian forests. For instance, Naka and Yoda (1984) found seeds of a single species (*Eurya japonica*) to be abundant in stands of varying ages in the Kasugayama Forest Reserve (Japan), while we found differing species to dominate in stands of different ages. Some studies (e.g. Roberts and Vankat 1991; Bakker et al. 1996) have shown that the species richness of a seed bank often decreases during secondary succession. This process is usually associated with a decline in the density of seeds (Kalamees and Zobel 1998). The higher seed bank densities found in earlier successional stages of forests may be due to the higher fecundity, earlier reproduction, and longer dormancy associated with the higher seed longevity of pioneer species that predominate in these sites (Janzen and Vázquez-Yanes 1990). The soil seed bank of pioneer species plays an important role in the post-disturbance regeneration and secondary succession of the tropical rain forest (Garwood 1989; Bazzaz 1991). In our subtropical study site, the density of germinated seedlings and the species richness in the soil seed banks also decrease with the increasing age of plant communities from the shrubland to the primary forest. The dominant canopy tree species of *Castanopsis, Lithocarpus, Machillus*, and *Schima* are absent from the seed bank, while a relatively high density of the somewhat light-demanding *Vaccinium duclouxii* and *Lyonia ovalifolia* is found in the soil seed bank of the primary forest. We have found the number of tree species to be relatively low in the soil seed banks of early successional communities of the shrubland and the *Populus bonatii* secondary forest (Li et al. 2010b). This may have several causes. Tree seeds may not be dormant. Trees usually have large seeds that cannot be dispersed easily over long distances. Thus, inputs of seed rain from adjacent natural forests to the soil seed bank of the early successional stage may be minimal (Hermy et al. 1999; Mengistua et al. 2005). And seeds may be lost through predation (Teketay 1998).

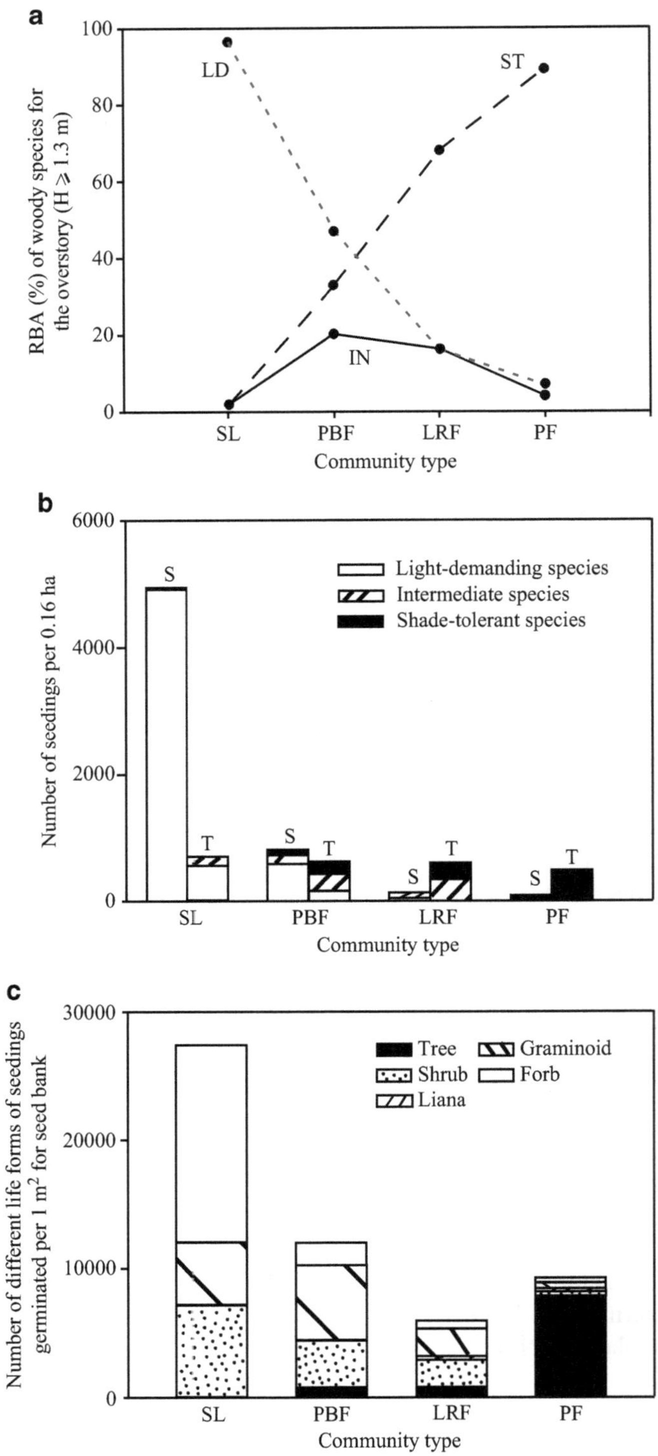

Fig. 2.21 Species with different ecological traits in the overstory (≥1.3 m in height) of the forest (**a**). Number of seedlings of woody species in the understory (**b**). Number of seedlings of different life forms germinated in soil seed bank (**c**). *SL* Shrubland; *PBF Populus bonatii* secondary forest; *LRF Lithocarpus* regrowth forest; *PF* primary forest; *LD* light-demanding; *IN* intermediate light demand; *ST* shade-tolerant; *S* shrub; *T* tree (Data source: Li et al. (2010b))

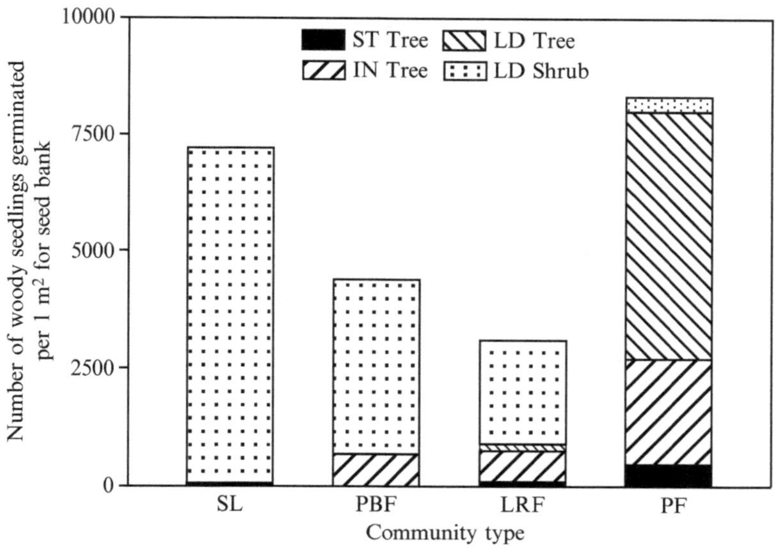

Fig. 2.22 Woody species with different ecological traits germinated in the soil seed bank. *SL* shrubland, *PBF Populus bonatii* secondary forest, *LRF Lithocarpus* regrowth forest, *PF* primary forest, *ST* shade-tolerant, *IN* intermediate light demand, *LD* light-demanding (Data source: Li et al. (2010b))

2.6 Threats of Habitat Fragmentation and Degradation

The most important factors in the loss of plant diversity are habitat fragmentation and degradation in the subtropical areas of SW China (Tang 2010). In Yunnan Province, 90 plant species were recorded in the national priority protection list/inventory (Li et al. 2003). Among them, 76 species exist in the subtropical EBLF zone. Three species are extinct, 45 are ranked as critically endangered, 27 as endangered, one as vulnerable. We find that the extinction of the species *Kingdonia uniflora*, *Ormosia henryi*, and *Diploknema yunnanensis* is 70 % due to habitat loss or degradation, and 30 % to overexploitation, for example collecting *Kingdonia uniflora* as an herb medicine. The rankings as critically endangered, endangered, and vulnerable are each over 65 % due to habitat loss/degradation. It is certain that the current status of the 76 species is largely attributable to habitat loss or degradation (Fig. 2.23).

Many genera of seed plants endemic to China are found in the boundary area of southeastern Yunnan, Guangxi and Viet Nam. Limestone outcrops in the area occur in a variety of configurations, from scattered small patches to extensive landscapes. The karst areas have an exceptionally high floristic richness and degree of endemism (Shui and Cheng 2003). However, intensive agricultural development began here in the 1950s and has resulted in widespread fragmentation of forests, leaving

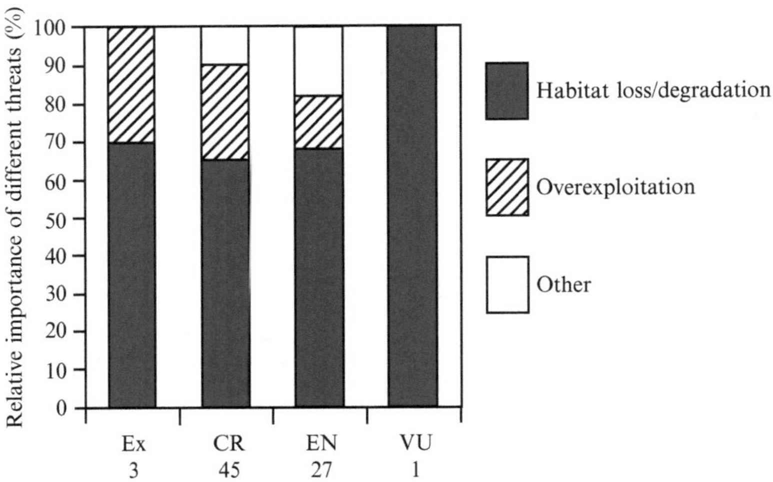

Fig. 2.23 Relative importance of different threats to apparently extinct (*EX*), critically endangered (*CR*), endangered (*EN*), vulnerable (*VU*) species; numbers below each column give numbers of species in each threat category (Data source: Tang (2010))

many small patches. Many sites were severely logged or partially cut 25–80 years ago as development proceeded. The extensive removal of native vegetation has resulted in fragmentation and small-grained patchiness in the remaining vegetation across the landscape. China, and Yunnan in particular, harbor a high number of Magnoliaceae species (Fig. 2.24), many of them in danger of extinction due to habitat destruction (Fu and Jin 1992; Cicuzza et al. 2007). One of the most threatened is the evergreen broad-leaved tree species *Michelia coriacea* Chang et B. L. Chen (Magnoliaceae), which is endangered by habitat destruction, fragmentation, and degradation.

M. coriacea grows only on limestone outcrop habitats in a few localities at 1,300–1,700 m in the Malipo and Xichou counties of Wenshan, southeastern Yunnan, near the border with Vietnam (Fig. 2.25). It is a shade-intolerant evergreen tree growing up to 22 m in height (Fig. 2.26a–c). Its timber is valuable. Its altitudinal range of 1,300–1,700 m is in an area originally belonging to the monsoon evergreen broad-leaved forest. However, the indigenous vegetation has been highly fragmented by human activities. There are four populations located 30 km or more from each other, all in unprotected areas (i.e. Daping, Shipen, Yangkaiping and Tiechang-Dongma) (Fig. 2.25). We are aware that one population is already extinct in Guangnan. Each population is small (no more than 50 individuals each). All of

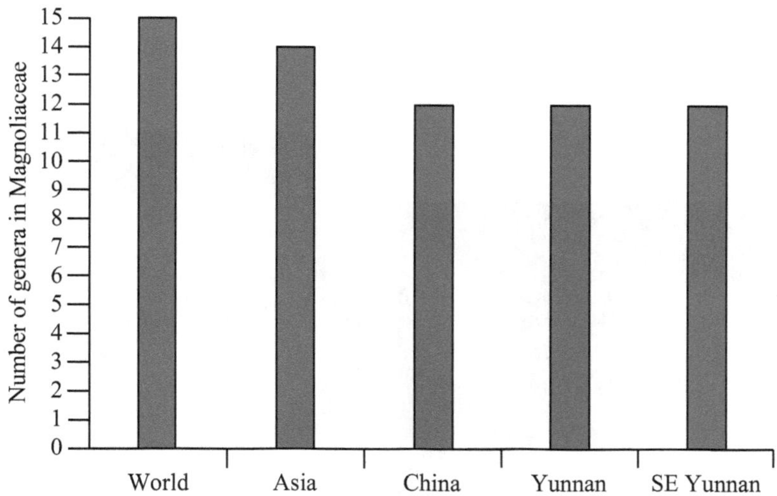

Fig. 2.24 Number of genera of Magnoliaceae in southeastern Yunnan compared with China and the world

Fig. 2.25 The location of *Michelia coriacea* populations

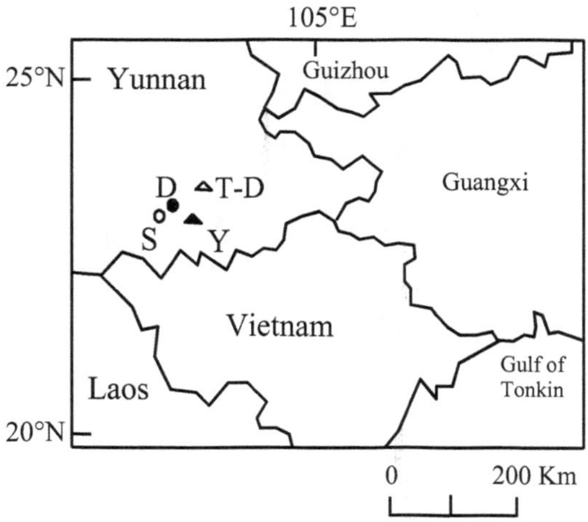

Fig. 2.26 *Michelia coriacea* tree (**a**), flower bud (**b**), fruits (**c**)

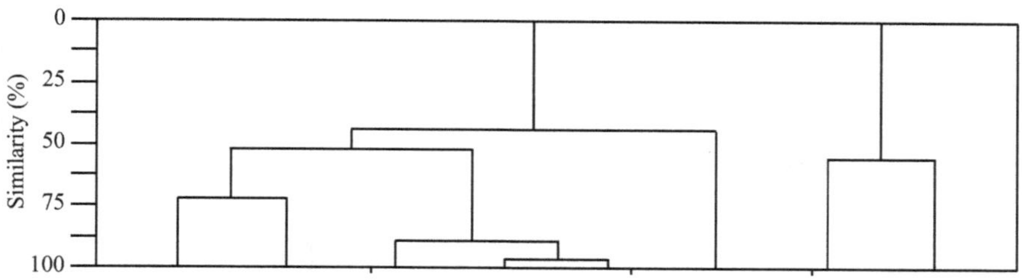

Plot	P1, P2	P6, P7, P8	P3	P4, P5
Site	Daping	Yangkaiping	Shipen	Tiechang-Dongma
Land use history	Partial-logging in 1950s-1980s	Clear-logging in 1930s, and partial-logging in 1970s-1980s	Partial-logging in 1950s-1980s	Clear-logging in the early 1970s
Score of degree of human disturbance	5-11	20	5-11	30
Max. age of *M. coriacea*	206	91	121	40
Dominant species	*Michelia coriacea* *Lithocarpus fenestratus* *Ormosia olivacea* *Ilex macrocarpa*	*Michelia coriacea* *Cyclobalanopsis myrsinaefolia*	*Michelia coriacea*	*Sapium chihsinianum* *Neocinnamomum delavayi* *Sloanea sinensis* *Michelia coriacea*
Mean coverage of invasive plants in the understory	20%	45.3%	25%	90.5%

Fig. 2.27 Stand characteristics of the four *Michelia coriacea* populations (Data source: Tang et al. (2011))

them are situated on sites originally occupied by forests, but now converted into secondary forests after clear or partial cutting during the 1930–1980s.

The stands in Daping, Shipen, and Yangkaiping are codominated by evergreen *M. coriacea*, *Lithocarpus fenestratus*, *Cyclobalanopsis myrsinifolia*, *Ilex macrocarpa*, and deciduous *Ormosia olivacea*. Deciduous broad-leaved *Celtis cerasifera* is also present in these stands. Larger numbers of deciduous broad-leaved species such as *Sapium chihsinianum*, *Carpinus tsaiana*, and *Cryptocarya chinensis* occur among the *M. coriacea* trees. Other evergreen broad-leaved species such as *Sloanea sinensis* and *Neocinnamomum delavayi* grow in the young secondary stand at Tiechang-Dongma. Notably, the coverage of the invasive plant *Eupatorium adenophorum* has significantly increased with greater degrees of human disturbance (Fig. 2.27).

Tang et al. (2011) find that the regeneration is poor (Fig. 2.28) and the annual mortality of seedlings of *M. coriacea* at early stages (individuals under 70 cm high) indicates very limited recruitment based on a census taken over 5 consecutive years from 2006 to 2010 (Fig. 2.29). The scanty seed production may be an aftereffect of

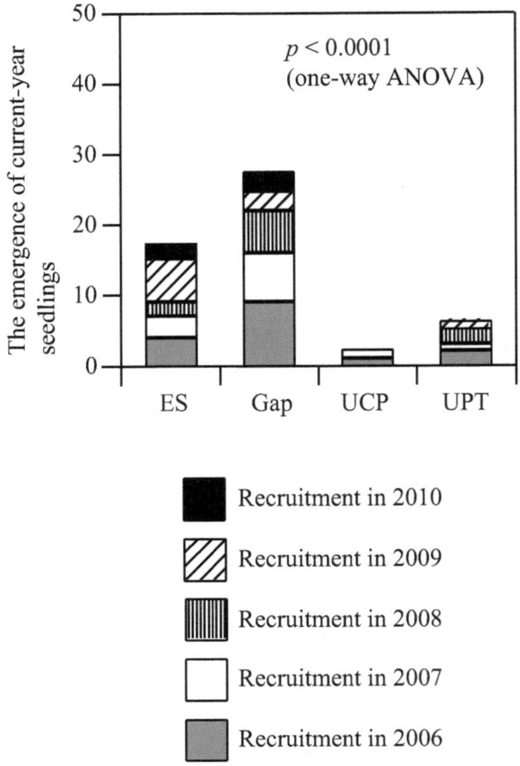

Fig. 2.28 The emergence of current-year seedlings and sprouts of *Michelia coriacea* for the four populations. *ES* edges of stands, *Gap* canopy gaps, *UCP* under canopy, *UPT* under parent tree (Data source: Tang et al. (2011))

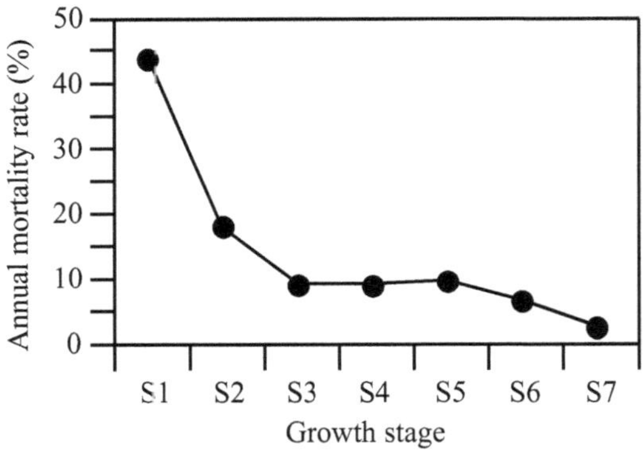

Fig. 2.29 Annual stem mortality in relation to growth stage. Growth stages (*S1–S7*) delimited for stems according to height (H) and DBH: *S1* 0 cm < H < 20 cm, *S2* 20 cm ≤ H < 40 cm, *S3* 40 cm ≤ H < 70 cm, *S4* 70 cm ≤ H < 100 cm, *S5* 100 cm ≤ H < 130 cm, *S6* H ≥ 1.3 m and DBH < 3 cm, *S7* H ≥ 1.3 cm and DBH ≥ 3 cm (Data source: Tang et al. (2011))

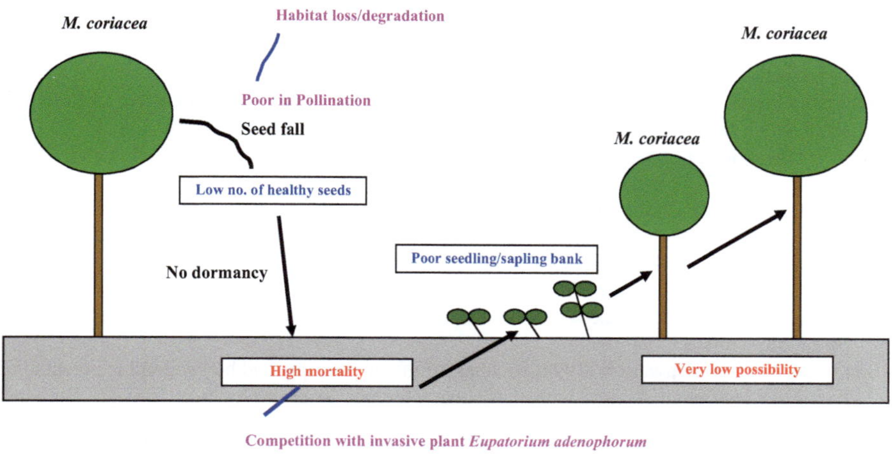

Fig. 2.30 A schematic diagram of the initial regenerative process of *Michelia coriacea*

earlier logging. In general, there are 2 to 6 embryos per fruit, but only about 1 filled seed per mature fruit. This may be due to abnormalities in sexual development (Zhao and Sun 2009). The growth and survival of this species may be superior where there is adequate sunlight. However, the invasive weed *Eupatorium adenophorum* invades aggressively in the habitats disturbed by humans and threatens *M. coriacea* habitats by reducing the availability of light, water, and nutrients. Competition for seedling establishment on sites with invasive weed species may have caused the high annual mortality in early recruitment, contributing to the decline of *M. coriacea* populations. A schematic diagram of the initial regenerative process of *M. coriacea* is shown in Fig. 2.30. Our data from the 5-year census show definitively that the limited recruitment in the early seedling stages is especially critical for *M. coriacea* and not adequately counteracted by its longevity. Threatening ecological factors include habitat damage, invasive plants, and low recruitment.

References

Auclair AN, Goff FG (1971) Diversity relations of upland forests in the western Great Lakes area. Am Nat 105:499–528

Bai Z-B, Zhang H-J, Ren G (2012) Quantitative classification and ordination of Simian Mountains, Chongqing. J Anhui Agric Sci 40(30):14839–14843

Bakker JP, Bakker ES, Rosen E, Verweij GL, Bekker RM (1996) Soil seed bank composition along a gradient from dry alvar grassland to Juniperus shrubland. J Veg Sci 7:165–176

Bao W, Liu Z (2002) Community features of primary and naturally secondary evergreen broad-leaved forests in Mt. Wawu in Sichuan, China. Chin J Environ Biol 8(2):120–126

Bazzaz FA (1979) The physiological ecology of plant succession. Annu Rev Ecol Syst 10:351–371

Bazzaz FA (1991) Regeneration of tropical forest: physiological responses of pioneer and secondary species. In: Gomez-Pompa A, Whitmore TC, Hadley M (eds) Rain forest regeneration and management. The Parthenon Publishing Group, Carnforth, pp 91–118

Binkley D (1992) Mixtures of nitrogen2-fixing and non-nitrogen2-fixing tree species. In: Cannell MGR, Malcom DC, Robertson PA (eds) The ecology of mixed-species stands of trees. Blackwell Science, Oxford, pp 92–123

Blatt SE, Crowder A, Harmsen R (2005) Secondary succession in two south-eastern Ontario old-fields. Plant Ecol 177:25–41

Bond WJ, Midgley JJ (2001) Ecology of sprouting in woody plants: the persistence niche. Trends Ecol Evol 16:45–51

Bond WJ, Midgley JJ (2003) The evolutionary ecology of sprouting. Int J Plant Sci 164:S103–S114

Box EO (1995) Climatic relations of East and Southeast Asia. In: Box EO, Peet RK, Masuzawa T, Yamada I, Fujiwara K, Maycock PF (eds) Vegetation science in forestry. Global perspective based on forest ecosystems of East and Southeast Asia. Papers from four symposia from the International Congress of Ecology, Yokohama 1990. Kluwer Academic Publishers, Dordrecht, pp 23–55

Box EQ, Chou CH, Fujiwara K (1998) Richness, climatic position, and biogeographic potential of East Asia laurophyll forests, with particular reference to examples from Taiwan. Bull Inst Environ Sci Technol (Yokohama National University) 24:61–95

Brown WH (1919) Vegetation of Philippine mountains. Bureau of Printing, Manila

Bruelheide H, Böhnke M, Both S, Fang T, Assmann T, Baruffol M, Bauhus JR, Buscot O, Chen X-Y, Ding B-Y, Durka W et al (2013) Community assembly during secondary forest succession in a Chinese subtropical forest. Ecol Monogr 81(1):25–41

Cadenasso ML, Pickett STA (2001) Effect of edge structure on the flux of species into forest interiors. Conserv Biol 15:91–97

Chen WL (1995) The present status of degraded ecosystems in China. In: Chen LZ, Chen WL (eds) Degraded ecosystems in China. China Science and Technology Press, Beijing, pp 16–23

Cicuzza D, Newton A, Oldfield S (2007) The Red List of Magnoliaceae. Fauna and Flora International, Cambridge, UK

del Moreal R, Wood DM (1993) Early primary succession on the volcano Mount St. Helens. J Veg Sci 4:223–234

Ding SY, Song YC (1997) Study on the synecological characteristics of the early succession stage of an evergreen broad-leaved forest on Tiantong National Forest Park, Zhejiang Province. Acta Phytoecologica Sin 23(2):197–207

Fu L-K, Jin J-M (eds) (1992) China plant red data book: rare and endangered plants, vol. 1. Science Press, Beijing

Garwood NC (1989) Tropical soil seed banks: a review. In: Leck M, Parker V, Simpson R (eds) Ecology of soil seed banks. Academic, San Diego, pp 149–209

Givnish TJ (1987) Comparative studies of leaf form–assessing the relative roles of selective pressures and phylogenetic constraints. New Phytol 106:131–160

González ME, Veblen TT, Donoso C, Valeria L (2002) Tree regeneration responses in a lowland *Nothofagus*-dominated forest after bamboo die-back in south-central Chile. Plant Ecol 161:59–73

Grime JP (2001) Plant strategies, vegetation processes, and ecosystem properties, 2nd edn. Wiley, New York

Grubb PJ (1977) Control of forest growth and distribution on wet tropical mountains: with special reference to mineral nutrition. Annu Rev Ecol Systematics 8:83–107

Grubb PJ, Lloyd JR, Pennington TD, Whitmore TC (1963) A comparison of montane and lowland rain forest in Ecuador I. The forest structure, physiognomy, and floristics. J Ecol 51:567–601

Guariguata MR, Rheingans R, Montagnini F (1995) Early woody invasion under tree plantations in Costa Rica: implications for forest restoration. Restoration Ecol 3(4):252–260

Guo QB, Liu YC, Li XG (1999) Dynamics of species diversity in secondary succession series of forest communities on Mt. Jinyun J Appl Ecol 10(5):521–524

Habeck JR (1968) Forest succession in the Glacier Park cedarhemlock forests. Ecology 49:872–880

Halpern CB, Spies TA (1995) Plant species diversity in natural and managed forests of the Pacific Northwest. Ecol Appl 5:913–934

Hattori T (1985) Synecological study on the lucidophyllous forest of *Castanopsis–Persea* type in Japan proper. Bull Kobe Geobot Soc 1:1–98

He M, Zhang J, Liu S, Li W (1996) A preliminary study on the natural vegetation of Black Bamboo Ravine in Ebian County of Sichuan. J Sichuan University (Nat Sci Ed) 33(3):347–350

Hermy M, Honnay O, Firbank L, Grashof-Bokdam C, Lawesson J (1999) An ecological comparison between ancient and other forest plant species of Europe, and the implications for forest conservation. Biol Conserv 91:9–22

Hill MO, Gauch HG (1980) Detrended correspondence analysis: an improved ordination technique. Vegetatio 42:47–58

Holz CA, Veblen TT (2006) Tree regeneration responses to *Chusquea montana* bamboo die-off in a subalpine *Nothofagus* forest in the southern Andes. J Veg Sci 17:19–28

Hu C-T (1999) Vegetation types of Xiaoheshan in the Daxue Mountain Nature Reserve, Longling. Forestry Invest Des 24(2):1–7

Huang W-L, Tu Y-L, Yang L (eds) (1988) The vegetation of Guizhou. Guizhong People's Press, Guiyang

Janzen DH, Vázquez-Yanes C (1990) Aspects of tropical seed ecology of relevance to management of tropical forested wildlands. In: Gómez-Pompa A, Whitmore TH, Hadley M (eds) Rain forest regeneration and management, MAB series. UNESCO, Paris, pp 137–154

Jia GM, Cao J, Wang CY, Wang G (2005) Microbial biomass and nutrients in soil at different stages of secondary forest succession in Ziwulin, northwest China. For Ecol Manage 217:117–125

Jiang F-L, Lu J-P (1991) Tropical forest ecosystem of Jianfengling, Hainan. Science and Technology Press, Beijing

Jose S, Andrew RG, George SJ, Kumar BM (1996) Vegetation responses along edge-to-interior gradients in a high altitude tropical forest in peninsular India. For Ecol Manage 87:51–62

Kalamees R, Zobel M (1998) Soil seed bank composition in different successional stages of a species rich wooded meadow in Laelatu, western Estonia. Acta Oecologica 19:175–180

Kelly DL, Tanner EVJ, Nic Lughadha EM, Kapos V (1994) Floristics and biogeography of a rain forest in the Venezuelan Andes. J Biogeogr 21:421–440

Kira T (1948) On the altitudinal arrangement of climax zones in Japan. Kanti-Nougaku 2:143–173

Kira T (1991) Forest ecosystems of east and southeast Asia in a global perspective. Ecol Res 6:185–200

Lancaster R (1989) Travels in China, a plantsman's paradise. Antique Collector's Club, Woodbridge

Lande R (1996) Statistics and partitioning of species diversity, and similarity among multiple communities. Oikos 76:5–13

Laurance WF, Perez-Salicrup D, Delamonica P, Fearnside PM, D'Angelo S, Jerozolinski A, Pohl L (2001) Rain forest fragmentation and the structure of Amazonian liana communities. Ecology 82:105–116

Letcher SG, Chazdon RL (2009) Lianas and self-supporting plants during tropical forest succession. For Ecol Manage 257:2150–2156

Li C (1997) Distribution of evergreen broad-leaved forests in East Asia. Nat Res 2:37–46

Li C-B, Yang Y-B, Wang J-X, Liang H-C (eds) (1990) The forest ecological study of Sichuan. Sichuan Science Press, Chengdu

Li H, Guo H, Dao Z (2000) Flora of the Gaoligong Mountains. Science Press, Beijing

Li YY, Sima YK, Fang B, Guo L, Jiang H, Zhao WS (2003) Current situation and evaluation of natural resources of the priority protection wild plants in Yunnan province of China. Acta Bot Yunnanica 25(2):181–191

Li D, Tang J-W, Luo C-K, Li J-S, Liu Z-A (2006) Community characteristics of the monsoon evergreen broad-leaved forest in Xishuangbanna, Yunnan. J Mountain Sci 24(3):257–267

Li X-S, Liu W-Y, Chen J-W, Tang CQ, Yuan C-M (2010a) Regeneration pattern of primary forest species across forest-field gradients in the subtropical mountains of southwestern China. J Plant Res 123(6):751–762

Li X, Liu W, Tang CQ (2010b) The role of the soil seed and seedling bank in the regeneration of diverse plant communities in the subtropical Ailao Mountains, Southwest China. Ecol Res 25:1171–1182

Lin WL, Hsieh CF, Chen YK, Lo HC (2003) A community study of the *Keteleeria davidiana* var. *formosana* forest stands in Taiwan. J Exp Forest (National Taiwan University) 17:33–47

Liu Y (1985) Numerical classification to the evergreen broadleaf forest in the Jinyun Mountain, Sichuan Province. Acta Phytoecologica et Geobotanica Sin 9(4):315–325

Liu ED, Peng H (2007) A preliminary floristic study on the mid-montane humid evergreen broad-leaved forest of Yongde Snow Mountain, SW Yunnan, China. Acta Bot Yunnanica 29(2):129–136

Liu Z-G, Qiu F-Y (1986) The main vegetation types and their distribution in the Gongga mountainous region. Acta Phytoecologica et Geobotanica Sin 10(1):26–34

Liu J-M, Zhong Z-C (2000) Nature of seed rain, the seed bank and regeneration of a *Castanopsis fargesii* community on Fanjing Mountain. Acta Phytoeologica Sin 24(4):402–407

Magurran AE (1988) Ecological diversity and its measurement. Princeton University Press, Princeton

Malhado ACM, Malhi Y, Whittaker RJ, Ladle RJ, ter Steege H, Phillips OL et al (2009) Spatial trends in leaf size of Amazonian rainforest trees. Biogeosciences 6:1563–1576

Mengistua T, Teketayb D, Hultenc D, Yemshaw Y (2005) The role of enclosures in the recovery of woody vegetation in degraded dryland hillsides of central and northern Ethiopia. J Arid Environ 60:25–281

Monk CD (1967) Tree species diversity in the eastern deciduous forest with particular reference to north central Florida. Am Nat 101:173–187

Naka K, Yoda K (1984) Community dynamics of evergreen broadleaf forests in southwestern Japan II. Species composition and density of seeds buried in the soil of a climax evergreen oak forest. Botanical Magazine Tokyo 97:61–79

Nakashizuka T (1987) Regeneration dynamics of beech forests in Japan. Vegetation 69:169–175

Nakashizuka T, Numata M (1982) Regeneration process of climax beech forest. I. Structure of beech forest with the undergrowth of *Sasa*. Jpn J Ecol 32:57–67

Ohsawa M (1990) An interpretation of latitudinal patterns of forest limits in south and east Asian mountains. J Ecol 78:326–339

Ohsawa M (1993a) Latitudinal pattern of mountain vegetation zonation in southern and eastern Asia. J Veg Sci 4:13–18

Ohsawa M (1993b) Forest pattern along hierarchical habitat gradients in east Asia. Biol Int Special Issue 29:30–36. In: Kawanabe H, Ohgushi T, Higashi M (eds) Symbiosphere: ecological complexity for promoting biodiversity

Ohsawa M (1995) Latitudinal comparison of altitudinal changes in forest structure, leaf-type and species richness in humid Asia. Vegetation 121:3–10

Ohsawa M, Nitta I (1997) Patterning of subtropical/warm-temperate evergreen broad-leaved forests in east Asian mountains with special reference to shoot phenology. Tropics 6(4):317–334

Ohsawa M, Ozaki K (1992) Hierarchical analysis of vegetation/environment pattern for east Asian extra-tropical evergreen broad-leaved forests. J Japan Biometeorol 29(Suppl):93–103

Ovington JD (1983) Temperate broad-leaved evergreen forests. Elsevier, Amsterdam

Peet RK (1978) Forest vegetation of the Colorado Front Range: patterns of species diversity. Vegetatio 43:131–140

Peng M, Dang C (1998) Studies on interspecific association of both *Castanopsis orthacantha* and *Castanopsis delavayi* communities on Jizu Mountain, Yunnan Province. Acta Ecologica Sin 18(2):159–166

Peng SL, Fang W (1994) Studies on the dynamics of the Castanopsis chinensis and Schima superba populations in forest succession in Dinghushan. Acta Phytoecologica Sin 19(4):311–318

Peng H, Wu Z-Y (2001) The floristic characteristics and its significance in conservation of semi-humid evergreen broad-leaved forests in Mt. Wuliangshan Acta Bot Yunnanica 23(30):278–286

Peng M-C, Wang C-Y, Dang C-L (2006) Studies on biodiversity and conservation in Yaoshan Nature Reserve. Science Press, Beijing

Pielou EC (1975) Ecological diversity. Wiley, New York

Raunkiaer C (1916) The use of leaf size in biological plant geography. Botanisk Tidsskrifit 34:368–378. In: Raunkiaer C (ed) (1977) Life forms of plants and statistical plant geography, Arno Press, New York

Richards PW (1996) The tropical rain forest, 2nd edn. Cambridge University Press, Cambridge

Roberts TL, Vankat JL (1991) Floristics of a chronosequence corresponding to old field-deciduous forest succession in southwestern Ohio. II. Seed banks. Bull Torrey Botanical Club 118:377–384

Roeder M, Hölscher D, Kossmann-Ferraz ID (2012) Traits and growth of liana regeneration in primary and secondary forests of Central Amazonia. Appl Veg Sci 15:108–118

Satoo T (1983) Temperate broad-leaved evergreen forests in Japan. In: Ovington JD (ed) Temperate broad-leaved evergreen forest, vol 10, Ecosystems of the World. Elsevier, Amsterdam, pp 169–189

Schnitzer SA, Dalling JW, Carson WP (2000) The impact of lianas on tree regeneration in tropical forest canopy gaps: evidence for an alternative pathway of gap-phase regeneration. J Ecol 88:655–666

Sharma EL, Ambasht RS (1984) Seasonal variation in nitrogen fixation by different ages of root nodules of Alnus nepalensis plantations in the eastern Himalayas. J Appl Ecol 21:265–270

Shi JP, Zhu H (2009) Tree species composition and diversity of tropical mountain cloud forest in the Yunnan, southwestern China. Ecol Res 24:83–92

Shi J-P, Chang Y-F, Xu C-D, Zhu H (2011) Synecological studies on the semi-humid evergreen broad-leaved forest in Huajiaoyuan Nautre Reserve, Chuxiong, Yunnan. Guihaia 31(2):204–207

Shimwell DW (1971) Description and classification of vegetation. Sidgwick & Jackson, London

Shui YM, Cheng YH (2003) The plants of Honghe Region, Yunnan. Science Press, Beijing

Singh JS, Sing SP (1987) Forest vegetation of the Himalaya. Botanical Rev 53(1):135–143

Song Y-C (1995) On the global position of the evergreen broad-leaved forests of China. In: Box EO, Peet RK, Masuzawa T, Yamada I, Fujiwara K, Maycock PF (eds) Vegetation science in forestry. Global perspective based on forest ecosystems of East and Southeast Asia. Papers from four symposia from the International Congress of Ecology, Yokohama 1990. Kluwer Academic Publishers, Dordrecht, pp 69–84

Song Y-C (ed) (2013) Evergreen broad-leaved forests in China. Classification-ecology-conservation. Science Press, Beijing

Song HX, Su ZX, Peng YY (2005) Relationships between soil fertility and secondary succession of plant community in Jinyun Mountain. Chin J Ecol 24(12):1531–1533

Sun H, Zhou ZK, Yu HY (1996) A preliminary probe into secondary succession series of tropical forests in the Big Bend Gorge of Yalu Tsangpo River in SE Tibet, eastern Himalayas. Acta Bot Yunnanica 18(3):308–316

SVEG (Sichuan Vegetation Editorial Group) (1980) The vegetation of Sichuan. Sichuan People's Press, Chengdu

Tagawa H (1995) Distribution of lucidophyll oak-laurel forest formation in Asia and other areas. Tropics 5:1–40

Tang CQ (2006) Forest vegetation as related to climate and soils at varied altitudes on a subtropical mountain, Mt. Emei, Sichuan, China. Ecol Res 21(2):174–180

Tang CQ (2010) Subtropical montane evergreen broad-leaved forests of Yunnan, China: diversity, succession dynamics, human influence. Front Earth Sci China 4(1):22–32

Tang CQ (2015) Distribution patterns of the subtropical evergreen broad-leaved forests of south-western China, as compared with those of the eastern Chinese subtropical regions. Collectanea Botanica (in press)

Tang CQ, He L-Y (2013) Population status of the Tertiary relict *Tetracentron sinense* in the subtropical Ailao Mountains, Yunnan, SW China, and proposed conservation efforts. Annu Rep Pro Natura Fund 21:141–150

Tang CQ, Ohsawa M (1997) Zonal transition of evergreen, deciduous and coniferous forests along the altitudinal gradient on a humid subtropical mountain, Mt. Emei, Sichuan, China. Plant Ecol 133:63–78

Tang CQ, Ohsawa M (1999) Altitudinal distribution of evergreen broad-leaved trees and their leaf-size pattern on a humid subtropical mountain, Mt. Emei, Sichuan, China. Plant Ecol 145:221–233

Tang CQ, Ohsawa M (2000) Transitional pattern of tree leaf-types from tropical to temperate forests on Mt. Emei, Sichuan, China. In: Zheng D et al (eds) Formation, evolution and sustainable development on the Qinghai-Tibet Plateau. Academy Press, Beijing, pp 397–407

Tang CQ, Ohsawa M (2002a) Coexistence mechanisms of evergreen, deciduous and coniferous trees in a mid-montane mixed forest on Mt. Emei, Sichuan, China. Plant Ecol 161:215–230

Tang CQ, Ohsawa M (2002b) Tertiary relic deciduous forests on a subtropical mountain, Mt. Emei, Sichuan, China. Folia Geobotanica 37:93–106

Tang CQ, Ohsawa M (2009) Ecology of subtropical evergreen broad-leaved forests of Yunnan, southwestern China as compared to those of southwestern Japan. J Plant Res 122:335–350

Tang CQ, Zhu X (2007) Structure and dynamics of mid-montane moist evergreen broad-leaved forest on subtropical mountains, Ailaoshan and Gaoligongshan, Yunnan, China, and the relationship between the forests and the people, with a view of future sustainability. Annu Rep Pro Natura Fund 16:153–175

Tang CQ, Li T, Zhu X (2007a) Structure and regeneration dynamics of three subtropical midmontane moist evergreen broad-leaved forests in southwestern China, with special reference to bamboo in the forest understories. Can J Forest Res 37:2701–2714

Tang CQ, Hou X, Gao K, Xia T, Duan C, Fu D (2007b) Man-made versus natural forests in mid-Yunnan, southwestern China: plant diversity and initial data on water and soil conservation. Mountain Res Dev 27(3):242–249

Tang CQ, Zhao M-H, Li X-S, Ohsawa M, Ou X-K (2010a) Secondary succession of plant communities in a subtropical mountainous region of SW China. Ecol Res 25(1):149–161

Tang CQ, Li Y-H, Zhang Z-Y (2010b) Species diversity patterns of natural secondary plant communities and man-made forests in a subtropical mountainous karst area, Yunnan, SW China. Mountain Res Dev 30(3):244–251

Tang CQ, He L-Y, Gao Z, Zhao XF, Sun WB, Ohsawa M (2011) Habitat fragmentation, degradation, and population status of endangered *Michelia coriacea* in Southeastern Yunnan, China. Mountain Res Dev 31(4):343–350

Tang CQ, Chiou C-R, Lin C-T, Lin J-R, Hsieh C-F, Tang J-W, Su W-H, Hou X (2013) Plant diversity patterns in subtropical evergreen broad-leaved forests of Yunnan and Taiwan. Ecol Res 28:81–92

Tang CQ, Li Y-H, Zhang Z-Y, Hou X, Hara K, Tomita M, He L-Y, Li X-S (2014) Effects of management on vegetation dynamics and associated nutrient cycling in a karst area, Yunnan, SW China. Landsc Ecol Eng. doi:10.1007/s11355-014-0258-7

Taylor AH, Qin Z (1988) Regeneration patterns in old-growth *Abies-Betula* forests in the Wolong natural reserve, Sichuan, China. J Ecol 76:1204–11218

Taylor AH, Qin Z (1993) Bamboo regeneration after flowering in the Wolong Giant panda Reserve, China. Biol Conserv 63:231–234

Taylor AH, Qin Z, Liu J (1995) Tree regeneration in an *Abies faxonianan* forest after bamboo dieback, Wang Lang Natural Reserve, China. Can J Forest Res 25:2034–2039

Teketay D (1998) Soil seed banks at an abandoned Afromontane arable site. Feddes Repert 109:161–174

Thompson J, Proctor J, Viana V, Milliken W, Ratter JA, Scott DA (1992) Ecological studies on a lowland evergreen rain forest on Maraca Island, Roraima, Brazil. I. Physical environment, forest structure and leaf chemistry. J Ecol 80:689–703

Turner IM (2001) The ecology of trees in the tropical rainforest. Cambridge University Press, Cambridge

Turner IM, Wong YK, Chew PT, Bin Ibrahim A (1997) Tree species richness in primary and old secondary tropical forest in Singapore. Biodivers Conserv 6:537–543

Vesk PA, Westoby M (2004) Sprouting ability across diverse distribution and vegetation types worldwide. J Ecol 92:310–320

Wang J, Ma Q-Y, Du F, Yang Y-M (eds) (2005) Altitudinal patterns of seed plants on Dawei Mountain, Yunnan Province, China. Acta Phytoecologia Sinica 29(6):894–900

Wang XH, Kent M, Fang XF (2007) Evergreen broad-leaved forest in Eastern China: its ecology and conservation and the importance of resprouting in forest restoration. For Ecol Manage 245:76–87

Wang C-Y, He Z-R, Peng M-C (2013) Studies on vegetation and plants in Dulongjiang (upper Irrawaddy River) watershed and adjacent area. Science Press, Beijing

Webb LJ (1959) A physiognomic classification of Australian rain forests. J Ecol 47:551–570

Whitmore TC (1975) Tropical rain forests of the Far East. Clarendon, Oxford

Wolfe JA (1979) Temperature parameters of humid to mesic forests of eastern Asia and relation to forests of other regions of the northern hemisphere and Australasia. United States Geological Survey, Professional Paper 1106

Wu Z-Y, Zhu Y-C, Jiang H-C (eds) (1987) The vegetation of Yunnan. Science Press, Beijing

Xie Z, Chen W (1998) The remaining evergreen broad-leaved forest and its significance in Three Gorges Reservoir Area. Acta Phytoecologia Sin 22(5):422–427

Xu D-Y (2005) Preliminary investigation on the vegetation of Dabashan Nature Reserve and its characteristics. J Anhui Agric University 32(2):332–335

Xu X, Qin Z-S, Liao Y-M, Gu X-D, Yang L, Zhao J (2004) Study on vegetation of Xiaozhaizigou-Piankou Natural Reserve. J China West Normal University (Nat Sci) 25(2):135–143

Yamamoto S (1992) Gap characteristics and gap regeneration in primary evergreen broad-leaved forests of western Japan. Botanical Magazine Tokyo 105:29–45

Yang L (1983) The structure and dynamics of *Cyclobalanopsis stewardiana* forest of the Fanjingshan Mountain. Acta Phytoecologica et Geobotanica Sin 7(3):204–214

Yang Y-M, Du F (eds) (2004) Integrated scientific studies of Nangunhe Nature Reserve, Yunnan. Yunnan Science Press, Kunming

Yang Y-M, Du F (eds) (2006) Integrated scientific studies of Tongbiguan Nature Reserve, Yunnan. Yunnan Science Press, Kunming

Yang Y-M, Li T-S (eds) (1999) Integrated scientific studies of Dawei Nature Reserve, Yunnan. Yunnan Science Press, Kunming

Yang X, Xie S (2012) A comparative study on surface spore-pollen in the mountaintop and mountainside of Mt. Jinfo J Mountain Sci 30(96):655–662

Yang Y, Wu Z-L, Wang D-S (eds) (2006) Nabanhe Watershed National Nature Reserve, Xishuangbanna. Yunnan Science Press, Kunming

Yang Y-M, Tian K, He S-J (eds) (2008) Integrated scientific studies of Wenshan National Nature Reserve, Yunnan. Science Press, Beijing

Yang C-H, Chen J-Y, Ding F-J (2012) Study on the plant diversity of Karst forest community in Tianlong Mountain of Puding County. Guizhou Forestry Sci Technol 40(2):1–7

Yuan C-M, Liu W-Y, Tang CQ, Li X-S (2009) Species composition, diversity, and abundance of lianas in different secondary and primary forests in a subtropical mountainous area, SW China. Ecol Res 24:1361–1370

Zhang Z, Hu G, Zhu J, Ni J (2013) Aggregated spatial distributions of species in a subtropical karst forest, southern China. J Plant Ecol 6(2):131–140

Zhao XF, Sun WB (2009) Abnormalities in sexual development and pollinator limitation in Michelia coriacea (Magnoliaceae), a critically endangered and endemic to southeastern Yunnan, China. Flora 204(6):463–470

Zhou ZX (ed) (1992) The forest of Guizhou. Guizhou Science and Technology Press, Guiyang, pp 73–77

Zhou XY, Huang ZL, Ouyang XJ, Li J, Guan LL, Xu GL, Zhang C (2005) Succession of the original Castanopsis chinensis-Cryptocarya chinensis-Schima superba community of monsoon evergreen broad-leaved forest in Dinghushan Nature Reserve. Acta Ecologica Sin 25(1):37–44

Zhu H (1997) Ecological and biogeographical studies on the tropical rain forest of south Yunnan, SW China with a special reference to its relation with rain forests of tropical Asia. J Biogeogr 24:647–662

Zhu H (2007) On the classification of forest vegetation in Xishuangbanna, southern Yunnan. Acta Bot Yunnanica 29(4):377–387

Zhu H, Yan LC (2009) Biogeographical affinities of the flora of southeastern Yunnan, China. Botanical Stud 50(4):467–475

Zimmerman JK, Pascarella JB, Aide TM (2000) Barriers to forest regeneration in an abandoned pasture in Puerto Rico. Restoration Ecol 8(4):350–360

Zuo J-F (1995) Investigation on *Michelia martini*. Chin J Ecol 14(1):27–32

Chapter 3
Evergreen Sclerophyllous *Quercus* Communities

Abstract Sixteen evergreen sclerophyllous *Quercus* species (*Quercus* Subgen. *Quercus* sect. *Brachylepides* and *Quercus* subgen. *Quercus* sect. *Engleriana*) are found within 23–32° N in China, mainly in Yunnan, western Sichuan, southeastern Tibet and western Guizhou. They exhibit exceptional environmental adaptations, ranging from the subtropical to the altitudinal subalpine/alpine zones. Twenty-two evergreen *Quercus* communities are recorded in SW China. Among them, two types are distinguished. The first type, the *Quercus* forests, grow in cold-dry or subhumid environments at high altitudes. The second type, the *Quercus* scrub communities, are found both on mountain slopes and in warm- or hot-dry valleys. The *Quercus* scrub communities are similar to the Maquis growing in areas of Mediterranean-type climate such as Spain and California. The *Quercus* forests have high dominance concentration (usually by a single species) and limited species diversity, with sparse ground vegetation including herbaceous species and climbing plants, and low seedling establishment that is complimented by sprouting. In all respects they are similar to the *Quercus agrifolia* forest of central coastal California (USA) and a *Quercus ilex* community of Catalonia (NE Spain).

3.1 Distribution and Community Types

The word "sclerophyllous" comes from the Greek roots *scler-* (hard) and *phyll-* (leaf). These species have drought-resistant foliage, often with dense hairs, thick cuticles, leaf margins with sharp spines or spinose-dentate leaf margins, and thick, rough barks. *Quercus* (Oak), the largest genus in the family Fagaceae, comprises two subgenera, subgenus *Quercus* (with deciduous and evergreen species) and subgenus *Cyclobanopsis* (having only evergreen species), based on whether the cupule is imbricate-scaled or lamellate (Camus 1936–1954). Various evergreen sclerophyllous *Quercus* species are widely distributed in the Mediterranean region, North and Central America, Colombia of South America and Asia. Figure 3.1 indicates the general geographic distribution of the two subgenera as well as the evergreen species of the subgenus *Quercus*. After the final uplift of the Himalayas, along with the gradual climate change in the Tethys area, which had become arid since the Neocene, the subtropical evergreen broad-leaved species have been gradually replaced under semiarid or arid conditions by a sclerophyllous flora, that ultimately became an

C.Q. Tang, *The Subtropical Vegetation of Southwestern China*, Plant and Vegetation 11, DOI 10.1007/978-94-017-9741-2_3

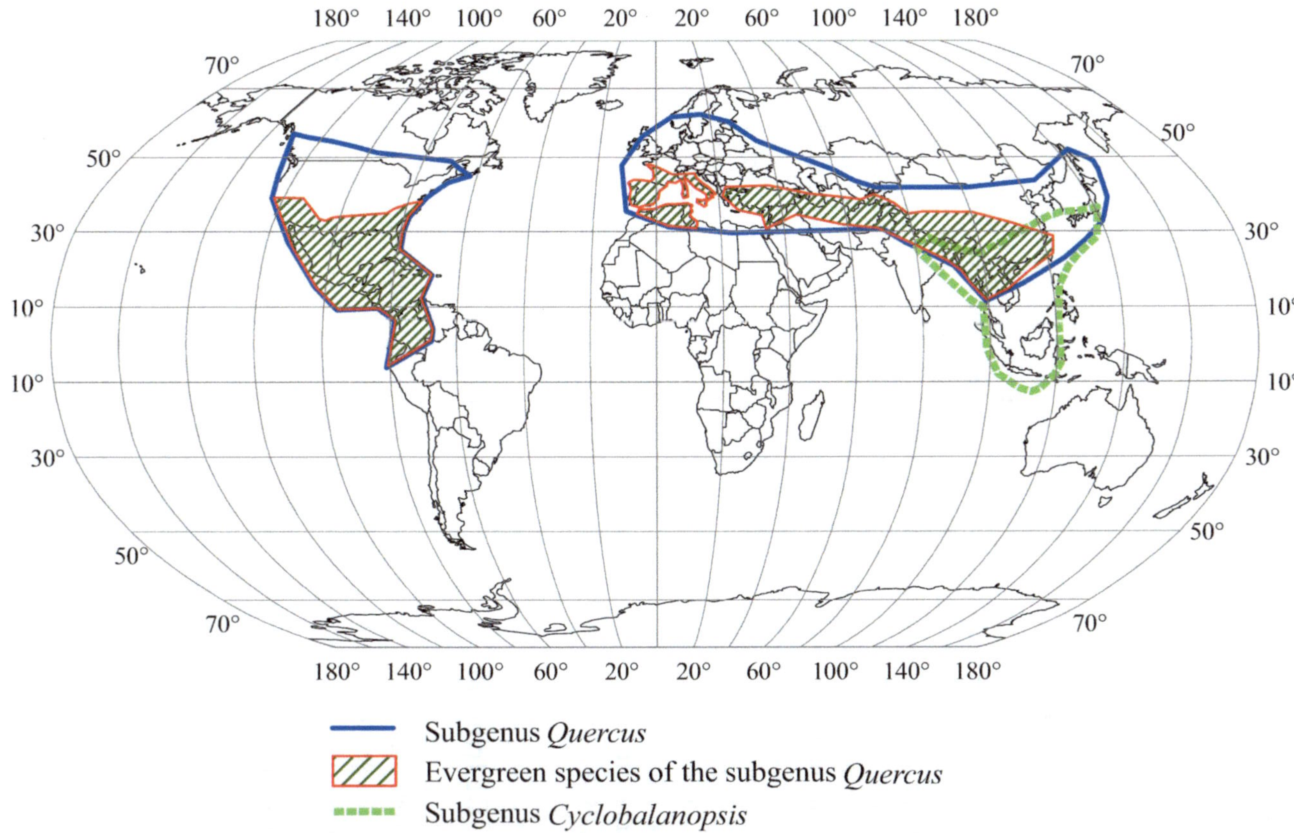

Fig. 3.1 General distribution of *Quercus*. Geographic areas of the two subgenera *Quercus* and *Cyclobalanopsis* and the evergreen species of the subgenus *Quercus* are indicated (Revised from Tang (2006))

important vegetation type (Tao and Kong 1973; Li and Guo 1976; Tao 1986; Zhou 1993, 1996; Zhou et al. 1995, 2003). Seventeen fossil species of evergreen *Quercus* are found in Tertiary geological strata of southern Tibet. Presumably the Hengduan Mountains are the origin of the evergreen sclerophyllous *Quercus* (Yang 1990; Zhou 1992, 1999). The present Chinese evergreen sclerophyllous *Quercus* forests have close affinities with paleo-Mediterranean subtropical vegetation (e.g. Jing 1981; Yang 1990; Zhou 1999; Tang 2000; Sun 2002; Sun and Li 2003). These oaks are morphologically similar to Mediterranean species such as *Q. suber* (cork oak) and *Q. ilex* (holm oak) (Li et al. 2000; Zhou et al. 2003).

Sixteen evergreen sclerophyllous *Quercus* (*Quercus* sect. *Brachylepides* and *Quercus* subgen. *Quercus* sect. *Engleriana*) species (e.g. Yang 1990; Huang et al. 1998; Zhou et al. 2003; Liu et al. 2008) are found between 23° (e.g. Yuanjiang, Yunnan) and 32° N (e.g. southern Minshan, Sichuan) in China, mainly in Yunnan, western Sichuan, southeastern Tibet and western Guizhou. In China, 81–88 % of the *Quercus* species are found from 23° to 29° N, with 90 % recorded around 27° N (Fig. 3.2a). The latitudinal distribution of *Q. spinosa* ranges between 23° and 35° N, while *Q. marlipoensis* and *Q. kongshanensis* are confined to small areas in Yunnan. More than 50 % of the *Quercus* species are located between 98° and 104° E, almost all between 99° and 103° E (Fig. 3.2b). *Q. engleriana* is widely distributed from 95° to 117° E. The *Quercus* are found at elevations of 500–4,800 m, and are most abundant between 2,500 and 3,000 m (Fig. 3.2c). *Q. aquifolioides* (1,900–4,800 m), *Q. monimotricha* (1,700–3,800 m) and *Q. dolicholepis* (500–2,700 m) have wide altitudinal distribution ranges (Fig. 3.2c). Among the 16 species, *Q. aquifolioides*, *Q. fimbriata*, *Q. guyavifolia* and *Q. monimotricha* are endemic to the Hengduan region. Several of the 16 species are also found in Bhutan, Nepal, India, Afghanistan, Pakistan, Myanmar, Thailand and Taiwan. The evergreen *Quercus* species exhibit exceptional environmental adaptations, corresponding to their ranges from the subtropical to the altitudinal subalpine/alpine zones of SW China. The tree height of a given *Quercus* species can change greatly with altitudes.

Two types of evergreen sclerophyllous *Quercus* communities in SW China are distinguished: The *Quercus* forests in cold-dry or subhumid environments at high altitudes, and the *Quercus* scrub communities on mountain slopes and in warm- or hot-dry valleys (Fig. 3.3a–c). The scrub communities (height c. 1–5 m) are similar to the Maquis in the areas of Mediterranean-type climate, such as Spain and California. *Q. cocciferoides*, *Q. franchetii* and *Q. rehderiana* are common components of the *Quercus* scrub communities. In western Sichuan, the evergreen *Quercus* forest and the scrub community together occupy a disproportionate amount of the province's total forest area (c. 15 %) (Wu 1980). *Quercus* species are numerous in NW Yunnan, though they cover less area than in western Sichuan. *Quercus* communities in southeastern Tibet and western Guizhou are scattered. The characteristics of the major evergreen sclerophyllous *Quercus* species in SW China are summarized in Table 3.1.

In Yunnan, *Q. aquifolioides* forest is mainly found at 1,900–3,900 m. It has a height of 15–25 m, and mainly includes *Populus szechuanica* and *Alnus nepalensis* at 2,400–2,800 m in Zhongdian, and is found along with *Larix potaninii, Picea*

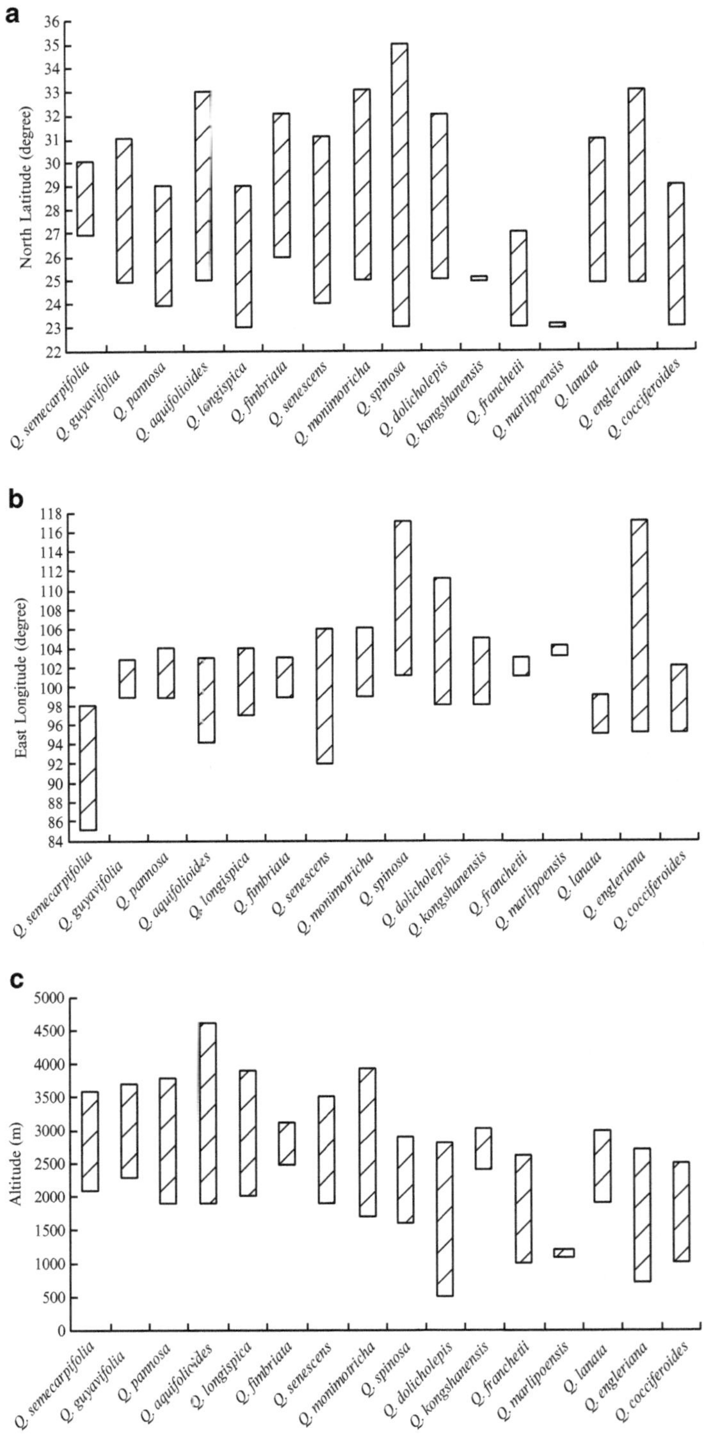

Fig. 3.2 Major evergreen *Quercus* distribution along latitudes (**a**), and longitudes (**b**), and altitudes (**c**) (After Liu et al. (2008))

a

b

c

Fig. 3.3 *Quercus pannosa* forest at 3,550 m in Bitahai (**a**), and *Quercus longispica* forest at 3,400 m in Adong (**b**), and *Quercus pannosa* scrubland at 3,200 m in the Baimaxue Mountains (**c**), NW Yunnan

Table 3.1 Characteristics of major evergreen *Quercus* species in SW China. For latitudinal, longitutinal and altitudinal ranges, see Fig. 3.2a–c

Species	Height (m)	DBH	Shade tolerance	Habitats
Quercus aquifolioides	Usually 8–15 m; may reach 25 m; shrub form 3–8 m	20–60 cm; may reach 100 cm	Intolerant	Sunny slopes and semi-sunny slopes
Quercus pannosa	Usually 15–20 m; may reach 25 m; shrub form 1–3.5 m	30–62 cm; may reach 80 cm	Somewhat tolerant	Widely distributed; sunny & semi-sunny slopes; mainly in areas of limestone as bedrock
Quercus rehderiana	Usually 5–16 m; shrub form 1–3 m	20–40 cm; may reach 55 cm	Somewhat tolerant	Semi-sunny slopes and mountain valleys; limestone areas
Quercus senescens	Usually 10–16 m; Shrub form 1–3 m	10–40 cm; may reach 50 cm	Somewhat tolerant	Many areas of limestone as bedrock
Quercus longispica	Usually 15–20 m; may reach 26 m; shrub form 3–8 m	19–65 cm; may reach 135 cm	Tolerant	Sheltered shady; semi-shady, semi-sunny slopes; rare in sunny slopes
Quercus guyavifolia	Usually 8–10 m; may reach 20 m; shrub form 3–8 m	20–30 cm; may reach 50 cm	Tolerant	Semi-shady slopes; semi-sunny slopes
Quercus cocciferoides	Usually 5–8 m; may rech 10 m; shrub form 1–1.5 m	10–20 cm; may reach 25 cm	Somewhat tolerant	Jinsha river bank & valleys; limestone areas; stony soils
Quercus franchetii	Usually 6–8 m; may reach 10 m; shrub form 1–2 m	15–20 cm; may reach 30 cm	Intolerant	Jinsha river bank and valleys
Quercus engleriana	Usually 4–15 m; may reach 25 m; shrub form 1–4 m	15–30 cm	Intolerant	Sunny slopes, limestone areas
Quercus kongshanensis	Usually 6–10 m	15–20 cm	Intolerant	Sunny slopes; forest edges
Quercus lanata	Usually 15–20 m; may rach 30 m	30–80 cm; may reach 2.6 m	Intolerant	Sunny slopes
Quercus fimbriata	usually 4–10 m	10–20 cm	Somewhat tolerant	Semi-shady slopes
Quercus dolicholepis	Usually 3–16 m	Usually 10–30 cm	Intolerant	Sunny slopes forest edges
Quercus marlipoensis	Usually 7–18 m	10–30 cm	Somewhat tolerant	Semi-shady slopes

(continued)

Table 3.1 (continued)

Species	Height (m)	DBH	Shade tolerance	Habitats
Quercus monimotricha	Only shub form 0.5–2 m	1–3.5 cm	Intolerant	Sunny slopes, ridges
Quercus spinosa	Usually 4–15 m; shrub form 1–4 m	10–20 cm; may reach 30 cm	Intolerant	Sunny slopes, ridges, cliffs, limestone areas

Partial data from Tang (2006), Yang (1990), Wu (1980), SVEG (1980), Wu et al. (1987)

likiangensis, Abies georgei var. *smithii,* and *Rhododendron* spp. at higher altitudes (3,500–3,900 m) in Bitahai, NW Yunnan. The *Q. pannosa* forest in Bitahai reaches 25 m in height. The companion species are *Picea likiangensis, Pinus armandii,* and some shrubs such as *Rhododendron* spp., *Sorbus* spp., *Rosa sweginzowii*. In Luquan, central Yunnan, the *Quercus pannosa* community reaches only about 3–9 m in height, along with deciduous *Sorbus pallescens, Acer robusta, Sorbus rufopilosa* and coniferous *Picea likiangensis* at 2,900–3,300 m. In Feilaisi, Yubeng and Adong of NW Yunnan, the *Q. longispica* forest reaches 26 m in height, and is accompanied by *Pinus densata* and some shrub species. In Yulongxueshan, *Q. guyavifolia-Q. longispica* forest with a height of 25 m, is found in limestone habitats at 2,600–3,200. *Pinus yunnanensis, Pinus armandii, Picea likiangensis* and *Q. rehderiana* are also found in this forest. On low mountain slopes and in warm or hot-dry valleys (e.g. along the Jinsha River and Pudu River in Luquan, central Yunnan) *Q. coccifer-oides* or *Q. rehderiana* in limestone habitats, and *Q. franchetii* in non-limestone areas, form scrub communities (3–8 m in height). *Pistacia weinmanniifolia* and *Engelhardia colebrookiana* are often present in the scrublands.

In western Sichuan, *Quercus aquifolioides* forms a forest with a height of 20–30 m on mountain slopes at elevations of 3,000–3,800 (4,700) m. The average diameter at breast height is 55 cm, with some reading 100 cm. Species of *Larix, Abies, Picea, Pinus, Betula, Populus* are commonly found in the forest. It becomes scrubby over 3,800 m. The scrubs extend over the timber line (4,500 m), mixed with *Sabina, Rhododendron* and other shrub species. In dry habitats with thin soils below 3,800 m *Q. aquifolioides* appears in valley scrubland along rivers, i.e. the Jinshajiang, Yalongjiang, Anninghe, Daduhe and their tributaries. *Q. pannosa* is widely distributed in the Xichang and Liangshan areas. It forms a forest of 10–25 m high on mountain slopes with fertile soils at 2,500–3,300 m, or appears as scrub ranging from 1 to 4 m in height in warm or hot-dry valleys with thin soils developed on limestone bedrocks. *Q. longispica* with a height of 8–15 m is found on sunny slopes at 2,300–3,400 m in areas of southwestern Sichuan including Derong, Xiangcheng, Daocheng, Yanbian and Muli. It is often accompanied by *Pinus yunnanensis* or *Pinus densata* to form a community. *Q. rehderiana* is mainly distributed on mountain slopes or in valleys at 2,000–3,200 m in Ganzi and Xichang. It mixes with *Pinus densata, Pinus armandii, Populus, Rhododendron, Acer, Caragana,* etc. (Wu 1980).

In Western Guizhou (e.g. Weining, Hezhang, Bijie, and Panxian), *Q. rehderiana* occurs in a *Carpinus pubescens* var. *firmifolia* community (10–16 m in height) at 2,200–2,500 m, *Quercus senescens* (13–16 m) along with *Corylus yunnanensis* and *Pinus yunnanensis* in habitats with many outcrops, with *Q. pannosa* (5–10 m) and deciduous *Q. aliena* at 2,300 m, with scrubby *Q. spinosa* (3–5 m) at 2,000–2,800 m, and *Q. aquifolioides* forest (10–15 m) at 2,700 m. *Q. engleriana* forest (c. 15 m) is found at 1,600–1,800 m in Fanjingshan (Huang et al. 1988).

3.2 Species Diversity and Dynamics of the Forests

In this section, three representative species, *Q. aquifolioides*, *Q. pannosa* and *Q. longispica* occurring at high altitudes (≥2,400 m) in the Zhongdian (Shangrila) and the Deqin regions of NW Yunnan, are chosen to demonstrate the ecological patterns in detail. NW Yunnan has warm rainy summers and cold dry winters. Average monthly temperatures range between −3.6 °C in December and 13.5 °C in July at the Zhongdian climate station (27°42′ N, 99°42′ E, 3,276 m asl) and between −5.5 °C in December and 11 °C in June at the Deqin station (28°26′ N, 99°11′ E, 3,485 m asl). The mean annual precipitation is 609 mm for Zhongdian and 651 mm for Deqin. In both localities rain falls mostly during the monsoon months, June to October.

Q. aquifolioides is shade-intolerant, *Q. pannosa* is somewhat tolerant, and *Q. longispica* is shade-tolerant. *Q. pannosa* is mainly found in areas of limestone bedrock. *Q. aquifolioides* and *Q. longispica* are plentiful, ranging from somewhat moist to dry slopes and valley floors. The woody species in the three forests are shown with their relative basal area in Table 3.2. The forests are named by their principal dominant species, being those with the highest values of relative basal area (RBA). In NW Yunnan, *Q. aquifolioides*, *Q. pannosa* and *Q. longispica* are clearly the most important species, accompanied by *Rhododendron rubiginosum*, *Rhododendron vernicosum*, *Rhododendron oreotrephes*, *Rhododendron decorum*, *Larix potaninii*, *Picea likiangensis*, *Pinus densata*, *Pinus armandii*, *Populus szechuanica*, *Sorbus rufopilosa* and *Sorbus rehderiana*.

The vegetation on the forest floor of each of the three forests is sparse, with occurrence of some herbaceous species and climbing plants. The herbaceous species include *Ainsliaea latifolia*, *Thalictrum finetii*, *Saxifraga ficicaulis*, *Leibnizia nepalensis*, *Adiantum fimbriatum* and *Carex ruchebraunii*, and the climbing plants *Clematis rannunculoides* and *Smilax mairei*. *Usnea longissima* often hangs from the branches of the *Quercus* trees.

Figure 3.4a,b shows the woody species richness and the reciprocal of the Berger-Parker index (1/d) (Magurran 1988) for the three forests. The mean levels of woody species richness are low (7–11 species) (Fig. 3.4a), but significantly higher in the *Q. longispica* communities (11 species) than in the other two. The low reciprocals of the Berger-Parker indices (1.01–1.33) show the high dominance concentration

Table 3.2 Woody species of three evergreen *Quercus* forests in northwestern Yunnan

		Northwestern Yunnan		
	Forest types	QAq	QP	QL
	Number of plots	6	4	6
	Plot area (m^2)	400	400	400
	Area of plots (m^2)	2,400	1,600	2,400
	Sites	Bitahai	Bitahai, Baima-xueshan	Feilaisi, Yubeng, Adong
	Elevation (m)	3,500–3,900 RBA (%)	3,550–3,600 RBA (%)	3,353–3,538 RBA (%)
Family	**Dominant species**			
Fagaceae	*Quercus aquifolioides*	74.3		
	Quercus pannosa		95.2	
	Quercus longispica			94.9
	Companion woody species			
Ericaceae	*Rhododendron rubiginosum*	9.9	1.1	
	Rhododendron spiciferum	0.2		
	Rhododendron vernicosum		0.9	
	Rhododendron oreotrephes		0.5	1.4
	Rhododendron decorum			0.8
	Lyonia doyonensis			0.3
	Rhododendron yunnanensis			0.2
Pinaceae	*Larix potaninii*	8		
	Picea likiangensis	5.9	0.8	0.19
	Abies georgei var. *smithii*	0.7		0.09
	Pinus densata			0.8
	Pinus armandii		0.8	
Rosaceae	*Sorbus rufopilosa*	1		
	Sorbus rehderiana		0.4	
	Rosa sweginzowii		0.03	
	Dichotomanthes tristaniaecarpa		0.03	0.04
	Cotoneaster franchetii			0.04
Caprifoliaceae	*Lonicera hispida*		0.1	0.05
	Viburnum glomeratum subsp.			
	rotundifolium			0.03
Hydrangeaceae	*Deutzia purpurascens*		0.04	0.03
Leguminosae	*Caragana franchetiana*		0.02	0.02
Grossulariaceae	*Ribes glaciale*		0.01	0.05
Salicaceae	*Populus szechuanica*			0.8
Betulaceae	*Betula utilis*			0.2
Rhamnaceae	*Berchemia yunnanensis*			0.02
Berberidaceae	*Berberis stiebritziana*			0.01
Labiatae	*Elsholtzia densa*			0.002

Data sources: Tang (2006) and Yu et al. (2003)

QAg Quercus aquifolioides forest, *QP Quercus pannosa* forest, *QL Quercus longispica* forest, *RBA* relative basal area

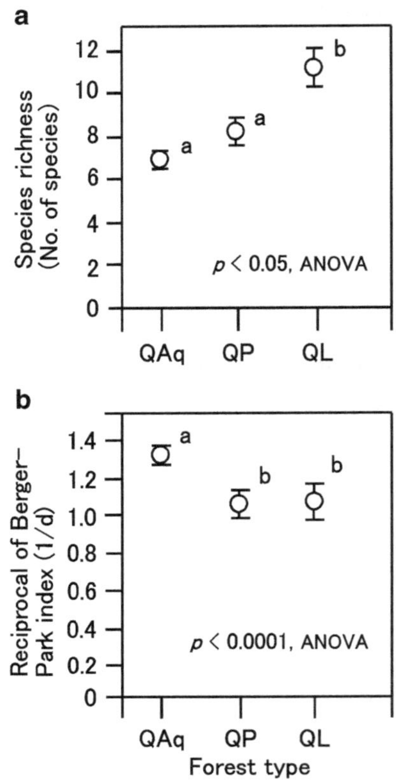

Fig. 3.4 Species richness (mean ± SE) (**a**) and the reciprocal of Berger-Park index (1/d) (mean ± SE) for the evergreen *Quercus* forests in NW Yunnan. Significantly different values are indicated by different letters. Forests sharing the same letters do not differ significantly by ANOVA, $p<0.05$ for (**a**), $p<0.001$ for (**b**). *QAg Quercus aquifolioides* forest, *QP Quercus pannosa* forest, *QL Quercus longispica* forest (Data source: Tang (2006))

and low diversity in the forest communities (Fig. 3.4b). *Q. aquifolioides* has a value higher (1.33) than the others. The tree height of the three *Quercus* species reaches over 20 m (Fig. 3.5a). The maximum DBH is 57 cm for *Q. aquifolioides*, 62 cm for *Q. pannosa*, and 75–82 cm for *Q. longispica* (Fig. 3.5b). The basal area of 43.5 m^2 ha^{-1} for *Q. aquifolioides* is the highest, while those for *Q. pannosa*, *Q. longispica* are low (<20 m^2 ha^{-1}) (Fig. 3.5c). *Q. pannosa* shows the lowest density with 918.8 stems ha^{-1} (Fig. 3.5d). Though the ages of the three forests differ, their structural features represent the oldest, most natural *Quercus* communities of their respective local regions. The *Q. aquifolioides* and *Q. pannosa* trees are c. 60–75 years old, and those of *Q. longispica* c. 60–90 years old.

These evergreen species of *Quercus* in the three forests show both sporadic and inverse-J types of frequency of size distributions (Fig. 3.6). The sporadic type, with highest values for the intermediate size classes, indicates the possibility of good regeneration for *Q. pannosa* and *Q. aquifolioides*. The inverse-J type, having the highest frequency in small dbh classes with a gradual decrease in the number of individuals towards the large size-classes, suggests that regeneration is active in *Q. longispica*. Sprouting accounts for most of the regeneration in these forests. *Q. longispica* reaches the highest ratio of sprouts to main stems at 0.94, followed by *Q. pannosa* (0.47) and *Q. aquifolioides* (0.35). The sprouts originate above or

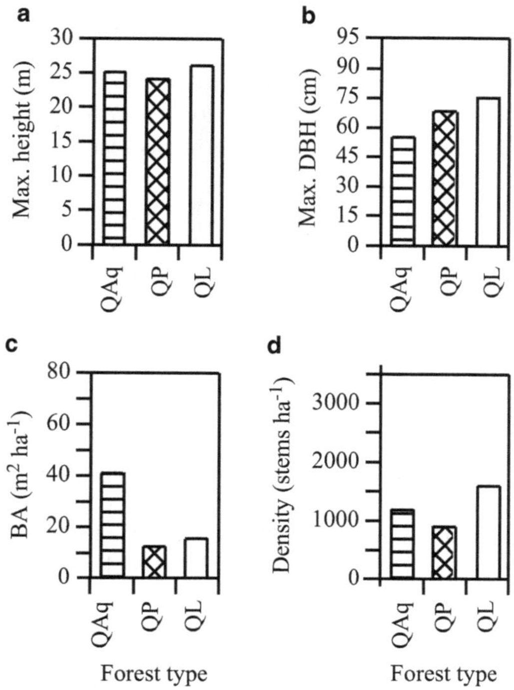

Fig. 3.5 (**a–d**) Structural features of the evergreen sclerophyllous *Quercus* forests. *QAq Quercus aquifolioides*, *QP Q. pannosa*, *QL Q. longispica* (Data source: Tang (2006))

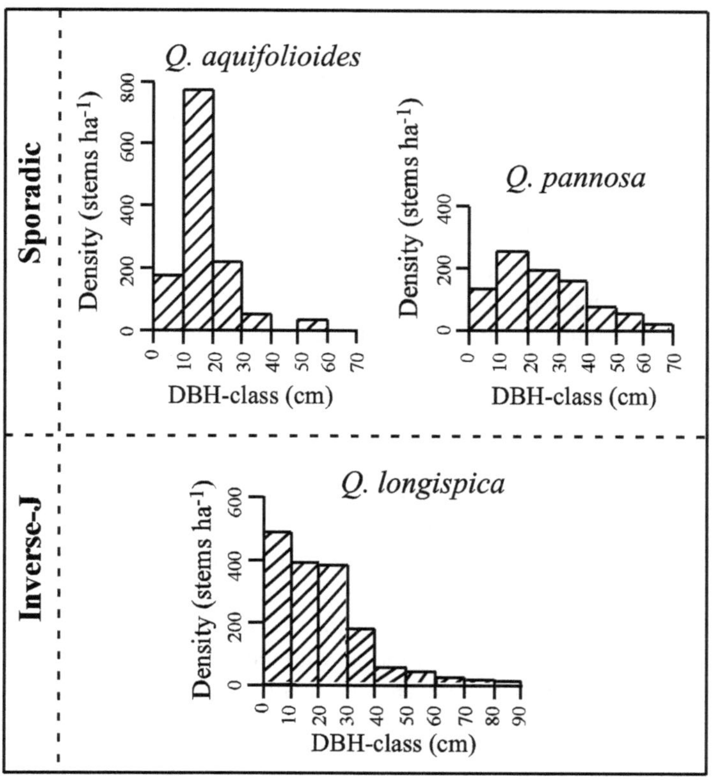

Fig. 3.6 Frequency of DBH-classes of the studied evergreen sclerophyllous *Quercus* species (Data source: Tang (2006))

below ground level. For *Q. pannosa* and *Q. longispica*, sprouting also occurs in seedlings and saplings. *Q. longispica* resprouts more vigorously after being cut. The ratio of sprouts to main stems reaches 4.4 in the coppices of *Q. longispica* in NW Yunnan.

In some species a sprout has a higher capacity to form adventitious roots than cuttings taken from shoots in the crown of the same tree. This indicates a high resilience after disturbance. Sprouting in seedlings promotes their survival under a variety of stressful conditions, including suppression by canopy trees, herbivory, site exposure, pathogens and desiccation. Although sprouting is more conspicuous in mature trees than in seedlings, it is less significant ecologically (del Tredici 2001). The *Quercus* species generally have strong sprouting ability, especially after disturbances.

Development of deep roots is thought to be one of the mechanisms for surviving drought. Data on rooting depth for evergreen *Quercus* of NW Yunnan are not available, though all the species have been observed by the author on sites with deep soils as well as on very shallow, rocky soils. And also some seedlings and a few saplings of *Q. longispica* were found on moderately dry slopes in Adong and Deqin.

According to Zhang et al. (2005), the optimum photosynthetic temperature (T*opt*) of *Q. pannosa* is significantly lower than those of Mediterranean oaks. This species was found to adapt well to low but not to high temperatures, which indicates that cold-tolerance seems an important determinant of the distribution of this oak in the NW Yunnan area.

The healthy survival of evergreen sclerophyllous *Quercus* forests in NW Yunnan is threatened by firewood production and livestock raising. Most of the evergreen *Quercus* forests have been turned into coppice and scrub communities by human activities. Some aged stumps left from cutting in relatively natural forests are still found in NW Yunnan.

Tang (2006) compared the forests of *Q. aquifolioides*, *Q. pannosa*, *Q. longispica* in NW Yunnan with *Q. agrifolia* forest in central coastal California (USA) and *Q. ilex* forest in Catalonia (NE Spain). Although NW Yunnan has warm rainy summers and cold dry winters, the reverse of the hot dry summers and cool rainy winters of the Mediterranean-type climate in California and Spain, and in spite of the long distances separating the communities, the general patterns of the evergreen *Quercus* forests are similar. These general patterns are defined by a similar high dominance concentration and low species diversity according to the reciprocal Berger-Park index (1/d), sparse ground vegetation with herbaceous species and climbing plants, and a low rate of seedling establishment that is compensated by sprouting. Though differing assemblages of species are involved in the three regions, the forests in NW Yunnan share species of the families Ericaceae, Rosaceae, Betulaceae, Grossulariaceae and Berberidaceae with the California forest, and of Ericaceae, Caprifoliaceae, Rhamnaceae and Rosaceae with the forest in NE Spain. Altogether, the *Quercus* forests of NW Yunnan are structurally more similar to the *Q. agrifolia* forest of California than to the *Q. ilex* forest in NE Spain. This may be due to long-time human influence on the Spanish forest. In contrast, the evergreen *Quercus* forests of NW Yunnan and California are relatively natural. The Chinese evergreen *Quercus* forests are thus precious remnants of a once widespread forest type which today is threatened wherever it survives.

References

Camus A (1936–1954) Les Chênes. Monographie du genre Quercus. Lechevalier, Paris

del Tredici P (2001) Sprouting in temperate trees: a morphological and ecological review. Bot Rev 67:121–140

Huang WL, TU YL, Yang L (eds) (1988) The vegetation of Guizhou. Guizhong People's Press, Guiyang

Huang C-J, Zhang Y-T, Xu Y-C (1998) Flora of China, vol 22. Science Press, Beijing, pp 213–332

Jing ZZ (1981) A mossy forest appearing in the sclerophyllous evergreen broad-leaf forests-Quercetum *pannosum nimborum*. Acta Bot Yunnanica 3:75–88

Li H, Guo S (1976) Miocene flora of Nanmulin in Tibet. Acta Palaeobot Sin 15:598–609

Li JQ, Guo YS, Romane F (2000) Environmental heterogeneity and population variability of sclerophyllous oaks (*Quercus* Sec. suber) in East Himalayan Region. For Stud China 2:1–15

Liu X-L, Liu S-R, He F, Yang D-S, Yang Y-P, Ma Q-Y (2008) Taxonomy and modern geographical distribution of species of sclerophyllous alpine oak plants in China. J Sichuan For Sci Technol 29(3):1–7

Magurran AE (1988) Ecological diversity and its measurement. Princeton University Press, Princeton

Sun H (2002) Tethys retreat. Himalaya and Hengduan Mountains and its significance for the origin and development of Sino-Himalayan elements and alpine flora. Acta Bot Yunnanica 24:273–288

Sun H, Li ZM (2003) The Qinghai-Tibet Plateau uplift and its impact on Tethys flora. Advanced Earth Sci 18:852–862

SVEG (Sichuan Vegetation Editorial Group) (1980) Vegetation of Sichuan. Sichuan People's Press, Chengdu

Tang YC (2000) On the affinities and the role of the Chinese flora. Acta Bot Yunnanica 22:1–26

Tang CQ (2006) Evergreen sclerophyllous *Quercus* forests in northwestern Yunnan, China as compared to the Mediterranean evergreen Quercus forest types in California, USA and northeastern Spain. Web Ecol 6:88–101

Tao JR (1986) Fossil flora and its significance in Lanping–the middle section of the Hengduan Range. In: Qinghai-Tibet Plateau Exploring Group (ed) The exploration of the Hengduan Range (II). Beijing People's Press, Beijing, pp 58–65

Tao JR, Kong SH (1973) Fossil flora and sporo-pollen assemblage in Eryuansanying coal formation, Yunnan. Acta Bot Sin 15:120–126

Wu ZY (ed) (1980) Vegetation of China. Science Press, Beijing

Wu ZY, Zhu YC, Jiang HQ (eds) (1987) The vegetation of Yunnan. Science Press, Beijing

Yang QZ (1990) The characteristics and classification of sclerophyllous oaks in the Himalayan region of China (in Chinese with English abstract). Acta Phytoecol Geobot Sin 14:197–211

Yu Y, Cao M, Liu WS (2003) Diversity of tree species of subalpine coniferous forests and *Quercus* sclerophyllous forests in Northwest Yunnan. J Mt Sci 21:568–575

Zhang SB, Zhou ZK, Hu H (2005) Photosynthetic performances of *Quercus pannosa* vary with altitude in the Hengduan Mountains, southwest China. For Ecol Manag 212:291–301

Zhou ZK (1992) Origin, phylogeny and dispersal of Quercus from China. Acta Bot Yunnanica 14(3):227–236

Zhou ZK (1993) The fossil history of Quercus. Acta Bot Yunnanica 15(1):21–33

Zhou ZK (1996) Study on Dryophyllum complex from China and its geological and systematic implication. Acta Bot Sin 38(8):666–671

Zhou ZK (1999) Fossils of the Fagaceae and their implications in systematics and biogeography. Acta Phytotaxon Sin 37:369–385

Zhou ZK, Sun H, Yu HY (1995) Distribution of Fagaceae in Tibet. Acta Bot Yunnanica 17(2):144–152

Zhou ZK, Pu CX, Chen WY (2003) Relationships between the distributions of *Quercus* sect heterobalanus (Fagaceae) and uplift of Himalayas. Advanced Earth Sci 18:884–890

Chapter 4
Warm Temperate Coniferous Forests

Abstract Warm temperate coniferous species, principally including *Pinus yunnanensis, Pinus kesiya, Pinus armandii, Pinus massoniana, Keteleeria evelyniana, Keteleeria davidiana, Cunninghamia lanceolata, Cryptomeria japonica* var. *sinensis, Cupressus funebris* and *Cupressus duclouxiana,* are often found in areas where the evergreen broad-leaved forest formerly existed in SW China. After destruction of the evergreen broad-leaved forest by human activities, these coniferous species grow in early or intermediate-successional stands in the subtropical evergreen broad-leaved forest zone. Topo-climax warm temperate coniferous forests (e.g. Tertiary relict forests consisting of so-called living fossil plants, such as coniferous *Taiwania cryptomerioides, Metasequoia glyptostroboides, Cathaya argyrophylla, Thuja sutchuenensis,* and gymnospermous *Ginkgo biloba*) also grow in the subtropical region.

In Yunnan, a large area is now dominated by the coniferous *Pinus yunnanensis* forest, which has naturally regenerated and represents an early intermediate stage in the reestablishment of the evergreen broad-leaved forest after human disturbance. Forest fires are frequent in Yunnan. The dynamics of the seral *Pinus yunnanensis* forest after a major fire indicate that the burned *Pinus* communities should recover naturally to their pre-fire condition over a period of between 20 and 35 years, assuming that regeneration does occur and the colonists are not predominantly exotic. Disturbance by fire has facilitated *Pinus* regeneration. Natural succession of plant communities after a major fire, as opposed to human intervention as in plantations, is the most efficient way to its reestablishment.

4.1 Distribution and Forest Types

Many warm temperate coniferous species, including *Pinus yunnanensis, Pinus kesiya, Pinus massoniana, Pinus armandii, Keteleeria evelyniana, Keteleeria davidiana, Cunninghamia lanceolata, Cryptomeria japonica* var. *sinensis, Cupressus funebris, Cupressus duclouxiana, Taiwania cryptomerioides, Metasequoia glyptostroboides, Cathaya argyrophylla*, and *Thuja sutchuenensis,* are present in the subtropical region where the evergreen broad-leaved forest is the zonal forest in SW China. The term "zonal forest" reflects the climate, rather than local effects of soil, topography, disturbance, and other non-climatic factors; where local influences are

C.Q. Tang, *The Subtropical Vegetation of Southwestern China*,
Plant and Vegetation 11, DOI 10.1007/978-94-017-9741-2_4

important, the forest is described as an "azonal forest". In the evolutionary process, warm temperate conifers such as pines appeared in the Jurassic. Before the Tertiary, species of *Pinus* had grown in the temperate Northern Hemisphere. They invaded more southern zones during the harsh conditions of the Quaternary, and also colonized north to boreal areas in the interglacial periods. Many other pine species adapted physiologically and ecologically to the subtropical environment (Nakagoshi 1995; Li 1999; Yu 1999). Fossils of *Metasequoia*, *Taiwania*, *Cathaya* and *Thuja* from the Mesozoic and Cenozoic are found in many parts of the Northern Hemisphere, but now their extant species grow only in southwestern and south-central China, except that *Taiwania* also occurs in Taiwan and Vietnam. Some other Chinese paleoendemic conifers (e.g. *Torreya fargesii*, *Taxus wallichiana* var. *chinensis*, *Cephalotaxus fortunei*) now appear scattered in the subtropical zone, but not as dominants in their stands.

Warm temperate coniferous forests in the subtropical zone most often represent secondary vegetation or seral forests developing in areas where the evergreen broad-leaved forest formerly existed. After destruction of the evergreen broad-leaved forest by humans, warm temperate coniferous forests (e.g. *Pinus yunnanensis*, *Pinus massoniana*, *Pinus armandii*, *Keteleeria evelyniana*) can be regarded as a seral stage of succession, whereas the late successional (climax) vegetation on these sites will be the evergreen broad-leaved forest (Fig. 4.1). It should be remarked, however,

Fig. 4.1 A naturally regenerated *Keteleeria evelyniana* and *Pinus yunnanensis* community after the natural evergreen broad-leaved forest was completely logged (clear-cut) 20 years ago in Luquan, central Yunnan

that on steep slopes, or in habitats with outcrops and soils poor in nutrients, or on limestone sites, primary forests of *Pinus yunnanensis* or *Pinus armandii* or *Pinus kesiya*, and topo-climax Tertiary relict coniferous forests consisting of so-called living fossil plants can be found. The species comprising the Tertiary relict forests are exemplified by coniferous *Taiwania cryptomerioides, Metasequoia glyptostroboides, Cathaya argyrophylla, Thuja sutchuenensis* and gymnospermous *Ginkgo biloba*. These species occur independently of each other, and all are dispersed in small specific areas. These warm temperate/subtropical coniferous forests are considered as azonal forests. The Tertiary relict forests will be detailed in Chap. 7.

Pinus yunnanensis grows in Yunnan, western Sichuan, western Guizhou, southeastern Tibet as well as in Guangxi (Figs. 4.2 and 4.3a–c). It has a wide altitudinal distribution range, 400–3,200 m, but grows mainly at 1,600–2,900 m. *Pinus yunnanensis* var. *tenuifolia* is found at 400–1,200 m, usually along river banks in southwestern Guizhou and western Guangxi. *Pinus yunnanensis* var. *pygmaea* forms a scrubby woodland can reach 3,100 m, but mainly from 1,600 to 2,800 m in central, northwestern Yunnan and southwestern Sichuan. *Pinus kesiya* is found at

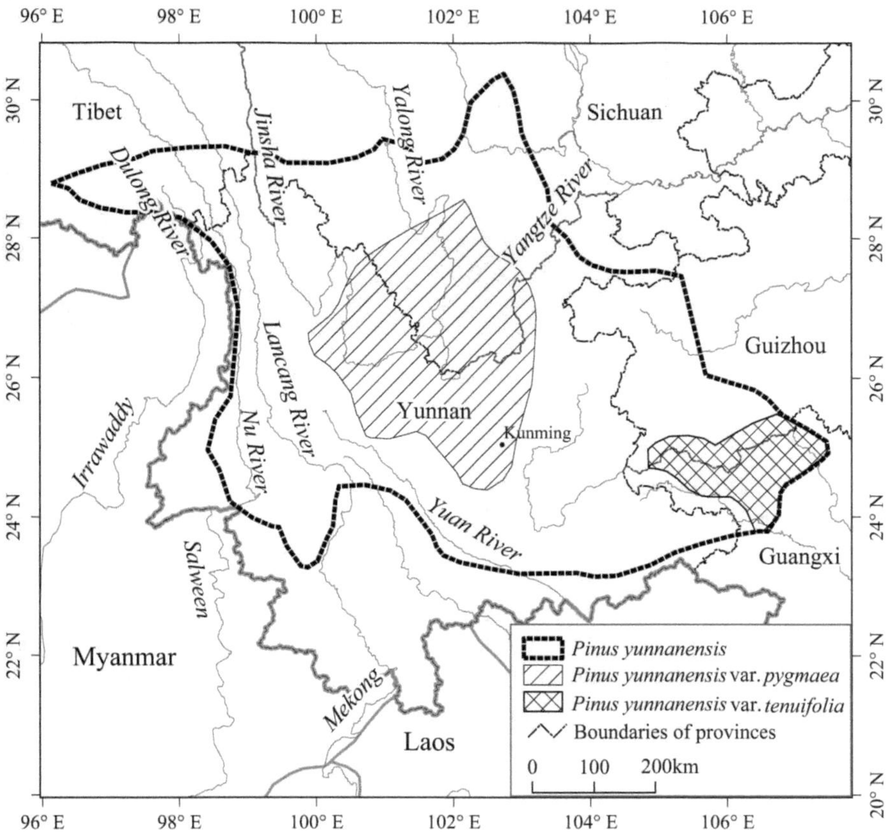

Fig. 4.2 Geographic distribution range of *Pinus yunnanensis*, *Pinus yunnanensis* var. *tenuifolia*, and *Pinus yunnanensis* var. *pygmaea* (Data source: Jin and Peng (2004))

Fig. 4.3 (**a**) The long needles and the seed cones of *Pinus yunnanensis*. (**b**) Needles 2 or 3 per bundle of *P. yunnanensis*. (**c**) A typical *P. yunnanensis* forest at 1,650 m in Yimen, central Yunnan

600–1,600 m in southern and southwestern Yunnan and southeastern Tibet, extending into NE India, Laos, Myanmar, Philippines, Thailand, and Vietnam. In southwestern Yunnan, *Pinus kesiya* grows mainly in valleys while *Pinus yunnanensis* occupies mountain slopes. *Pinus armandii* is widely distributed in southern Gansu, central and NW Guizhou, Hainan, SW Henan, western Hubei, southern Shaanxi, southern Shanxi, Sichuan, SE Tibet, Yunnan and northern Myanmar, with altitudes ranging from 1,000 to 3,300 m. *Pinus armandii* has two further varieties. *Pinus armandii* var. *mastersiana* is found at 1,800–3,300 m in central Taiwan. *Pinus armandii* var. *amamiana* is restricted to the Yaku-shima and Tanega-shima Islands, southwestern Japan. *Pinus massoniana* is distributed below 1,500 m in Guizhou, Sichuan, Chongqing, eastern and southern China. *Keteleeria evelyniana* grows at 700–2,800 m, but mainly at 1,600–2,200 m in Yunnan, western Sichuan and western Guizhou. *Keteleeria davidiana* is found at 600–1,500 m in Gansu, Hunan, Sichuan,

Guizhou, Shanxi and Hubei, but not in Yunnan. *Cunninghamia lanceolata* is distributed at 300–2,800 m in a vast area of subtropical China, but its native distribution range is uncertain due to widespread plantations of this species. *Cryptomeria japonica* var. *sinensis* is usually found below 1,100 m in Fujian, Jiangxi, Sichuan, Yunnan, and northwestern Zhejiang. *Cupressus duclouxiana* is distributed at 1,400–3,300 m in central and northwestern Yunnan, Guizhou, southwestern Sichuan, and southeastern Tibet. *Cupressus funebris* is found below 2,000 m in the subtropics, especially in limestone areas. However, all the species except *Pinus yunnanensis* var. *tenuifolia, Pinus yunnanensis* var. *pygmaea*, and *Keteleeria davidiana,* are extensively cultivated in SW China. Sometimes it is very difficult to tell whether they regenerated naturally or grew from seeds dispersed by airplanes.

The seral warm-temperate coniferous forest's height usually ranges from 18 to 25 m, but may reach 30 m. The vertical strata of the forests are rather simple, usually two layers – the canopy and the shrub layers. Species composition of five typical seral warm temperate coniferous forests (*Pinus yunnanensis*, *Pinus kesiya, Pinus armandii, Keteleeria evelyniana* and *Pinus massoniana* forests) is represented in Table 4.1. *Pinus yunnanensis* grows along with evergreen *Lithocarpus mairei, Rhododendron microphyton, Michelia yunnanensis, Myrica nana,* and *Alnus nepalensis* in Qiongzhusi in the Kunming area of Yunnan. *Pinus kesiya* grows with *Cyclobalanopsis delavayi* and *Schima wallichii* etc. in Sanguozhuang in the Ailao Mountains, Yunnan. *Pinus armandii* is mainly accompanied by *Castanopsis orthacantha* and *Alnus nepalensis* in Xizishan, Chuxioang, Yunnan. *Keteleeria evelyniana* grows along with *Cyclobalanopsis glaucoides*, *Quercus aliena,* etc. in Xishan of the Kunming area. The *Pinus massoniana* forest is found with *Castanopsis hystrix*, *Symplocos laurina*, and *Eurya semiserrulata* in the low altitudes of Mt. Emei, Sichuan, and with *Meliosma beaniana, Dalbergia hupeana, Populus adenopoda, Cinnamomum camphora, Cyclobalanopsis glauca* below 1,000 m in Guizhou Province. Around altitude 1,500 m in northwestern Guizhou grows a *Pinus yunnanensis* and *Pinus massoniana* mixed forest, including *Myrica nana, Rhododendron* spp. and *Quercus fabri*.

4.2 Dynamics of the *Pinus Yunnanensis* Forest after a Major Fire

Post-disturbance succession is a key factor in vegetation dynamics, a feature inherent to forests that favors species diversity and persistence (Faliński 1986; Knapp 1986). Differences in disturbance strongly influence successional trends. In tropical forests, ecological species groups are characterized by their successional status (pioneer or non-pioneer) which in part relates to the light requirements of various species (Swaine and Whitmore 1988). Among disturbances, fire is one of the most far-reaching in both space and time (Bond and Keeley 2005). Turner et al. (1997) demonstrated that species responses vary with fire size and pattern. Both naturally occurring and human-induced fires have been essential components of the

Table 4.1 Woody species (≥1.3 m at height) composition for typical seral warm temperate coniferous forests

Site	QZSY	XZSY	XSKY	SASY	ESS
Altitude (m)	2,193	2,400	1,985	1,540	655
Each plot area (m^2)	400	400	400	400	400
Number of plots	5	5	5	5	5
Forest type	PY	PA	KE	PK	PM
Species	RBA (%)	RBA (%)	RBA (%)	RBA (%)	RBA (%)
Pinus yunnanensis	**50.8**				
Lithocarpus mairei	7.6				
Rhododendron microphyton	6.5				
Michelia yunnanensis	4.4		0.004		
Lyonia ovalifolia	4.1				
Alnus nepalensis	3.9	2.3		0.004	
Myrica nana	3.6				
Rhododendron decorum	3.4				
Ternstroemia gymnanthera	2.5	0.1	0.02		
Myrsine africana	1.9		0.03		
Camellia reticulata	1.6				
Castanopsis orthacantha	1.4	24.1			
Rhododendron spiciferum	1.4				
Rhododendron siderophyllum	1.2				
Cupressus funebris	1.1				
Hypericum patulum	0.9				
Lithocarpus dealbatus	0.8	0.9	1.7		
Viburnum cylindricum	0.8				
Populus adenopoda	0.7				
Cotoneaster microphyllus	0.5				
Rhododendron spinuliferum	0.3		0.01		
Vaccinium bracteatum	0.07		0.04		
Elaeagnus stellipila	0.04				
Keteleeria evelyniana	0.01		**60.6**		
Buddleja asiatica	0.003				
Dichotomanthus tristaniaecarpa	0.003				
Cyclobalanopsis glaucoides	0.002		8.3		
Elsholtzia rugulosa	0.002				
Pyracantha fortuneana	0.002				
Pinus armandii		**70.4**			
Craibiodendron yunnanensis		1.4			
Eurya nitida		0.5			
Schima argentea		0.3			
Viburnum erubescens		0.02			
Hypericum monogynum		0.005			
Quercus aliena		0.005	29.1		

(continued)

Table 4.1 (continued)

Site	QZSY	XZSY	XSKY	SASY	ESS
Altitude (m)	2,193	2,400	1,985	1,540	655
Each plot area (m^2)	400	400	400	400	400
Number of plots	5	5	5	5	5
Forest type	PY	PA	KE	PK	PM
Species	RBA (%)	RBA (%)	RBA (%)	RBA (%)	RBA (%)
Daphne feddei		0.004			
Styrax gradiflora		0.004			
Lindera communis			0.2		
Castanopsis delavayi			0.04		
Camellia assamica			0.01		
Schoepfia jasminodora			0.01		
Symplocos chinensis			0.01		
Pyrus pashia			0.005		
Nothopanax delavayi			0.003		
Rhamnus leptophyllus			0.0004		
Pinus kesiya				**79.3**	
Photinia sp.				6.7	
Schima wallichii				4.9	
Cyclobalanopsis delavayi				4.1	
Rapanea neriifolia				2.3	
Wendlandia uvariifolia				1.7	
Diospyros blancoi				0.5	
Phyllanthus emblica				0.4	
Engelhardia colebrookiana				0.1	
Glochidion hirsutum				0.04	
Cyclobalanopsis kerrii				0.03	
Pinus massoniana					**67.4**
Castanopis hystrix					21.6
Symplocos laurina					4.1
Eurya semiserrulata					3.5
Cryptomeria japonica var. *sinensis*					2.8
Mallotus tenuifolius					0.4
Lindera nacusua					0.1
Vaccinium carlesii					0.01

Partial data from Tang et al. (2010)
QZSY Qiongzhusi of Kunming, Yunnan, *XZSY* Xizishan of Chuxiong, Yunnan; *XSKY* Xishan of Kunming, Yunnan, *SASY* Sanguozhuang in Ailaoshan, Yunnan, *ESS* Emeishan, Sichuan, *PY* Pinus yunnanensis, *PA* Pinus armandii, *PK* Pinus kesiya, *KE* Keteleeria evelyniana, *PM* Pinus massoniana, *RBA* relative basal area".

disturbance regimes influencing forest succession (Mooney et al. 1980; Russell 1983; Delcourt and Delcourt 1998). Forest fires are frequent in Yunnan, but a major fire, of a square kilometer or more, is relatively rare in the subtropical parts of the province (Zheng et al. 2007). In Yunnan, the precipitation is particularly low in winter and spring, and the fire season is in March and April. In 2006, an unusually large forest fire (it remains unknown whether it was wild or caused by human accident), composed of a mixture of crown and surface fires, took place in the Anning–Tuanjiexiang area of central Yunnan from 29 March to 12 April, burning approximately 18.49 km^2 of naturally regenerated seral *Pinus yunnanensis* forest and producing a heterogeneous mosaic of burn severities and islands of unburned vegetation across the landscape (Fig. 4.4).

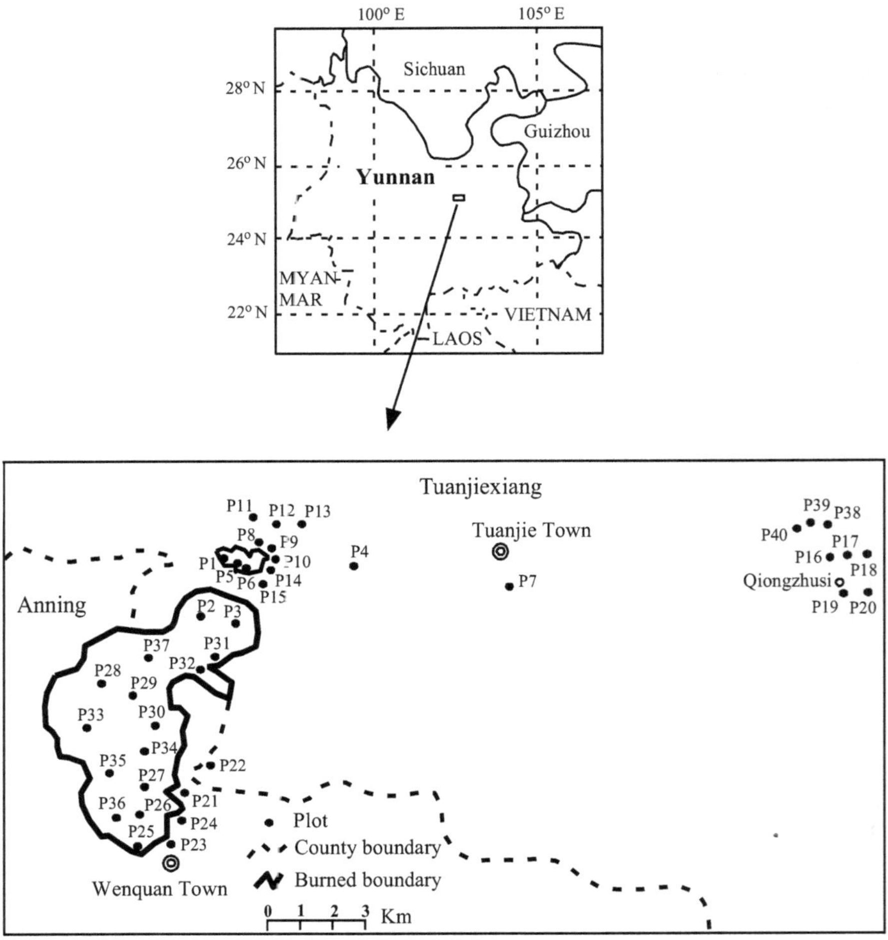

Fig. 4.4 The burned area and the location of 40 study plots (Tang et al. 2013)

4.2.1 Community Structure and Regeneration

Five years after the forest fire, patches of plant communities differed between the burned and unburned areas in terms of vegetation structure and regeneration pattern. A quick recovery of the *Pinus yunnanensis* community has been observed after the fire, involving the regeneration of woody species and the floristic composition. We documented the responses of the *Pinus* community to the disturbance of the major fire (Tang et al. 2013).

Based on the floristic composition of the 37 plots and using Sørensen's similarity index and group average clustering, five groups were identified at a similarity of 21 % (Fig. 4.5).

1. Group 1: a *Pinus* shrubland (PS) regenerated after the forest mega-fire. According to burn severity, this group can be divided into two subgroups. One (12 plots) was burned severely (SBPS), the other (six plots) was burned moderately

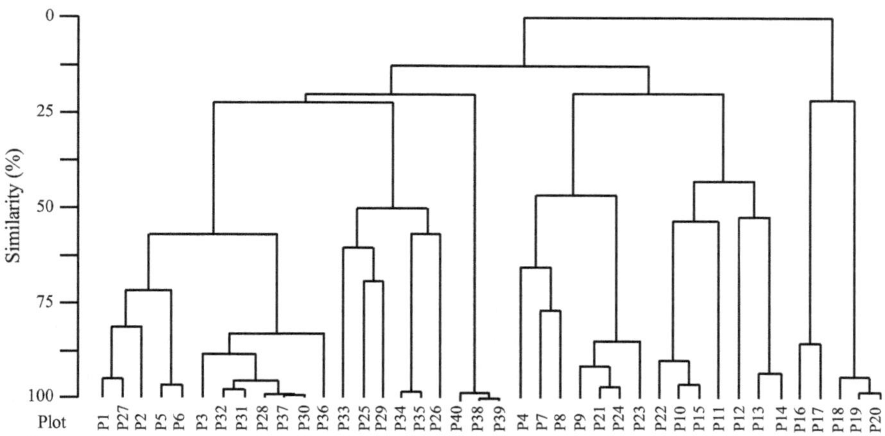

Group	Group 1		Group 2	Group 3	Group 4	Group 5
Burn severity	Severe	Moderate	Notburned	Not burned	Not burned	Not burned
Degree of disturbances (score)	40	30	15	10	5	1
Community age (year)	5	5-15	16	20-25	30-40	80-145
Community type	*Pinus* shrubland (SBPS)	*Pinus* shrubland (MBPS)	*Pinus* plantation (PP)	*Pinus* forest (PF)	Coniferous and ever-green broad-leaved mixed forest (PY-LM)	Mature EBLF (EBLF)
Dominant	*Pinus yunnanensis*		*Pinus yunnanensis*	*Pinus yunnanensis*	*Pinus yunnanensis* *Lithocarpus mairei*	LIDE CAOR

Fig. 4.5 Similarity dendrogram using relative Sørensen's similarity index and group average clustering for the sampled plots in the burned and unburned patches of a *Pinus yunnanensis* forest, and the unburned mature evergreen broad-leaved forest. *SBPS* severely burned *Pinus* shrubland, *MBPS* moderately burned *Pinus* shrubland, *PY–LM* mixed coniferous and evergreen broad-leaved forest, *EBLF* mature evergreen broad-leaved forest, *LIDE Lithocarpus dealbatus*, *CAOR Castanopsis orthacantha*

(MBPS). Both of the subgroups were dominated by *P. yunnanensis* in the understory and shrub layer; no trees more than 8 m tall were found 5 years after the fire. The latter subgroup contained abundant resprouts of some evergreen broad-leaved plants. No plots were classified as burned slightly.

2. Group 2: a *Pinus* plantation (PP) without fire occurrence. It was planted as a restoration measure when the evergreen broad-leaved forest was clear cut 16 years ago.
3. Group 3: a *Pinus* forest (PF) without fire occurrence. The canopy layer was exclusively dominated by *P. yunnanensis*. This forest was in a natural process of secondary succession after the original evergreen broad-leaved forest was destroyed 20–25 years ago.
4. Group 4: a coniferous and evergreen broad-leaved mixed forest (PY-LM) with an age of 30–40 years, dominated by *P. yunnanensis* and *Lithocarpus mairei* and without fire occurrence. It was at a middle successional stage in the reestablishment of the evergreen broad-leaved forest.
5. Group 5: a mature evergreen broad-leaved forest (EBLF) with an age of 80–145 years, dominated by *Castanopsis orthacantha* and *Lithocarpus dealbatus* and without fire occurrence. It represents the original natural vegetation of central Yunnan.

In the severely burned *Pinus* shrubland (SBPS), crown and surface fires occurred: as a result there was extensive tree mortality, and the soil organic layer was completely consumed. No individuals more than 4 m tall were found. In the moderately burned *Pinus* shrubland (MBPS), surface burns occurred (the soil organic layer was completely consumed), and canopy trees were generally killed, but some individuals between 4 and 8 m tall survived. A great number of seedlings of *Pinus yunnanensis* were found (Fig. 4.6a, c). Saplings and resprouts appeared in the burned area. The number of seedlings and saplings of *Pinus yunnanensis* was lower (7,431 ha^{-1}) in the severely burned *Pinus* shrubland (SBPS) than in the moderately burned *Pinus* shrubland (MBPS) (10,075 ha^{-1}). The frequency distribution at height-classes showed an inverse-J pattern, indicating strong regeneration in both severely and moderately burned sites. This shows that disturbance of fire has facilitated *Pinus yunnanensis* regeneration. Some small individuals of *Pinus yunnanensis,* 7 m in height and 9–12 cm DBH, were found in the sites, having survived moderate burning, but no larger trees survived. This coniferous species displayed a resprouting capacity after the forest fire. Evergreen broad-leaved *Lithocarpus mairei* resprouted vigorously in the shrubland, especially in the sites burned moderately (Fig. 4.6b, d). The plantation was exclusively dominated by a single species *Pinus yunnanensis*, with a large number of stems of 4–8 m in height, with DBH ranging from 6 to 15 cm. The absence of seedlings (<1.3 m high) in the plantation indicated poor regeneration (Fig. 4.6e). As the *Pinus yunnanensis* forest without fire occurrence had a larger number of stems in the height-class 1.3–8 m, its canopy layer had reached 16 m, corresponding to a DBH of 30 cm (Fig. 4.6f). *Lithocarpus mairei* had resprouted frequently after cutting for firewood 20 years ago (Fig. 4.6g). This *Pinus yunnanensis* forest reflects the state of the plant community before the occurrence

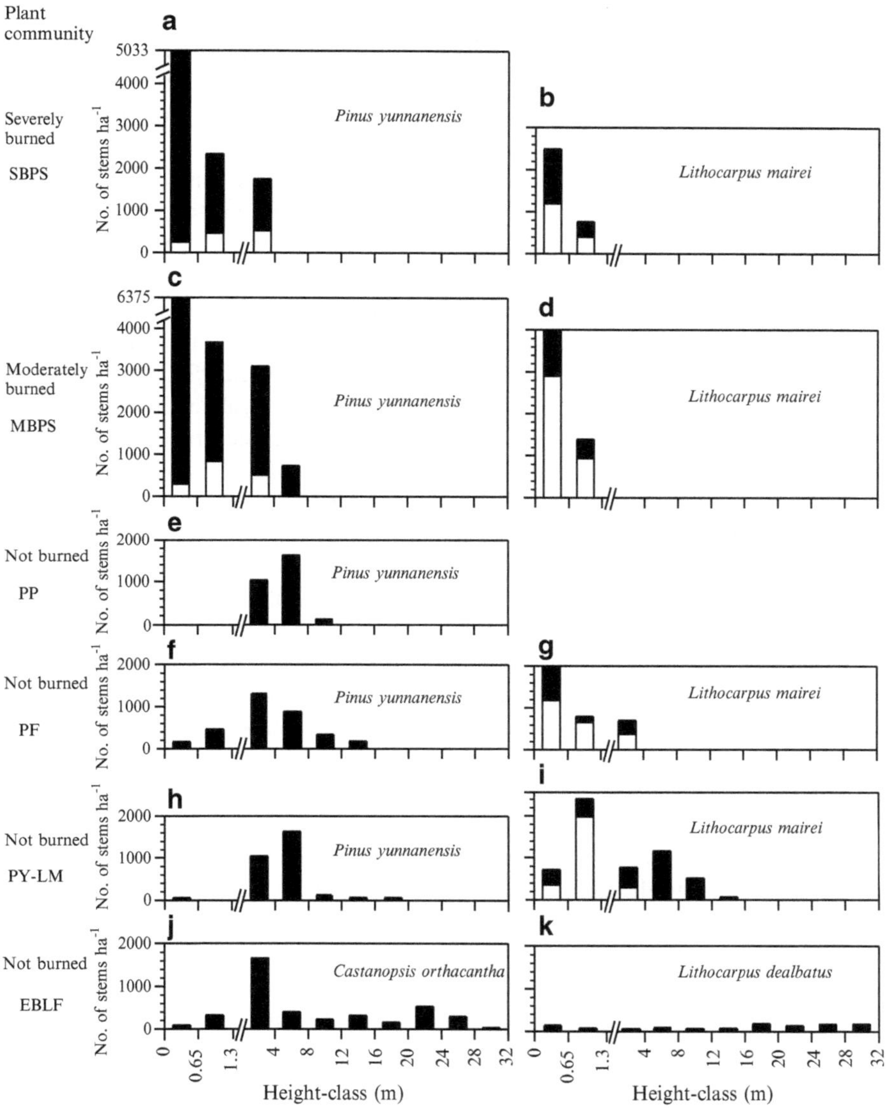

Fig. 4.6 The frequency distribution by height class for the representative species of each plant community in the burned and unburned patches of a *Pinus yunnanensis* forest, and the unburned mature evergreen broad-leaved forest (**a**–**k**). Main stems are indicated by shaded bars, resprouts by empty bars. *SBPS* severely burned *Pinus* shrubland, *MBPS* moderately burned *Pinus* shrubland, *PP Pinus* plantation, *PF Pinus* forest, *PY–LM* mixed coniferous and evergreen broad-leaved forest, *EBLF* mature evergreen broad-leaved forest (Data source: Tang et al. (2013))

of fire. In the mixed coniferous and evergreen broad-leaved forest, the height of *Pinus yunnanensis* stems reached 20 m with 40 cm DBH, and *Lithocarpus mairei* reached 15 m with 30 cm DBH. Many individuals were found in the shrub layer and understory (Fig. 4.6h, i). In the mature evergreen broad-leaved forest *Castanopsis orthacantha* and *Lithocarpus dealbatus* were the dominants, with a maximum height reaching 32 m with 100 cm DBH, and individuals occurring in each height-class (Fig. 4.6j, k). Figure 4.7a–e shows the *Pinus* community before the fire, and 2 years and 5 years after the fire.

The species composition of the communities that are representative of the mosaics of patchy vegetation in the local landscape is shown in Table 4.2. When the data for all the plots in each plant community are combined, the Sørensen's similarity index (beta diversity: $b = 2C/(S1 + S2)$, where $S1 =$ the total number of species recorded in the first plant community, $S2 =$ the total number of species recorded in the second plant community, and $C =$ the number of species common to both communities) (Sørensen 1948) between the severely burned (SBPS) and moderately burned *Pinus* shrubland (MBPS) is 63.4 %; between the shrubland (combining the SBPS and the MBPS) and the *Pinus* forest without fire occurrence it is 66.7 %; between the shrubland and mixed *Pinus–Lithocarpus forest* it is 60.6 %; between the shrubland and the mature EBLF, it is 31.3 %; and between the *Pinus* plantation and the mature EBLF, it is just 5 %. The Shannon–Wiener species diversity H' and evenness J' indices show the *Pinus* shrubland to be increasingly similar to the natural secondary and mature forests, while the *Pinus* plantation has significantly lower values on the two diversity indices than do the other plant communities (Fig. 4.8a, b). The light-demanding species (*Pinus yunnanensis, Lyonia ovalifolia, Rhododendron decorum, Myrica nana, Populus adenopoda* and *Alnus nepalensis*) appear with importance values (IV = (relative basal area + relative density)/2) of 92.1 % in the SBPS, and 86.7 % in the MBPS. The IV of shade-tolerant species (e.g. *Castanopsis orthacantha, Lithocarpus dealbatus, Cyclobalanopsis glaucoides, Castanopsis delavayi*) increases while the IV of light-demanding species decreases from the SBPS and the MBPS to the *Pinus* forest and the mixed *Pinus–Lithocarpus* forest, and finally to the EBLF along the past disturbance gradient, except for the plantation (Fig. 4.9). Intermediate (somewhat light-demanding, somewhat shade-tolerant) species (e.g. *Rhododendron* ssp., *Ternstroemia gymnanthera, Camellia reticulata, Photinia serrulata, Eurya nitida*) have IV values ranging from 7.5 to 36.8 %. The coverage of herbaceous species is 30–50 % in the burned area, while it is 50–80 % in the unburned area. This may be due to a large number of regenerating pine seedlings which occupy the available space and thus suppress the growth of herbaceous species in the burned area. Though invasive herbaceous *Eupatorium adenophorum* and *Senecio scandens* occur in places across the local landscape, they do not predominate in the understory of the plant communities in either the burned or unburned areas.

4.2.2 Succession

The effects of fire are great not only on the structure and species composition of the plant community, but also on soil fertility. The CCA ordination (canonical correspondence analysis) indicates the relationships between the communities and the

Fig. 4.7 **(a)** The shortly pedunculated seed cones of *Pinus yunnanensis* (Photograph courtesy of Longyuan He). **(b)** A naturally regenerated *P. yunnanensis* forest before the fire. **(c)** Burned *P. yunnanensis*. **(d)** A *P. yunnanensis* seedling 2 years after the fire. **(e)** A *P. yunnanesnsis* shrubland 5 years after the fire

Table 4.2 The species composition of the communities in the burned and unburned patches of a *Pinus yunnanensis* forest, and the unburned mature evergreen broad-leaved forest in the Anning-Tuanjiexiang area, central Yunnan

Community type		SBPS	MBPS	PP	PF	PY-LM	EBLF
Community age (year)		5	5–12	16	20–25	30–40	80–145
Number of plots (each plot size: 20 m×20 m)		12	6	3	7	7	5
Fire history		SB	MB	NB	NB	NB	NB
Degree of disturbance influence (score)		40	30	15	10	5	1
Number of species		15	26	4	35	38	36
Species	Traits	IV (%)	IV (%)	IV (%)	IV (%)	IV (%)	IV (%)
Pinus yunnanensis	LD	**79.1**	**56.3**	**96.2**	**51.5**	**31.2**	
Lyonia ovalifolia	LD	7.5	13.1		3.7	6.5	0.6
Ternstroemia gymnanthera	IN	3.1	5.9		2.5	3.7	0.4
Rhododendron siderophyllum	IN	2.6	4.4		1.3	6.9	0.001
Rhododendron decorum	LD	2.2	3.5	1.7	2.6	0.3	
Myrica nana	LD	1.9	3.3		3.5	1.9	0.001
Michelia yunnanensis	IN	1.8	2.4	1.2	4.6	0.4	0.3
Populus adenopoda	LD	1.2	1		0.8	0.3	
Lithocarpus mairei	IN	0.2	1.2		7.5	**24.3**	
Alnus nepalensis	LD	0.1	0.7		3.3		0.1
Elsholtzia rugulosa	LD	0.05	0.002		0.001	0.1	
Myrsine africana	LD	0.04	0.005		1.9	0.8	
Craspedolobium schochii	LD	0.005				0.1	
Vaccinium bracteatum	LD	0.002	0.002		0.07		1.3
Rhus chinensis	LD	0.001			0.001	0.08	2
Rhododendron microphyton	LD		3.4		5.9	1.8	
Campylotropis hirtella	LD		3.2				
Albizia mollis	LD		1.1			0.2	
Camellia reticulata	IN		0.3		1.7	0.3	2.2
Aralia chinensis	LD		0.2	0.9		0.2	
Dichotomanthus tristaniaecarpa	LD		0.05		0.003		
Castanopsis orthacantha	ST		0.04		1.6	4.5	**45.7**
Keteleeria evelyniana	LD		0.02		0.005	1	
Pyrus pashia	LD		0.01		0.001	0.1	
Quercus variabilis	LD		0.005		0.001		
Osyris wightiana	LD		0.004				
Quercus pannosa	IN		0.002				
Sabina gaussenii	LD		0.002				
Rhododendron spiciferum	IN				1.5	0.06	0.6
Cupressus funebris	LD				1.3		
Hypericum patulum	LD				1.1	0.03	

(continued)

Table 4.2 (continued)

Community type		SBPS	MBPS	PP	PF	PY-LM	EBLF
Community age (year)		5	5–12	16	20–25	30–40	80–145
Number of plots (each plot size: 20 m × 20 m)		12	6	3	7	7	5
Fire history		SB	MB	NB	NB	NB	NB
Degree of disturbance influence (score)		40	30	15	10	5	1
Number of species		15	26	4	35	38	36
Species	Traits	IV (%)	IV (%)	IV (%)	IV (%)	IV (%)	IV (%)
Lithocarpus dealbatus	ST				1.1	3.1	**22.3**
Viburnum cylindricum	IN				0.9	1.4	0.001
Viburnum foetidum var. *ceanothoides*	IN				0.6	0.5	0.2
Cotoneaster microphyllus	LD				0.5	0.5	
Rhododendron spinuliferum	IN				0.5	0.3	2.4
Elaeagnus stellipila	LD				0.04		
Buddleja asiatica	LD				0.003		
Cyclobalanopsis glauca	ST				0.002		
Pyracantha fortuneana	LD				0.002		0.1
Coriaria nepalensis	LD				0.001		
Campylotropis polyantha	LD				0.001		
Cyclobalanopsis glaucoides	ST					6.8	6.1
Lithocarpus variolosus	ST					1.4	0.001
Spiraea salicifolia	LD					0.6	
Docynia delavayi	LD					0.4	
Tripterygium hypoglaucum	LD					0.2	
Vaccinium sprengelii	IN					0.1	3.9
Rubus flosculosus	IN					0.005	
Pistacia chinensis	LD					0.005	
Photinia serrulata	IN					0.003	1.8
Celtis kunmingensis	LD					0.001	0.4
Ilex macrocarpa	IN					0.001	0.2
Lindera communis	LD						3.3
Eurya nitida	IN						1.6
Schima argentea	LD						1.6
Rhododendron trichostomum var. *radinum*	IN						0.9
Pieris formosa	LD						0.8
Diospyros lotus	LD						0.4
Styrax grandiflorus	LD						0.4
Castanopsis delavayi	ST						0.2
Cornus paucinervis	LD						0.1
Millettia dielsiana	LD						0.08

(continued)

Table 4.2 (continued)

Community type		SBPS	MBPS	PP	PF	PY-LM	EBLF
Community age (year)		5	5–12	16	20–25	30–40	80–145
Number of plots (each plot size: 20 m×20 m)		12	6	3	7	7	5
Fire history		SB	MB	NB	NB	NB	NB
Degree of disturbance influence (score)		40	30	15	10	5	1
Number of species		15	26	4	35	38	36
Species	Traits	IV (%)	IV (%)	IV (%)	IV (%)	IV (%)	IV (%)
Rhamnus leptophyllus	LD						0.07
Symplocos chinensis	IN						0.07
Schoepfia jasminodora	IN						0.05
Vaccinium dunalianum	IN						0.05

Data source: Tang et al. (2013)
IV (relative basal area+density)/2, *SBPS* severely burned *Pinus* shrubland, *MBPS* moderately burned *Pinus* shrubland, *PP Pinus* plantation, *PF Pinus* forest, *PY–LM* mixed coniferous and evergreen broad-leaved forest, *EBLF* mature evergreen broad-leaved forest, *SB* severely burned, *MB* moderately burned, *NB* not burned, *LD* light-demanding, *IN* intermediate, *ST* shade-tolerant

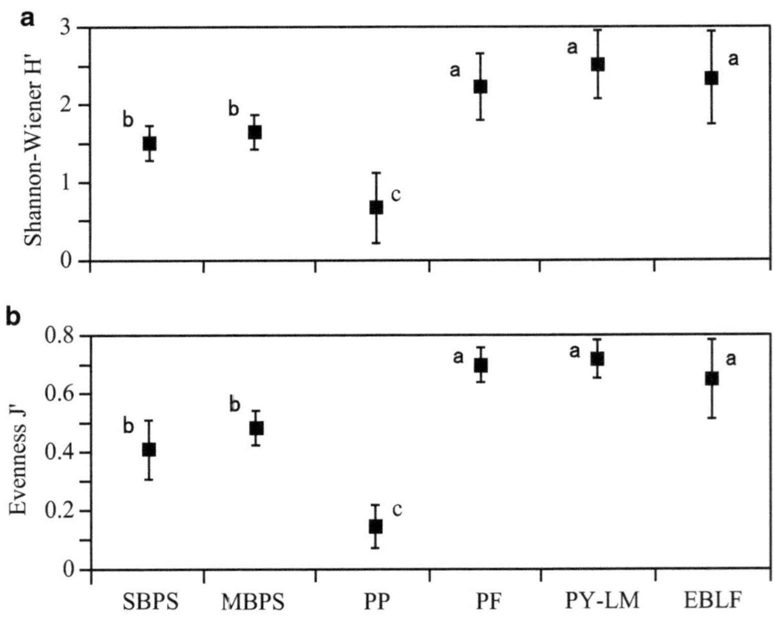

Fig. 4.8 (**a**) The Shannon–Wiener species diversity H′ and (**b**) evenness J′ indices in the various plant communities in the burned and unburned patches of a *Pinus yunnanensis* forest, and the unburned mature evergreen broad-leaved forest. Plant communities having different letters differ significantly ($p<0.05$). *Bars* represent standard deviations. *SBPS* severely burned *Pinus* shrubland, *MBPS* moderately burned *Pinus* shrubland, *PP Pinus* plantation, *PF Pinus* forest, *PY–LM* mixed coniferous and evergreen broad-leaved forest, *EBLF* mature evergreen broad-leaved forest (Data source: Tang et al. (2013))

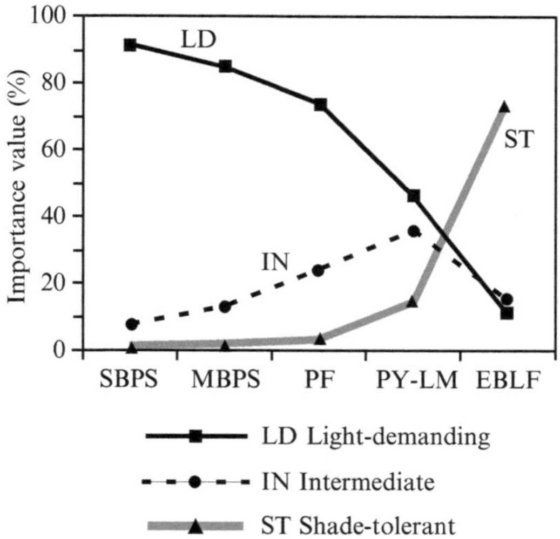

Fig. 4.9 Changes in the importance values (IV) for the plant communities in the burned and unburned patches of a *Pinus yunnanensis* forest, and the unburned mature evergreen broad-leaved forest (Data source: Tang et al. (2013)). *SBPS* severely burned *Pinus* shrubland, *MBPS* moderately burned *Pinus* shrubland, PF *Pinus* forest, *PY–LM* mixed coniferous and evergreen broad-leaved forest, *EBLF* mature evergreen broad-leaved forest

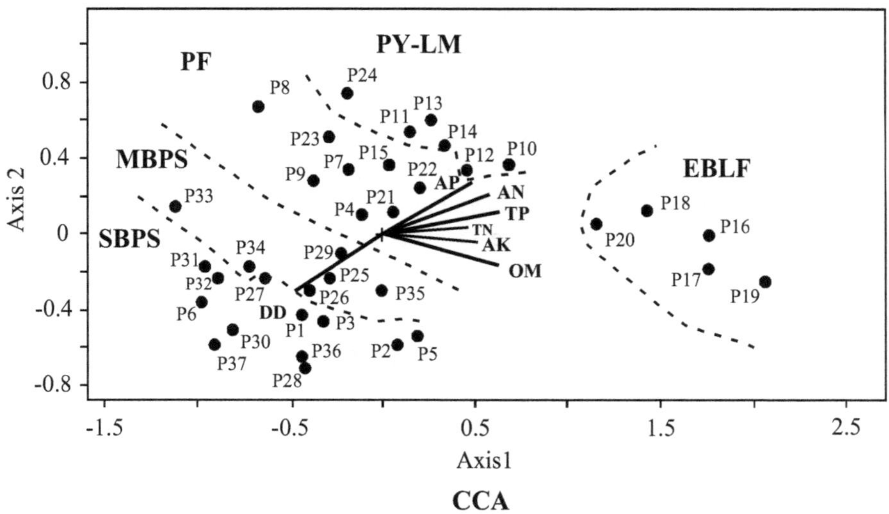

Fig. 4.10 CCA ordination for the sampled plots in the burned and unburned patches of a *Pinus yunnanensis* forest, and the unburned mature evergreen broad-leaved forest, excluding the three plots in plantations. A *dot* indicates a plot. *SBPS* severely burned *Pinus* shrubland, *MBPS* moderately burned *Pinus* shrubland, *PF Pinus* forest, *PY–LM* mixed coniferous and evergreen broad-leaved forest, *EBLF* mature evergreen broad-leaved forest, *DD* degree of disturbance, *OM* organic matter, *AK* available potassium, *TN* total nitrogen, *TP* total phosphorus, *AN* available nitrogen, *AP* available phosphorous (Data source: Tang et al. (2013))

soil properties as well as the degree of disturbances (Fig. 4.10) (Tang et al. 2013). The Monte-Carlo tests for the eigenvalues are significant for both axes ($p=0.01$ for axis 1, $p=0.04$ for axis 2). The tests for the species-environment correlations for each axis are significant ($p=0.01$ for axis 1 and $p=0.02$ for axis 2). The first ordination axis is strongly correlated with the degree of disturbance (DD) (r=–0.68), organic matter (r=0.76), total P (r=0.76), available N (r=0.73), available K (r=0.6), available P (r=0.59) and total N (r=0.57). According to the Indicator Species Analysis (Dufrêne and Legendre 1997), the severely burned SBPS can be defined by the pioneer (light-demanding) *Pinus yunnanensis*. The moderately burned MBPS is characterized by the presence of *Pinus yunnanensis* and pioneer *Lyonia ovalifolia*. These two communities have a higher overall degree of disturbance and grow in very nutrient-poor soils. The *Pinus* forest (PF) is characterized by the presence of *Pinus yunnanensis* in the overstory, and *Lithocarpus mairei* (seral, intermediate light-demanding), as well as pioneer light-demanding *Lyonia ovalifolia*, *Alnus nepalensis*, *Populus adenopoda* and *Keteleeria evelyniana* in the shrub layer. This community indicates that the plots are intermediate in degree of disturbance, and characterized by nutrient-poor soils. In the mixed coniferous and evergreen broad-leaved forest (PY–LM), *Pinus yunnanensis* and *Lithocarpus mairei* co-dominate the overstory, and the presence of shade-tolerant *Cyclobalanopsis glaucoides*, *Castanopsis orthacantha* and *Lithocarpus dealbatus* characterize the shrub layer. This mixed forest has a low degree of disturbance, and high values for available P and available N in some plots. The evergreen broad-leaved forest (EBLF) is almost free of disturbance and grows in fertile soils. It is characterized by the presence of late successional shade-tolerant *Castanopsis orthacantha* and *Lithocarpus dealbatus*. Axis 1 of the CCA, from the left side to the right, indicates that the successional direction is from the severely burned SBPS to the moderately burned MBPS, then to the early seral *Pinus* forest (PF), to the late seral mixed coniferous *Pinus* and evergreen broad-leaved *Lithocarpus* forest (PY–LM), finally to the late successional evergreen broad-leaved forest (EBLF). Though the boundary between the severely burned SBPS and the moderately burned MBPS is not distinctive, the MBPS is clearly at a somewhat more advanced stage in the successional sequence.

4.2.3 A Model on Successional Pathways

Tang et al. 2013 provided a model to compare the effects of fire, human restoration effort (*Pinus* plantation), selective cutting, clear cutting and abandonment of farmland on successional pathways in the central area of Yunnan (Fig. 4.11a–d). Both successional pace and species composition differ according to disturbance type and severity, though the eventual successional direction is toward the evergreen broad-leaved forest. The moderately burned *Pinus* shrublands (MBPS) should return naturally to their pre-fire condition over a period of 25–30 years after the fire, assuming that regeneration does occur and the colonists are not predominantly exotic, because

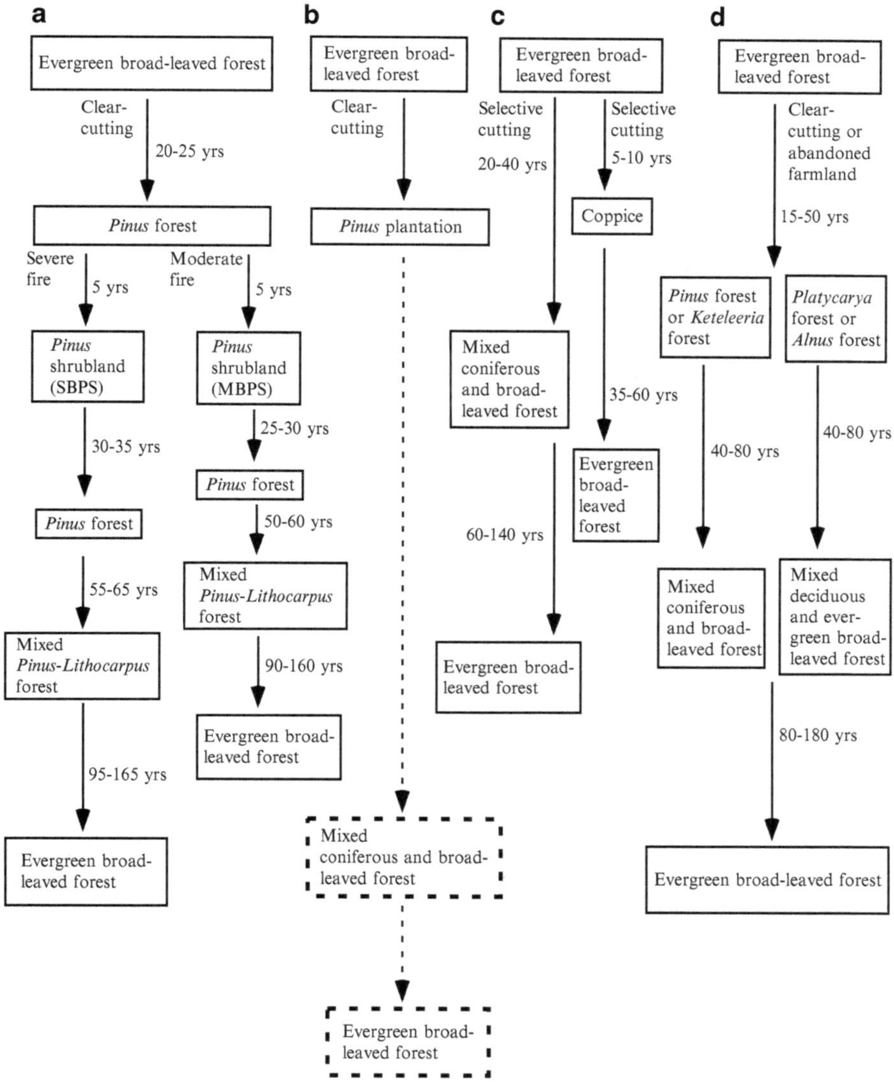

Fig. 4.11 A comparison of the secondary successional pathways after fire occurrence and human restoration efforts (pine plantations), as well as other disturbances in the area of central Yunnan (**a**–**d**). An *arrow* indicates the successional sequence. *SBPS* severely burned *Pinus* shrubland, *MBPS* moderately burned *Pinus* shrubland (Data source: Tang et al. (2013))

19 out of 35 species in the 20–25 year-old natural *Pinus* forest are already present in the MBPS (Table 4.2), and all these pioneer species grow very fast. Fourteen species existing in the severely burned *Pinus* shrubland (SBPS) have been found in the *Pinus* forest. The SBPS may take roughly 35–40 years after the fire to reach the status of *Pinus* forest (Fig. 4.11a). Though the successional pace in the SBPS is

slower than in the MBPS, it does not appear that the regeneration and growth of *Pinus yunnanensis* have been a problem in the SBPS (Fig. 4.6a). Tang et al. (2013), judging from tree ring analysis and the typical lifespan of *Pinus yunnanensis*, estimated that canopy gaps will begin to form, allowing invasion of evergreen broad-leaved species, when the *Pinus* forest has aged for 50–160 years. According to the data from Tang et al. (2010), it would take c. 25–30 years for a *Pinus* forest to become a mixed coniferous and broad-leaved forest, then another 40–100 years to establish as an evergreen broad-leaved forest.

As a human restoration effort, the *Pinus* plantation resulted in poor species diversity because the densely planted *Pinus yunnanensis* occupied the available space, hindering the regeneration of other species. In contrast, a number of light-demanding species colonized the burned area along with pioneer *Pinus yunnanensis*, contributing to higher species diversity. Though the time length is unpredictable for a *Pinus* plantation to succeed to the mixed coniferous and broad-leaved forest, it is certain that the successive stages will be long delayed (Fig. 4.11b).

After selective cutting for fuelwood, two typical plant communities have formed in the area. One is the mixed coniferous and broad-leaved forest; the other is the evergreen broad-leaved coppice with many sprouts. It takes 20–40 years for the former to take shape, and 5–10 years to establish the latter after selective cutting. The coppice community takes another 30–50 years to develop into a mature evergreen broad-leaved forest (Fig. 4.11c). In central Yunnan, the most common secondary succession follows after clear cutting or abandonment of farmland. From the early successional to the late-successional stage of the evergreen broad-leaved forest through the seral mixed coniferous and broad-leaved forest, or the mixed deciduous and evergreen broad-leaved forest, 80–180 years may elapse (Fig. 4.11d). At the early successional stage, there are diverse plant communities including the naturally regenerated *Pinus yunnanensis* community, the *Keteleeria evelyniana* community, the deciduous broad-leaved *Platycarya strobilacea* community, and the *Alnus nepalensis* community (Tang et al. 2010). Toward a future succession, the pioneer tree species of *Pinus, Populus, Alnus, Quercus* and *Keteleeria* in the early successional plant communities will be replaced by more shade-tolerant tree species of *Castanopsis* and *Lithocarpus*.

In conclusion, vegetation and plant diversity recover more efficiently by letting a burned *Pinus yunnanensis* community grow back naturally than by establishing plantations. Efforts to improve or accelerate reestablishment of a burned forest might most wisely be based on detailed observation of the natural processes.

References

Bond WJ, Keeley JE (2005) Fire as a global 'herbivore': the ecology and evolution of flammable ecosystems. Trends Ecol Evol 20:387–394

Delcourt PA, Delcourt HR (1998) The influence of prehistoric human-set fires on oak chestnut forests in the southern Appalachians. Castanea 63:337–345

Dufrêne M, Legendre P (1997) Species assemblages and indicator species: the need for a flexible asymmetrical approach. Ecol Monogr 67:345–366

Faliński JB (1986) Vegetation dynamics in temperate lowland primeval forest. Junk Publishers, Dordrecht

Jin ZZ, Peng J (2004) *Pinus yunnanensis*. Yunnan Science and Technology Press, Kunming

Knapp R (1986) Vegetation dynamics. Science Press, Beijing

Li N (1999) The geographical distribution, origin and diffusion of Pinaceae. In: Lu A (ed) The geography of spermatophytic families and genera. Science Press, Beijing, pp 17–39

Mooney HA, Bonnicksen JM, Christensen NL, Lotan JE, Reiners WA (eds) (1980) Fire regimes and ecosystem properties, Forest Service General Technical Report. United States Department of Agriculture, Washington, DC

Nakagoshi N (1995) Pine forests in East Asia. In: Box EO, Peet RK, Masuzawa T, Yamada I, Fujiwara K, Maycock PF (eds) Vegetation science in forestry. Kluwer Academic Publishers, Dordrecht, pp 85–104

Russell EWB (1983) Indian-set fires in the forest of the northern United States. Ecology 64:78–88

Sørensen TA (1948) A method of establishing groups of equal amplitude in plant sociology based on similarity of species content, and its application to analyses of the vegetation on Danish commons. K Dan Vidensk Selsk Biol Skr 5:1–34

Swaine MD, Whitmore TC (1988) On the definition of ecological species groups in tropical rain forests. Plant Ecol 75:81–86

Tang CQ, Zhao MH, Li XS, Ohsawa M, Ou XK (2010) Secondary succession of plant communities in a subtropical mountainous region of SW China. Ecol Res 25(1):149–161

Tang CQ, He L-Y, Su W-H, Zhang G-F, Wang H-C, Peng M-C, Wu Z-L, Wang C-Y (2013) Regeneration, recovery and succession of a *Pinus yunnanensis* community five years after a mega-fire in central Yunnan, China. For Ecol Manage 294:188–196

Turner MG, Romme WH, Gardner RH, Hargrove WW (1997) Effects of fire size and pattern on early succession in Yellowstone National Park. Ecol Monogr 67(4):411–433

Yu YF (1999) The origin, evolution and distribution of Taxodiaceae. In: Lu A (ed) The geography of spermatophytic families and genera. Science Press, Beijing, pp 40–62

Zheng HB, Zhang NQ, Fang Y, Zhang SY, Zhang YS (2007) The status of the prevention and countermeasure of forest fires in Yunnan Province. For Fire 4:19–22

Chapter 5
Man-Made Forests

Abstract In recent decades, afforestation and reforestation have led to an increase in forest and tree cover; however, in most areas of SW China, establishment of plantations has not achieved ecological restoration. Several case studies done in subtropical Yunnan have concluded that planting fast-growing monospecific *Eucalyptus* or *Pinus* plantation forests results in poor recovery of vegetation and plant diversity, great soil surface runoff and erosion, low litter nutrient input, and poor soil nutrient storage. In contrast, natural secondary succession facilitates regeneration of secondary forests, promotes recovery of vegetation and plant diversity and replenishes litter-soil nutrients, and, as a result, yields distinctive ecological benefits. Consideration of various individual case studies and evaluation of emerging patterns can help to explain the results of plantation establishment in the subtropical areas.

5.1 Controversial Opinions

In China, the total plantation area has reached 5.33×10^7 ha, accounting for 30 % of the country's forested area (SFA 2007) and 29 % of the world's total plantation area (FAO 2007). The plantation area of subtropical China accounts for 63 % of the nation's total (SFA 2007). Plantations of *Eucalyptus*, *Pinus*, *Keteleeria*, *Cunninghamia*, *Cryptomeria*, *Larix*, *Populus* and *Betula*, but especially *Eucalyptus* and *Pinus*, are widespread in this region, because of their rapid growth and high yield. Since plantation forestry contributes to economic growth, these fast-growing tree species are today being planted more and more; they produce wood, oil, pulp, and fuel, and are justified as a form of restoration of vegetation, and soil and water conservation. More recently, carbon sequestration and other environmental values are also cited. With rapid development, vast monocultures have been established after deforestation all over China. In the southwest, fast tree growth and inexpensive land and labor combine to make industrial production especially popular. The pressure on the man-made forests to deliver economic, social, and environmental services has reached unsustainable levels in SW China.

C.Q. Tang, *The Subtropical Vegetation of Southwestern China*, Plant and Vegetation 11, DOI 10.1007/978-94-017-9741-2_5

The legacies of land use in SW China have strongly influenced the plant communities' structure, dynamics, species diversity, litter nutrient inputs, water use efficiency, and soil chemical properties. The environmental impact of fast-growing monospecies plantations has been controversial. Fast-growing pine forests are inferior in the decomposition of leaf litter and nutrient cycling as compared with natural forests and mixed-species forests in central Yunnan, and the simple structure of the single-species pine plantation has resulted in a higher intensity of pluvial throughfall, with increased nutrient loss (Liu et al. 2000, 2003). The poor nutrient availability of the soil in the plantations of *Eucalyptus* or *Pinus* has a negative effect on plant species distribution and has diminished understory growth, leading to a decrease in biodiversity and further soil erosion according to Tang et al. (2007, 2010b) and Hou et al. (2010). Contrary opinions maintain that plantations of *Eucalyptus* or *Pinus* restore soil fertility and improve hydrological processes on degraded land in subtropical/tropical China (Peng and Fang 1995; Peng 2003).

As we have seen, the original vegetation in central Yunnan was a semi-humid evergreen broad-leaved forest dominated by evergreen species of *Cyclobalanopsis*, *Castanopsis* and *Lithocarpus* of the Fagaceae. Most of this forest has previously been utilized as coppices for fuelwood or turned into pastures, but since the 1980s some of the pastures and bare land have been replaced by plantations of *Pinus yunnanensis* and *Keteleeria evelyniana* or exotic *Eucalyptus smithii* and *Eucalyptus globulus* (Fig. 5.1a–c). Other remnant coppices and pastures were abandoned in the 1980s and have since regenerated naturally, with secondary stands consisting of evergreen and deciduous broad-leaved trees or shrubs. These secondary woodlands and plantations have been unmanaged, since they were abandoned or first planted. Their landscapes appear as dynamic habitats comprised of mosaics of diverse stands in various stages of vegetation recovery. Comparative studies on the influence of single-species plantations and natural successional plant communities on ecosystem restorative processes were done in order to make recommendations regarding restoration of the fragile mountain ecosystem in SW China.

5.2 Plant Diversity and Regeneration Dynamics

As compared to naturally regenerated plant communities, species diversity in monospecific plantations of *Eucalyptus* and *Pinus* is particularly low in woody and herbaceous species, as has become evident in the red-earth Mouding (Tang et al. 2007) and karst Shilin areas of central Yunnan (Tang et al. 2010b). In Mouding, at the time of surveying, there were six typical plant communities (Fig. 5.2): (1) 4-year-old shrubland (SL) that had formed since a degraded pasture was abandoned in 2001; (2) 10–14-year-old planted *Eucalyptus smithii* forest (EF) which was planted in a degraded pasture in 1991, and partially cut for fuel during the year 1995; (3) 35–45-year-old planted *Pinus yunnanensis* forest (PF) which was sowed by an airplane early in 1960, with the *Pinus* trees subsequently partially cut for industrial wood around 1970; (4) 35–45-year-old semi-natural *Keteleeria evelyniana* and

Fig. 5.1 (**a**) A *Eucalyptus smithii* plantation in Jinghong, southern Yunnan. (**b**) A *Eucalyptus globulus* plantation in Xinping, central Yunnan. (**c**) A *Pinus yunnanensis* plantation in Luquan, central Yunnan

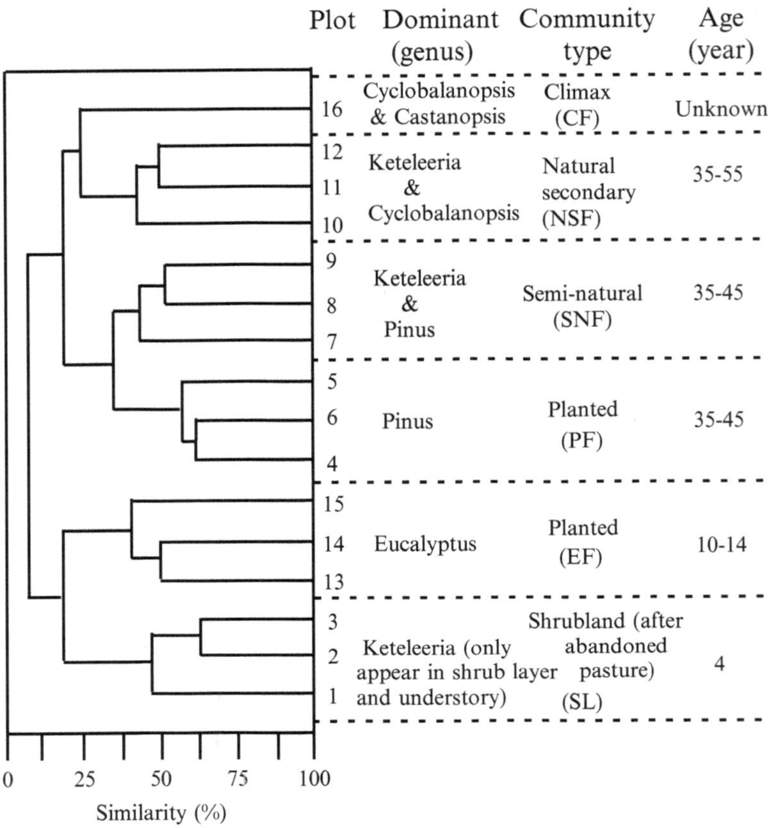

Fig. 5.2 Dendrogram of 16 study plots using Søerensen's similarity index, and group average clustering in Mouding, central Yunnan. *CF* climax forest, *NSF* natural secondary forest, *SNF* semi-natural forest, *PF Pinus* forest, *EF Eucalyptus* forest, *SL* shrubland (Data source: Tang et al. (2007))

Pinus yunnanensis forest (SNF); (5) 35–55-year-old natural secondary forest of *Keteleeria evelyniana* and *Cyclobalanopsis glaucoides* (NSF) which in a natural process appeared after the evergreen broad-leaved forest was clear cut in 1950 but then was partially cut again for fuel from 1965 to 1970; (6) a climax forest (CF) of *Cyclobalanopsis glaucoides*, *Castanopsis orthacantha*, and *Castanopsis delavayi*, being the original subtropical semi-humid evergreen broad-leaved forest of the area not affected by human activity.

Figure 5.3a shows that the *Eucalyptus* (10–14 years) (H′ = 0.25) and *Pinus* plantations (35–45 years) (H′ = 0.52) in the area have very low values on the Shannon-Wiener index (H′), while values for the secondary forest (35–55 years) reach 1.88 and for the natural mature forest (>80 years) 1.94. The secondary forest has more species (31) than the natural mature forest (26) because of its larger number of pioneer deciduous species (Fig. 5.3b). The Jaccard index (beta diversity) shows the

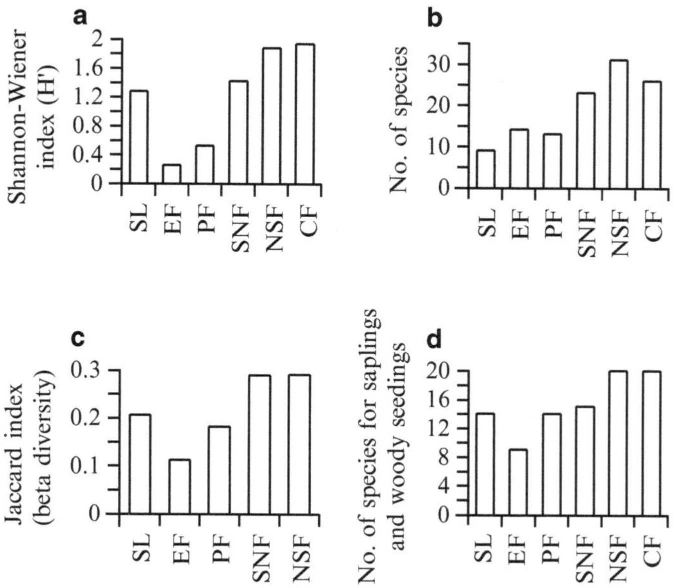

Fig. 5.3 Seven features of the plant communities in Mouding, central Yunnan (**a**–**d**). *SL* shrubland, *EF Eucalyptus* forest, *PF Pinus* forest, *SNF* semi-natural forest, *NSF* natural secondary forest, *CF* climax forest. Diversity index H′ according to Shannon-Wiener, Jacard index (beta diversity) according to $Cj = j/(a + b - j)$, where *Cj* is the Jaccard index, *a is* the number of species recorded in the first community, and *b* is the number of species recorded in the second community, *j* is the number of common species among the two communities (Data source: Tang et al. (2007))

differentiation in diversity values between the natural mature forest and the other plant communities: the *Eucalyptus* plantation is the lowest (0.11), which indicates the least similarity with the natural mature forest; on the other hand, the semi-natural (35–45 years) and secondary forests, with the same index of 0.29, more closely resemble the natural mature forest (Fig. 5.3c). The number of species of saplings and woody seedlings is lowest (9) in the *Eucalyptus* plantation, and next-lowest in the 4 year old shrubland that naturally regenerated from an abandoned pasture (14 species) and the *Pinus* plantation (14 species) (Fig. 5.3d).

In the local landscape of the karst Shilin area, there were four types of plant communities namely the *Pinus* plantation (30 years), the early successional shrubland (20 years) that had grown naturally from an abandoned pasture, the seral secondary forest (40–50 years), and the late successional maturing evergreen broad-leaved forest (hereafter the natural maturing forest) (80–120 years) (Fig. 5.4). Regarding the number of species, number of genera, and number of families, the natural maturing forest has values significantly higher than the shrubland and the secondary forest (Fig. 5.5a–c). And the secondary forest has a significantly higher number of species than the shrubland, which, in turn, has significantly higher values than the *Pinus* plantation. The number of either woody or herbaceous species, Shannon's H′, and Fishers' alpha, show the secondary forest to be increasingly similar to the natural

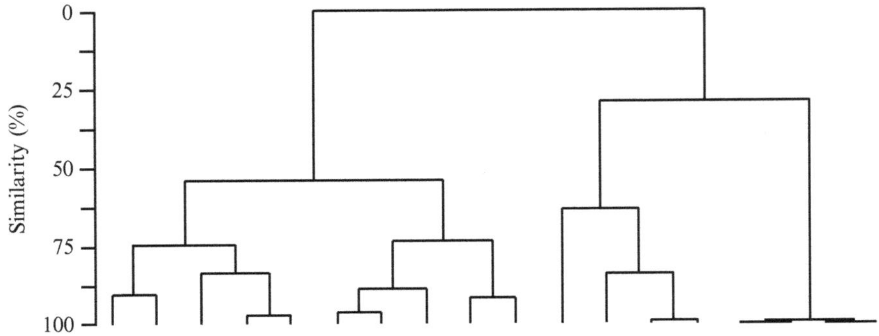

Transect	T1 T4 T2 T5 T14	T3 T10 T17 T6 T12	T7 T9 T15 T18	T11 T8 T13 T16
Area (m^2)	2000 1600 2000 1200 1600	1600 1600 800 1000 1000	800 1200 1200 800	1600 1200 1200 800
Community type	Natural maturing forest (MF)	Secondary forest (SF)	Shrubland (SL)	*Pinus* plantation (PP)
Age (year)	80-120	40-50	20	30
Dominants	*Cyclobanalopsis glaucoides* *Neolitsea homilantha* *Olea yunnanensis* *Ilex macrocarpa* *Pistacia weinmanniifolia* *Carpinus mobeigiana*	*Cyclobanalopsis glaucoides* *Olea yunnanensis* *Pistacia weinmanniifolia*	*Neolitsea homilantha*	*Pinus yunnanensis*
Land-use history	Limited cutting for fuelwood before 1985, protected from human activity since 1985	Cutting for fuelwood ceased in 1990	Used as pasture ceased in late 1980s	Grew after seed dispersal by airplanes in 1980

Fig. 5.4 Similarity dendrogram using Sørensen's relative similarity index and group average clustering in the Shilin karst area, central Yunnan (Data source: Tang et al. (2010b))

maturing forest with the passage of time (Fig. 5.5d–i). In the shrubland, the number of woody species, Shannon's H′ and Fisher's alpha for woody species, have significantly higher values than those of the *Pinus* plantation, while the numbers of herbaceous species, Shannon's H′ and Fisher's alpha for herbs in the shrubland, are not significantly higher than those for the *Pinus* plantation. To ascertain the diversity structure in relation to species traits (i.e. light-demanding, intermediate and shade-tolerant species) of all the four communities, dominance-diversity curves are drawn in Fig. 5.6. In the secondary forest and the natural maturing forest, many species occupy similar relative basal areas (RBA), and there are several co-dominant species. Hence, the diversity is high in the secondary forest and maturing forest. In the shrubland, the patterns become steeper. In the *Pinus* plantation, the top dominant (*Pinus yunnanensis*) becomes conspicuous. The diversity curves can represent the relative importance of the species in the community as well as illustrate the role played by certain species in determining community structure. The woody species ($\geq$1.3 m in height) composition in the various plant communities is shown in Table 5.1.

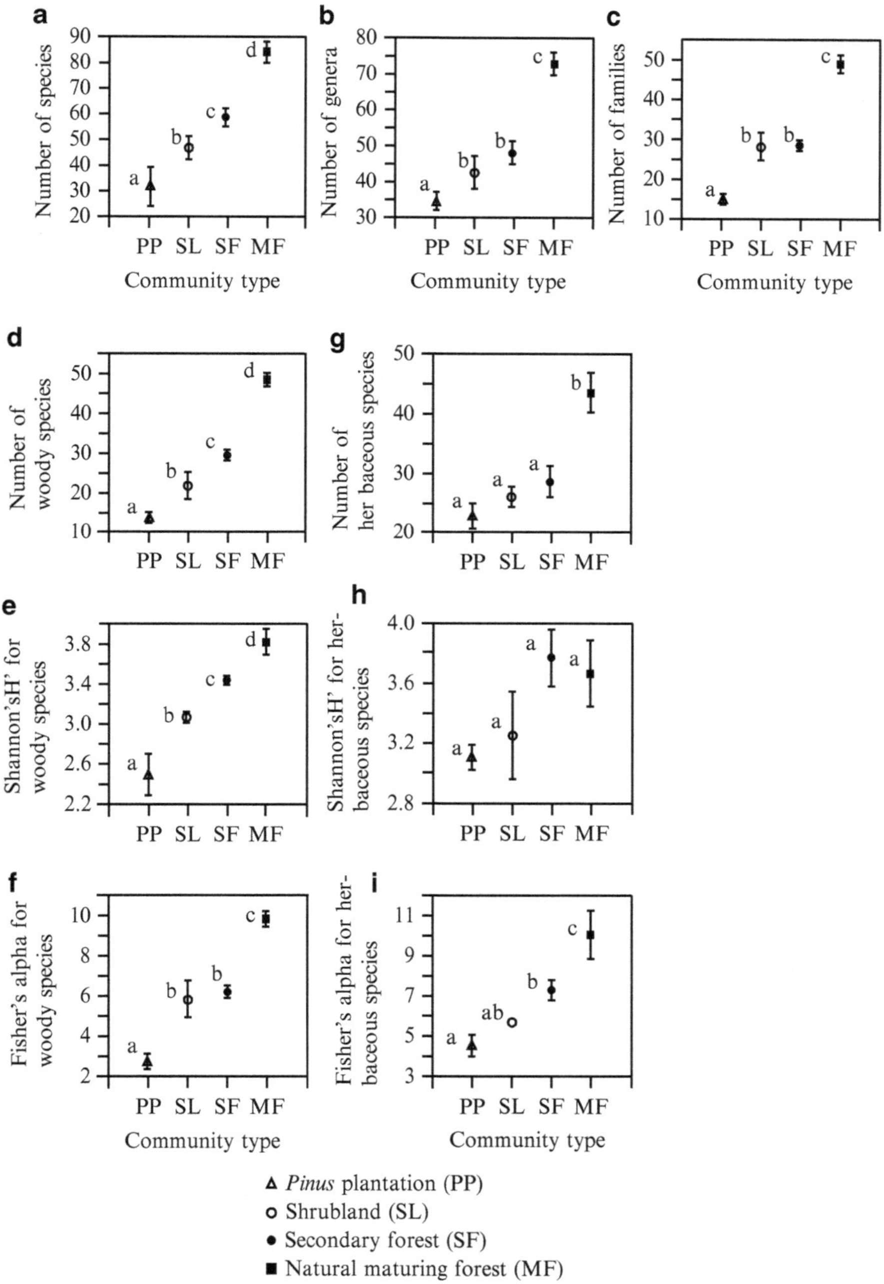

Fig. 5.5 Changes in diversity indices for woody and herbaceous species among the *Pinus* plantation and natural secondary plant communities (**a–i**) in the Shilin karst area, central Yunnan. Significantly different values are indicated by different letters, $p \leq 0.05$. *Bars* represent standard errors (Data source: Tang et al. (2010b))

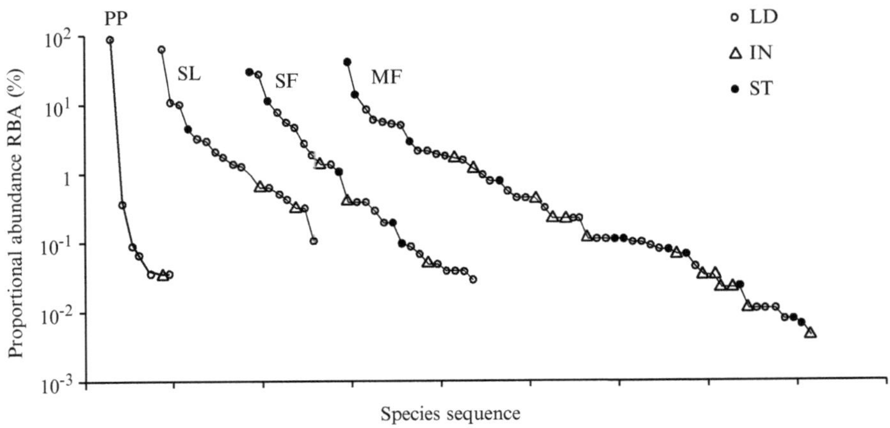

Fig. 5.6 Dominance-diversity curves of the four plant communities in the Shilin karst area. Species traits are distinguished with marks: *LD* light-demanding, *IN* intermediate, *ST* shade-tolerant. *PP* Pinus plantation, *SL* shrubland, *SF* secondary forest, *MF* natural maturing forest. The abundance of each species is plotted on a logarithmic scale from the most to the least abundant species (Data source: Tang et al. (2014))

Table 5.1 Woody species (H ≥ 1.3 m in height) composition in the overstory of each plant community in the Shilin area, central Yunnan

Species	Traits	PP	SL	SF	MF
Pinus yunnanensis	LD	**99.3**	9.6		
Albizia mollis	LD	0.41		8	1.6
Pistacia weinmanniifolia	LD	0.1	1.2	**28.1**	**8.1**
Pistacia chinensis	LD	0.07	1.9	5.7	4.7
Rhamnus leptophylus	LD	0.04		0.3	0.9
Dichotomanthes tristaniaecarpa	IN	0.04	0.3	1.4	0.1
Cotoneaster franchetii	LD	0.04		0.2	
Neolitsea homilantha	LD		**60.5**	4.8	4.4
Diospyros mollifolia	LD		10.1		0.1
Olea yunnanensis	ST		4.1	**11.7**	**12.5**
Rhamnella martini	LD		3	1.9	0.01
Mallotus philippinensis	LD		2.8		
Dalbergia mimosoides	LD		1.6	1.4	0.1
Prunus zippeliana	LD		1.3		1.9
Toxicodendron succedaneum	LD		0.9	0.04	0.7
Morus australis	IN		0.6		0.2
Sageretia theezans	LD		0.6		0.09
Celtis bungeana	LD		0.5		0.09
Nothapodytes tomentosa	LD		0.4		0.08
Pyrus pashia	LD		0.3		0.01
Prunus persica	LD		0.1		
Cyclobalanopsis glaucoides	ST			**30.7**	**39.8**
Toxicodendron griffithii	LD			2.8	1.9

(continued)

Table 5.1 (continued)

Species	Traits	PP	SL	SF	MF
Machilus yunnanensis	ST			1.1	2.6
Osyris wightiana	IN			0.4	
Juniperus formosana	LD			0.4	
Keteleeria evelyniana	LD			0.4	
Photinia glomerata	ST			0.2	0.02
Photinia sp.	ST			0.1	0.1
Rhus chinensis	LD			0.09	0.04
Campylotropis polyantha	LD			0.07	
Pentapanax henryi	IN			0.05	0.2
Quercus dentata	LD			0.05	
Celtis tetrandra	LD			0.04	1.4
Rosa banksiae	LD			0.04	
Cornus paucinervis	LD			0.03	
Ilex macrocarpa	LD				**5**
Carpinus mobeigiana	LD				4.5
Xylosma racemosum	LD				1.7
Lindera communis	IN				1.5
Jasminum humile	IN				1.1
Pittosporum brevicalyx	ST				0.7
Zanthoxylum scandens	LD				0.5
Distyliopsis laurifolia	LD				0.4
Quercus variabilis	LD				0.4
Ficus virens	IN				0.4
Milletia dielsiana	LD				0.3
Dalbergia collettii	LD				0.2
Milletia reticulata	LD				0.2
Myrsine semiserrata	ST				0.1
Schoepfia jasminodora	LD				0.07
Ficus pandurata	ST				0.07
Lonicera japonica	IN				0.06
Lithocarpus confinis	ST				0.06
Fortunella sagittifolia	IN				0.03
Indigofera cinerascens	IN				0.03
Leptodermis potaninii	IN				0.02
Paederia scandens	IN				0.02
Elaeagnus lanceolata	IN				0.01
Debregeasia edulis	LD				0.01
Crataegus scabrifolia	LD				0.007
Euonymus fortunei	ST				0.007
Ficus chartacea	ST				0.006
Buddleja officinalis	IN				0.004

Data source: Tang et al. (2014)
Values are relative basal area (RBA) of each species. Dominant species are shown in boldface
PP Pinus plantation, *SL* shrubland, *SF* secondary forest, *MF* natural maturing forest, *LD* light-demanding, *IN* intermediate, *ST* shade-tolerant

5.3 Hydrological Characteristics and Nutrient Loss

Natural forests and mixed stands have high canopy interceptions of rainfall as a result of their more complex vertical structure, compared to single-species. In the Mouding area, rain throughfall values are higher and rain interception and stemflow values are lower in *Pinus* and *Eucalyptus* forests than in the semi-natural and natural secondary forests (Fig. 5.7). This results in large surface runoff and leaching loss of soil nutrients. The *Pinus* and *Eucalyptus* plantation forests had the highest rates of surface runoff, with 1,607.2 and 1,193.8 m^3 ha^{-1}, respectively, whereas the secondary forest has the lowest value, 103.6 m^3 ha^{-1}. Soil erosion under *Eucalyptus* forest shows the highest values (876.5 kg ha^{-1}), 16 times greater than under the secondary forest and almost five times more than in the shrubland. This forest also shows the greatest leaching loss of total P (348.2 g ha^{-1}), about six times higher than the secondary forest, and about five times more than the shrubland. Leaching loss of total N and total K have patterns similar to those of surface runoff (Fig. 5.8).

5.4 Nutrient Cycling

5.4.1 Litter Production and Nutrient Inputs

Litter production, decomposition, and nutrient inputs are fundamental ecosystem processes. Tang et al. (2014) conducted a detailed study on these ecological processes as seen in Shilin. There, in the late successional maturing forest and the

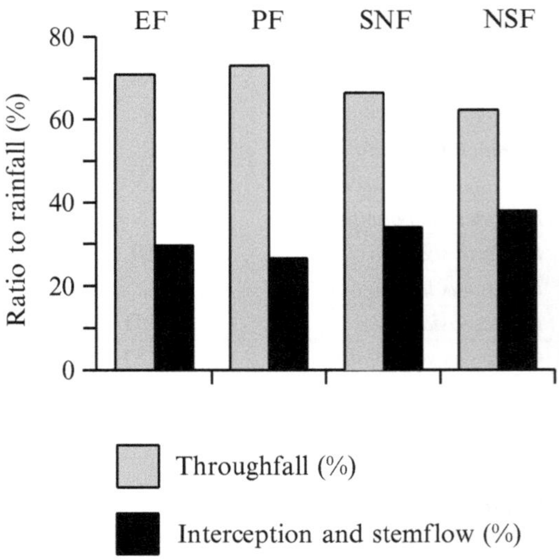

Fig. 5.7 Ratios of throughfall to rainfall and interception and stemflow to rainfall for *Eucalyptus* forest (*EF*), *Pinus* forest (*PF*), semi-natural forest (*SNF*), and natural secondary forest (*NSF*) in Mouding, central Yunnan (Data source: Tang et al. (2007))

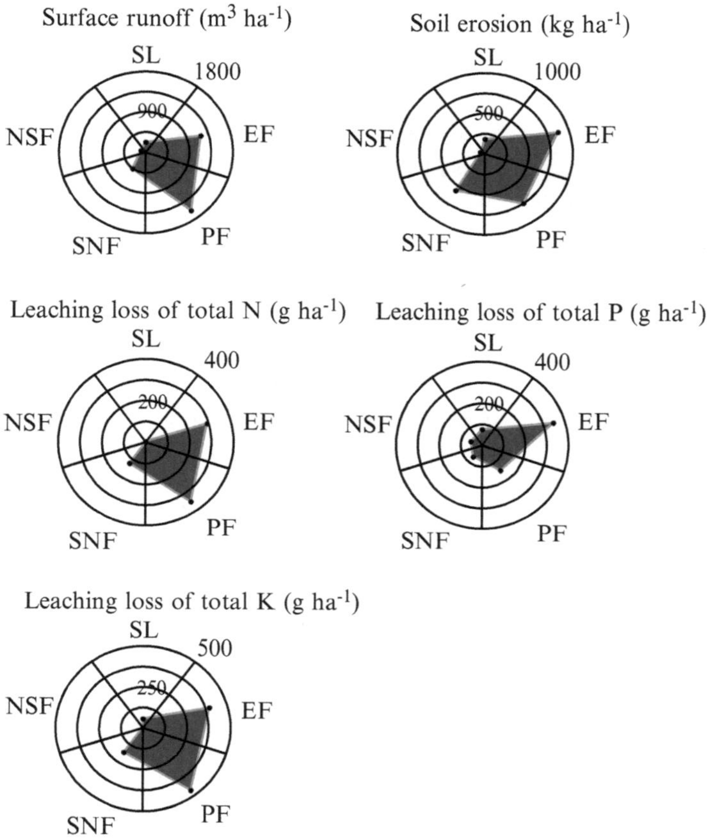

Fig. 5.8 Radar-diagram of soil surface runoff, soil erosion, and leaching loss of total N, P, and K in the representative local plant communities in Mouding, central Yunnan. *SL* shrubland, *EF Eucalyptus* forest, *PF Pinus* forest, *SNF* semi-natural forest, *NSF* natural secondary forest (Data source: Tang et al. (2007))

seral successional secondary forest, we found that values for the litter production (including leaves, flowers, fruits, seeds, bark, and twigs (<2 cm diameter) that have fallen to the ground) and forest floor mass (including two forest floor layers, first a litter (L) layer, fresh or slightly discolored, with little or no breaking up, and the other a fermenting (F) layer consisting of partially decomposed material located above the mineral soil surface and below the L layer) are significantly higher than those found either in an early successional shrubland that naturally regenerated in an abandoned field, or in a *Pinus* plantation (Table 5.2). The litter turnover coefficient (k) in the maturing forest (0.38 year^{-1}), the secondary forest (0.36 year^{-1}) and the shrubland (0.32 year^{-1}) has significantly higher values than in the *Pinus* plantation (0.24 year^{-1}). The concentrations of leaf litter nitrogen (ranging from 10.76 to 12.83 g kg^{-1}), phosphorus (from 1.24 to 1.53 g kg^{-1}) and potassium

Table 5.2 Litterfall, forest floor mass, litter turnover coefficient, turnover time, and nutrient concentrations and input of leaf litter in the four plant communities (±SD) in the Shilin karst area, central Yunnan

	PP	SL	SF	MF
Litterfall (t ha^{-1} year^{-1})	1.49±0.52[c]	1.6±0.69[c]	3.31±0.82[b]	4.59±1.04[a]
Forest floor mass (t ha^{-1})	6.48±1.17[c]	4.91±1.06[c]	9.25±1.07[b]	12.29±2.03[a]
Litter turnover coefficient (k) (year^{-1})	0.24±0.12[c]	0.32±0.09[b]	0.36±0.06[a]	0.38±0.1[a]
Turnover time (1/k) (year)	4.76±2.43[a]	3.34±0.87[b]	2.88±0.53[c]	2.76±0.68[c]
Leaf litter N (g kg^{-1})	6.64±0.98[c]	10.76±1.84[b]	11.64±2.59[a,b]	12.83±1.2[a]
Leaf litter P (g kg^{-1})	1.09±0.45[c]	1.24±0.26[b]	1.41±0.38[a,b]	1.53±0.42[a]
Leaf litter K (g kg^{-1})	3.15±2.32[c]	7.51±4.44[a]	5.55±2.74[b]	6.12±3.29[b]
Leaf litter Ca (g kg^{-1})	3±2.85[c]	2.89±1.04[c]	5.03±2.23[b]	7.85±1.53[a]
Leaf litter Mg (g kg^{-1})	4.83±1.24[b]	3.57±0.73[b]	9.13±2.26[a]	11.06±4.24[a]
Leaf litter N input (kg ha^{-1} year^{-1})	9.91±2.75[d]	17.22±7.81[c]	38.53±9.56[b]	58.88±13.03[a]
Leaf litter P input (kg ha^{-1} year^{-1})	1.62±0.56[c]	1.98±0.86[c]	4.67±1.12[b]	7.01±1.31[a]
Leaf litter K input (kg ha^{-1} year^{-1})	4.69±1.38[d]	12.02±5.04[c]	18.36±4.56[b]	28.08±6.33[a]
Leaf litter Ca input (kg ha^{-1} year^{-1})	4.47±1.55[c]	4.62±2[c]	16.64±4.13[b]	36±8.54[a]
Leaf litter Mg input (kg ha^{-1} year^{-1})	7.19±2.31[b]	5.71±2.76[b]	30.2±8.51[a]	50.81±11.5[a]

Data source: Tang et al. (2014)
Values followed by different letters are significantly different at $p<0.05$ by the non-parametric Kruskal–Wallis all-pairwise comparisons test
PP Pinus Plantation, *SL* shrubland, *SF* secondary forest, *MF* natural maturing forest

(from 5.55 to 7.51 g kg^{-1}) in all the secondary successional plant communities showed significantly higher values than those of the *Pinus* plantation. Leaf litter N (9.91 kg ha^{-1} year^{-1}), P (1.62 kg ha^{-1} year^{-1}), and K (4.69 kg ha^{-1} year^{-1}) inputs in the communities also show significantly higher values than those of the plantation. Leaf litter in the plantation has the lowest values as compared with the other plant communities. Leaf litter Ca and Mg inputs in both the *Pinus* plantation and the shrubland are low and do not significantly differ from each other.

5.4.2 Soil Chemical Properties

Plantations can greatly affect the way that soil physical and chemical properties influence nutrient cycling. Table 5.3 shows soil chemical properties among the four plant communities in the Shilin area. The soil pH values vary from 5.1 to 7 among

Table 5.3 Soil properties in each plant community (±SD)

	PP	SL	SF	MF
pH	6.0 ± 0.47^{b}	7.0 ± 0.2^{a}	$6.5\pm0.4^{a,b}$	5.1 ± 0.18^{c}
OM (%)	7.38 ± 1.26^{b}	7.57 ± 2.04^{b}	10.66 ± 4.1^{a}	13.73 ± 0.9^{a}
OC (%)	4.28 ± 0.75^{b}	4.71 ± 1.17^{b}	6.31 ± 2.27^{a}	7.96 ± 0.53^{a}
TN (%)	0.23 ± 0.09^{c}	0.35 ± 0.08^{b}	0.45 ± 0.18^{a}	0.55 ± 0.03^{a}
C/N	19.41 ± 6.52^{a}	13.44 ± 1.82^{b}	15.12 ± 7.25^{b}	14.56 ± 0.32^{b}
TP (%)	0.03 ± 0.01^{c}	$0.07\pm0.02^{a,b}$	0.06 ± 0.03^{b}	0.08 ± 0.01^{a}
TK (%)	0.58 ± 0.2^{b}	0.3 ± 0.06^{c}	0.73 ± 0.14^{a}	0.28 ± 0.03^{c}
TCa (%)	0.69 ± 0.14^{b}	1.02 ± 0.23^{a}	1.1 ± 0.27^{a}	0.5 ± 0.1^{b}
TMg (%)	0.39 ± 0.17^{b}	0.63 ± 0.05^{a}	0.57 ± 0.08^{a}	0.36 ± 0.05^{b}
AN (mg kg^{-1})	150.3 ± 72.05^{d}	268 ± 73.2^{c}	313.6 ± 160.5^{b}	436 ± 74.98^{a}
AP (mg kg^{-1})	3.42 ± 1.04^{c}	3.88 ± 1.67^{b}	4.51 ± 1.73^{a}	$4.18\pm0.76^{a,b}$
AK (mg kg^{-1})	103 ± 32.77^{b}	105.17 ± 25.18^{b}	144 ± 46.04^{a}	129.67 ± 24.93^{a}
ECa [cmol (1/2 Ca^{2+}) kg^{-1}]	23 ± 5.71^{b}	32.5 ± 6.24^{a}	30.03 ± 8.1^{a}	18.9 ± 5.59^{c}
EMg [cmol (1/2 Mg^{2+}) kg^{-1}]	1.82 ± 0.41^{b}	4.79 ± 1.05^{a}	1.73 ± 0.74^{b}	1.12 ± 0.42^{b}

Data source: Tang et al. (2014)
In each row values with different letters differ significantly ($p<0.05$) by the non-parametric Kruskal–Wallis all-pairwise comparisons test
PP Pinus plantation, *SL* shrubland, *SF* secondary forest, *MF* natural maturing forest, *OM* organic matter, *OC* organic carbon, *TN* total nitrogen, *AN* available nitrogen, *TP* total phosphorus, *AK* available potassium, *TK* total potassium, *TCa* total calcium, *ECa* exchangeable calcium, *TMg* total magnesium, *EMg* exchangeable magnesium

these communities. In the natural maturing forest and the secondary forest, the organic matter (ranging from 10.66 to 13.73 %), organic C (ranging from 6.31 to 7.96 %), total N (ranging from 0.45 to 0.55 %), available N (ranging from 313.6 to 436 mg kg^{-1}), available P (ranging from 4.18 to 4.5 mg kg^{-1}), and available K (ranging from 129.67 to 144 mg kg^{-1}) are higher than those of both the shrubland and the *Pinus* plantation. The shrubland has significantly higher concentrations of total N (0.35 %), total P (0.07 %), available N (268 mg kg^{-1}), and available P (3.88 mg kg^{-1}) than the *Pinus* plantation. The concentrations of total Ca and exchangeable Ca tend to be higher in the secondary forest and the shrubland than in the other plant communities. The ratio of organic C to total N in the *Pinus* plantation is the highest (19.41) among all the plant communities, indicating a slow litter decomposition rate.

Age-related patterns of leaf litter nutrient inputs and soil available nutrients are observed across the natural secondary successional plant communities (Fig. 5.9a–f). The concentrations of leaf litter N, P and K inputs and soil available N, P and K increase with increasing plant community ages. In contrast, the 30-year-old *Pinus* plantation, with its low leaf litter nutrient input, has delayed soil development.

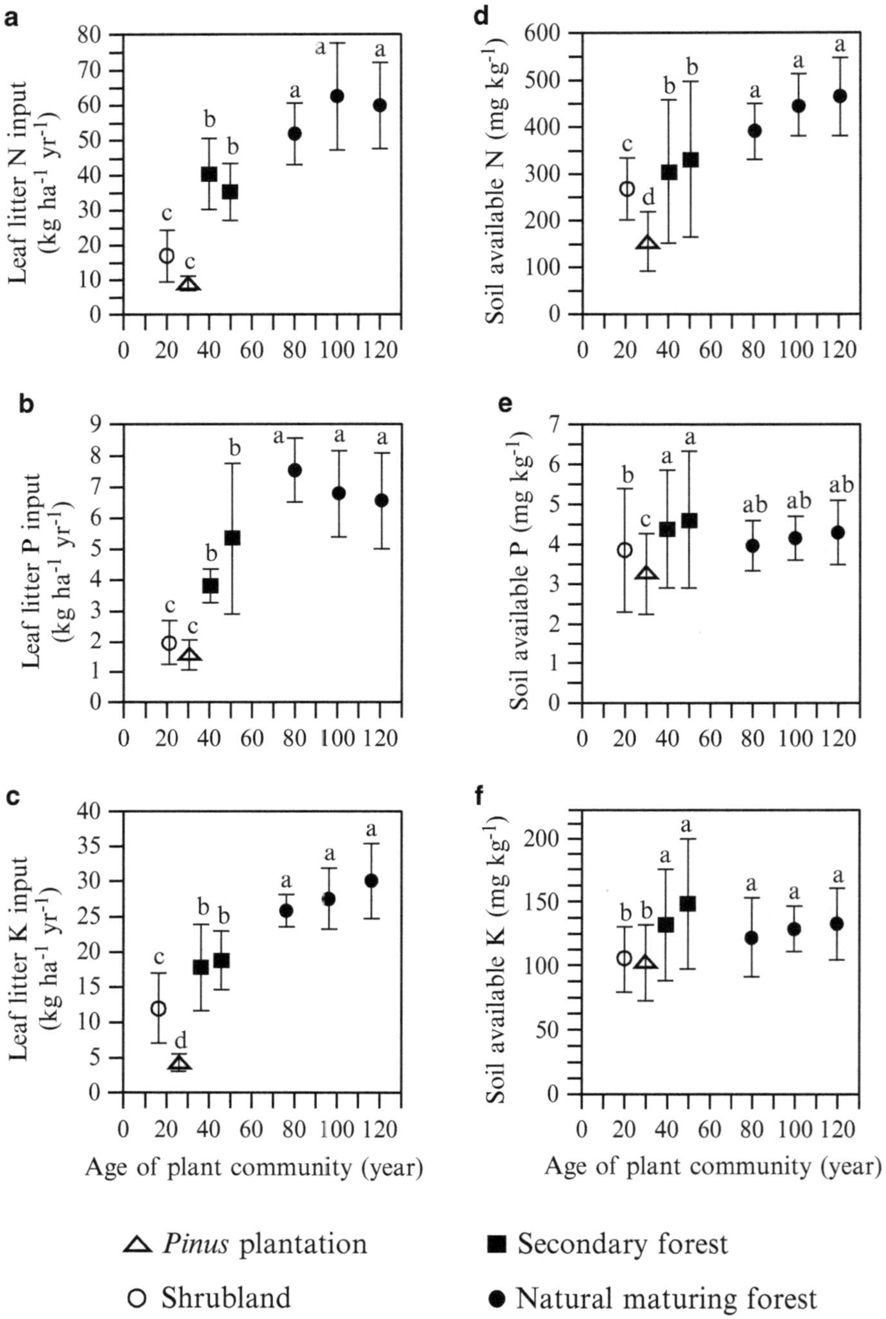

Fig. 5.9 Changes in leaf litter nutrient inputs and soil available nutrients with different ages of successional plant communities and the *Pinus* plantation in the Shilin karst area, central Yunnan. Different letters indicate significant differences ($p<0.05$) by the non-parametric Kruskal–Wallis all-pairwise comparisons test; *vertical bars* indicate standard deviations (Data source: Tang et al. (2014))

5.5 Ecosystem Processes

Litterfall production in a forest ecosystem is determined by climatic conditions, species composition, and successional stage (Vogt et al. 1986; Haase 1999; Sundarapandian and Swamy 1999). Litter accumulation during succession has traditionally been considered an ecosystem process with significant impact on carbon and nutrient cycles (Odum 1960; Mellinger and McNaughton 1975; Holland and Coleman 1987). Litterfall and decomposition are critical processes for transferring nutrients from the aboveground forest biomass to soils (Golley et al. 1975; Swift et al. 1979). Our study in the Shilin area shows that broad-leaved trees tended to have a higher litterfall production than coniferous trees in the subtropics of China (Deng et al. 1993; Xu and Hirata 2002). Litter nutrient inputs vary with differing successional plant communities, as changes in species assemblages would be expected to alter nutrient cycling. With increasing age in the various successional communities, leaf litter N, P, K, Ca and Mg inputs increase, showing a relationship between leaf litter nutrient inputs and soil nutrient storage. In the *Pinus yunnanensis* plantation, those inputs have been very low, due to the low concentration of nutrients in the pine needles and their low litter decomposition rate (showing in the high ratio of organic C to total N in the soil of the *Pinus* plantation) on the forest floor. Considering the N dynamics under all the plant communities, the lowest N concentration is found in pine litter, as is the lowest value of soil-available N in the *Pinus* plantation. However, in the *Pinus* plantation in the Shilin area, the soil-available N concentration (150.3 mg kg^{-1}) is much higher than in a *Pinus* plantation of Mouding (70 mg kg^{-1}, from Hou et al. 2010), and Luquan (35.8 mg kg^{-1}, from Tang et al. 2010a), both in central Yunnan. This is because *Albizia mollis* is present in the *Pinus* plantation in the Shilin area, and this species has a nitrification function. As for the degraded ecosystem, its recovery relies on an increase in species diversity, litterfall production, litter nutrient input, and on recovery of soil fertility. The *Eucalyptus* and *Pinus* plantations, considered as human restoration efforts, have resulted in poor species diversity and regeneration of the species. *Eucalyptus* plantations use more water and nutrients than do forests of other tree species. Like many introduced exotic plants, they may thrive at the expense of their environment. It is clear that *Eucalyptus* is not suitable for planting in large areas as a means to restore forest cover in the subtropical area of Yunnan.

The studies demonstrate that after abandonment of pastures or agricultural fields in Yunnan, habitat recovery generally appears sufficiently robust to achieve restoration of the native plant community through natural succession without human intervention. Plant diversity, litter nutrient input, and soil nutrient storage are much more efficiently recovered by letting plant communities grow back naturally than by establishing mono-species plantations.

References

Deng CZ, Hou JP, Li SC, Zhao HK, Fu Y (1993) Researches on litterfall distributed in seven forests at varied altitudes, on Ailao Mountain, Yunnan. Acta Phytoecologica et Geobotanica Sinica 17:364–370

FAO (2007) State of the world's forests 2007. Food and Agriculture Organization of the United Nations, Rome

Golley FB, Mcginnis JT, Clements RG, Child GI, Duever MJ (1975) Mineral cycling in a tropical moist forest ecosystem. University of Georgia Press, Athens

Haase R (1999) Litterfall and nutrient return in seasonally flooded and non-flooded forest of the Pantanal, Mato Grosso, Brazil. For Ecol Manage 117:129–147

Holland EA, Coleman DC (1987) Litter placement effects on microbial and organic matter dynamics in an agroecosystem. Ecology 68:425–433

Hou X, Duan C, Tang CQ, Fu D (2010) Nutrient relocation, hydrological functions, and soil chemistry in plantations as compared to natural forests in central Yunnan, China. Ecol Res 25:139–148

Liu WY, Liu LH, Jing GF, He AJ (2000) Decomposition of leaf litter in *Pinus yunnanensis* forest and evergreen broad-leaved forest in central Yunnan. Acta Bot Yunnanica 22(3):298–306

Liu SR, Sun PS, Wen YG (2003) Comparative analysis of hydrological functions of major forest ecosystems in China. Acta Phytoecologica Sinica 27(1):16–22

Mellinger MV, McNaughton SJ (1975) Structure and function of successional vascular plant communities in central New York. Ecol Monogr 45:161–182

Odum EP (1960) Organic production and turnover in old field succession. Ecology 41:34–49

Peng SL (ed) (2003) Study and application of restoration ecology in tropical and subtropical China. Science Press, Beijing

Peng SL, Fang W (1995) Development of species diversity in restoration process of tropical man-made forest ecosystems in China. Acta Ecologica Sinica 15(Suppl A):18–30

SFA (State Forestry Administration) (2007) China's forestry 1999–2005. China Forestry Publishing House, Beijing

Sundarapandian SM, Swamy PS (1999) Litter production and leaflitter decomposition of selected tree species in tropical forests at Kodayar in the Western Ghats, India. For Ecol Manage 123:231–244

Swift MJ, Heal OW, Anderson JM (1979) Decomposition in terrestrial ecosystems, vol 5, Studies in ecology. University of California Press, Berkeley

Tang CQ, Hou X, Gao K, Xia T, Duan C, Fu D (2007) Man-made versus natural forests in mid-Yunnan, southwestern China: plant diversity and initial data on water and soil conservation. Mt Res Dev 27(3):242–249

Tang CQ, Zhao M-H, Li XS, Ohsawa M, Ou XK (2010a) Secondary succession of plant communities in a subtropical mountainous region of SW China. Ecol Res 25(1):149–161

Tang CQ, Li YH, Zhang ZY (2010b) Species diversity patterns of natural secondary plant communities and man-made forests in a subtropical mountainous karst area, Yunnan, SW China. Mt Res Dev 30(3):244–251

Tang CQ, Li YH, Zhang ZY, Hou X, Hara K, Tomita M, He LY, Li XS (2014) Effects of management on vegetation dynamics and associated nutrient cycling in a karst area, Yunnan, SW China. Landsc Ecol Eng. doi:10.1007/s11355-014-0258-7

Vogt KA, Grier CC, Vogt DJ (1986) Production, turnover and nutrient dynamics of the above- and below-ground detritus of world forests. Adv Ecol Res 15:303–377

Xu XN, Hirata E (2002) Forest floor mass and litterfall in Pinus luchuensis plantations with and without broad-leaved trees. For Ecol Manage 157:165–173

Chapter 6
Semi-Savannas in Hot-Dry Valleys

Abstract The hot-dry valleys are deeply incised gorges created by major rivers in SW China. A foehn effect caused by the rain shadow on the leeward slopes of mountains accounts for their hot-dry climate. Many plant genera have tropical affinities. The most representative vegetation types of these semi-savannas are the sparse woody grassland and the thorny succulent shrubland, similar to those of the savannas of tropical India and tropical Africa. Here I describe the major communities of the hot-dry valleys along four rivers, Yuanjiang, Jinshajiang, Nujiang and Lancangjiang, concentrating on the floristic composition of Yuanjiang.

6.1 Characteristics of Hot-Dry Valleys

SW China's plant life is exceptionally diverse and has developed various types of vegetation because of its varied climates, topographies, and soils. A traveller will see striking changes going from north to south, from west to east, and from moist mountains to the arid, rain-shadowed, hot-dry river valleys of the region under study, particularly the deeply incised valleys along the Nu River (Nujiang, upper reaches of the Salween, below 1,200 m asl), the Lancang (Lancangjiang, upper reach of the Mekong, below 1,000 m), the Jinsha (Jinshajiang, the upper section of the Yangtze, below 800 and 1,200 m), the Yuan (Yuanjiang, the upper stretch of the Red River, below 1,000–1,400 m), and the Nanpan (Nanpanjiang, below 1,000 m) in Yunnan, and in the down-stream stretch of the Jinsha river, as well as its tributaries including the Yalong (Yalongjiang), Anning (Anninghe), Dadu (Daduhe), Min (Minjiang), Baishui (Baishuihe) rivers of western Sichuan (SVEG 1980; Zhang 1992; He et al. 2000; Jin 2002) (Figs. 6.1 and 6.2a–c). Some hot-dry valleys are also found in the Southeast (Wushan, Fengjie, Xuyong, Gulan and Pingshan in Chongqing Municipality) along the Yangtze and Chishui rivers of the Sichuan Basin (SVEG 1980). In Guizhou, hot-dry valleys are confined to the southwestern part of the province, such as in the Nanpanjiang and Beipanjiang valleys where the plant community is dominated by *Pinus yunnanensis* var. *tenuifolia*.

The usual temperature and precipitation values, as compared to those in other regions in the world with typical savannas, are shown in Table 6.1. Precipitation is low, around 600–810 mm per year, falling mostly from May to October; the annual

C.Q. Tang, *The Subtropical Vegetation of Southwestern China*,
Plant and Vegetation 11, DOI 10.1007/978-94-017-9741-2_6

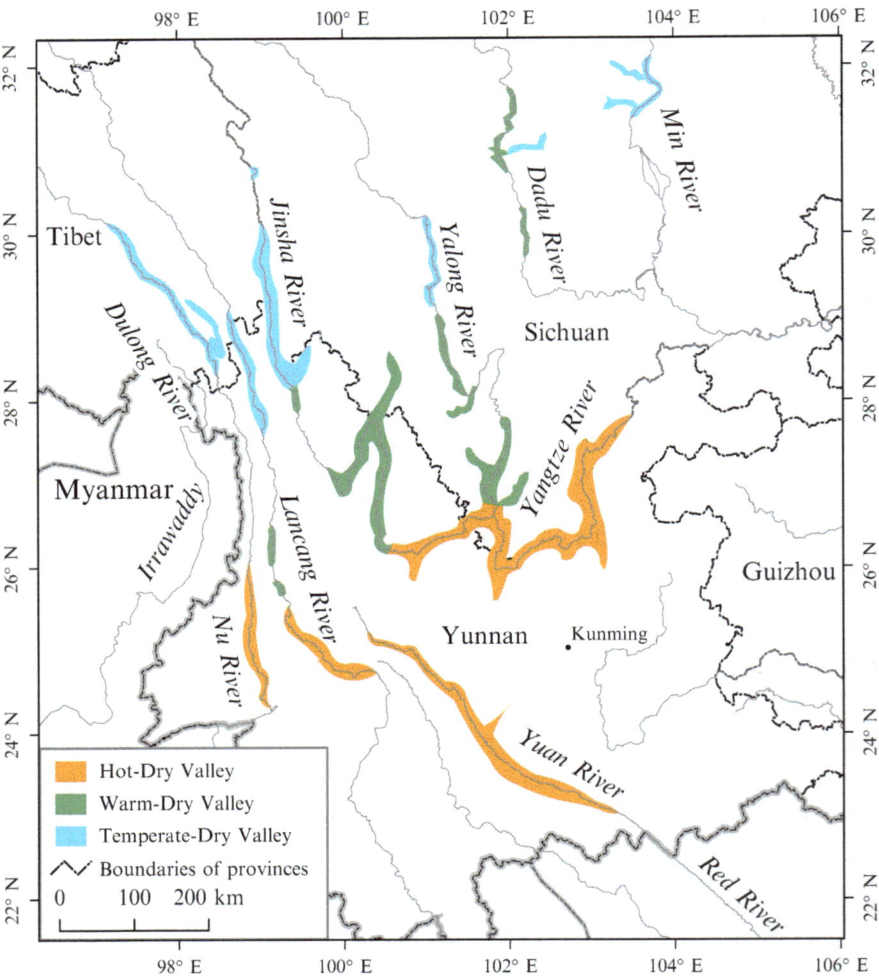

Fig. 6.1 Locations of major dry valleys in SW China (Revised from Zhang (1992) and Jin (2002))

evaporation is 2,300–3,200 mm. The weather is extremely dry from November to May, leaving water in short supply for a long period. The mean temperature during the coldest month is above 12 °C, and during the warmest month, from 24 to 28 °C. For 350 days of each year temperatures are above 10 °C. The climatic conditions compared to Dodoma (Tanzania) and Ouagadougou (Burkina Faso) are shown in Fig. 6.3.

Temperature and rainfall are the chief determinants of the type of vegetation found in an area, though soil, topography, sun exposure and geology are also important abiotic factors. The climate results from complex interactions among ocean currents, as well as latitude, altitude, continental terrain and local topography. The Hengduan Mountains to the north block the humid air from the Bay of Bengal, creating a rain shadow. Most of the river valleys are alternately affected by monsoon and foehn winds coming from the

Fig. 6.2 The hot-dry valleys along Lancangjiang (**a**), along Yuanjiang (**b**), along Jinshajiang (**c**)

Table 6.1 The climate of the Yuanjiang and Yuanmou hot-dry valleys as compared with typical savanna regions of the world

	Yuanjiang	Yuanmou	Typical savanna regions
Annual solar radiation (Kcal/year/cm^2)	128	153	140–190
Annual precipitation (mm)	805.1	634	1,000–1,500
Precipitation during rainy season per year	649.1	583.8	>600
Precipitation during dry season per year	155.1	50.2	<50
Proportion of precipitation during rainy season/year	81	92	>75
Mean annual temperature (°C)	23.7	21.8	>24
Mean temperature of the coldest month (°C)	16.7	14	13–18

Data sources: Jin and Ou (2000), SWFC-KZI-YJCG (2009)

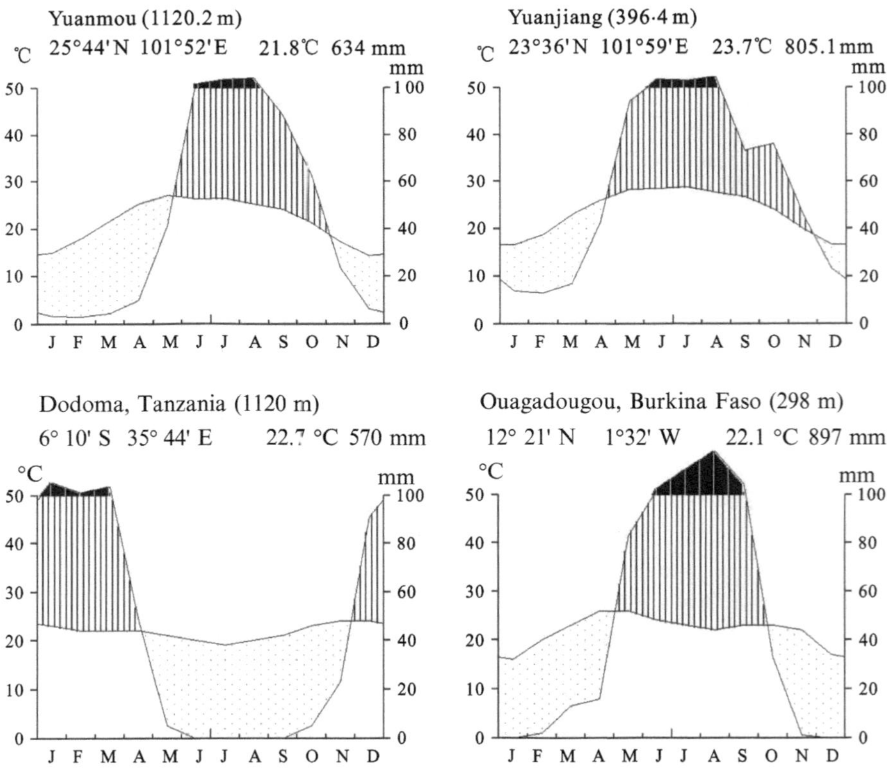

Fig. 6.3 The climatic conditions of Yuanmou and Yuanjiang compared to Dodoma (Tanzania) and Ouagadougou (Burkina Faso). Climatic diagrams are explained in Fig. 1.18a and its legend

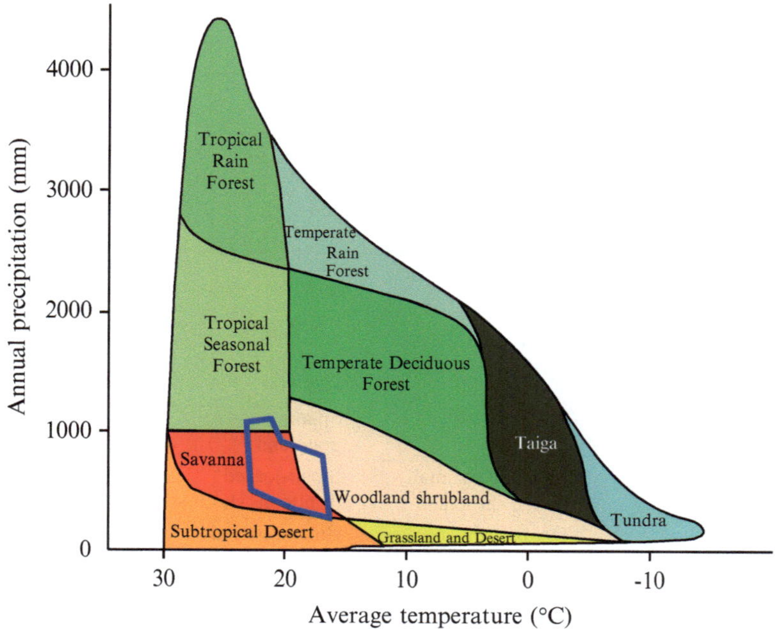

Fig. 6.4 The climate range (shown in *blue*) of hot- and warm-dry valleys in SW China corresponding to the savanna vegetation as found in the Whittaker Biome Diagram (Whittaker 1975)

Indian Ocean. The foehn creates a hot-dry season lasting 7 months; scorching dry spells also afflict the relatively wet seasons. Figure 6.4 illustrates the climatic range of the hot- and warm-dry valleys of SW China, corresponding largely to the savanna and shrubland vegetation as described in Whittaker's scheme (Whittaker 1975).

In these valleys the loss of water by evaporation exceeds the annual rainfall. The dry valleys can be divided into hot-dry, warm-dry and temperate-dry (Zhang 1992) (Table 6.2). In the warm-dry and temperate-dry valleys, the evaporation is 1–3 times higher than the annual rainfall, while in the hot-dry valleys it is 3–6 times higher. The warm-dry and temperate-dry valleys are found in areas to the north or at higher altitudes. There the vegetation is mainly small-leaved scrubland similar to the Maquis in areas with a Mediterranean-type climate (Jin 2002). Although some small-leaved scrubland (of e.g. *Pinus*, *Quercus*, *Terminalia*, *Sophora* or *Desmodium*) (community height usually below 10 m) does occur also in the transitional area from hot-dry to warm-dry valleys, in this chapter I focus only on the sparse woody grassland and thorny succulent shrubland of the semi-savanna in hot-dry valleys.

6.2 Semi-Savannas

The vegetation of hot-dry valleys, as characterized by a grassy ground layer and a sparse but distinct upper layer of dwarf woody plants with a height lower than 10 m, is called sparse woody grassland. In hot and dry habitats with thin soils and many

Table 6.2 Types of dry valleys in China

Types	Hot-dry	Warm-dry	Temperate-dry
Coldest monthly mean temperature (°C)	>12	12–5	5–0
Warmest monthly mean temperature (°C)	28–24	24–(20) 22	22–16
Days with mean daily temperature ≥10 °C	>350	350–251	250–151
Mean annual temperature (°C)	≥20	14	10
Annual accumulated temperature ≥10 °C	>7,000	5,500–7,000	N/A
Annual rainfall (mm)	800–900	800–900	600–700
Vegetation	Sparse woody grassland	Sparse woody grassland	Thorny small-leaved deciduous shrubland
	Medium- to small-leaved shrubland	Small-leaved deciduous shrubland	
Soil	Dry red soil	Cinnamon-red soil	Cinnamon soil

Partial data from Zhang (1992)

outcrops or on rocky slopes, thorny succulent shrubland occurs. Here plant growth is scattered and thus is in strong contrast with the high plant density of the evergreen broad-leaved forests of humid and semi-humid mountain slopes. The plants that grow in the hot-dry valleys are especially adapted to withstand or avoid water stress (see below). These plants have affinities with the xeric flora of tropical Asia and tropical Africa, and in their origin, are considered to have a connection with the flora of Gondwana (Zhu 1990; Jin 2002). The best examples are *Woodfordia*, *Calotropis*, and *Adinandra*, which are disjunctly distributed in areas of tropical Asia (including India, Malaysia and southern Yunnan) and tropical Africa. Ecologically the vegetation in the hot-dry valleys is comparable to that of savannas of India and Africa. However, these valleys are densely inhabited by humans because their high temperatures favor good crops as well as grasses for cattle grazing. Human activities have greatly modified the vegetation and changed these into what we now call semi-savannas (semi-natural savannas) or secondary savannas.

The earliest description of the vegetation of hot-dry valleys is found in the *Diary on touring in Yunnan Province* written by Xu Xiake during the Ming Dynasty (Xu 1985) as cited by Ma and McConchie (2001). He arrived in the Yuanmou Basin on 6 December 1638. He described the Earth Forest (now a tourist attraction because the soil has been deeply eroded into columnar shapes resembling standing tree trunks) as follows:

> Bambax [sic] trees grow to more than 1 zhang and the leaves can last for 2–3 years. There is a dry ravine from the west with a precipice stretching along it; the sand at the base of the ravine is a foot deep. There is no water in the bottom of the ravine. The soil was eroded into columnar shapes of different sizes and looks like a ghost on a moonlight night. The sand is white, which is not likely to be salt but frost from the ground……

Photos taken in 2010 depict what Xu Xiake described (Fig. 6.5a, b).

Fig. 6.5 The Earth Forest (Tulin) at Yuanmou in the Jinshajiang valley, Yunnan (**a**, **b**)

In this environment, plants have developed special biological features in response to the stresses produced by intense radiation, summer heat, high evaporation, and persistently dry conditions. The three main adaptive strategies are drought tolerance, drought evasion and succulence. In the dry season, grasses wither into straw, and many tree species shed their leaves. Other xerophytes usually have small hairy leaves with sunken stomas. They often are dwarfish, cushion-like, thorny, clumped,

Table 6.3 Traits enabling reduction of transpiration area in various species of the hot-dry valleys

Traits	Species
Dwarf (H<5 m)	*Buchanania latifolia*, *Grewia biloba* var. *parvifolia*
Distorted stem	*Grewia celtidifolia*, *Phyllanthus emblica*, *Engelhardia spicata*
Deciduous plant	*Buchanania latifolia*, *Terminthia paniculata*, *Polyalthia cerasoides*
Thick rough bark	*Lannea coromandelica*, *Polyalthia cerasoides*, *Eriolaena spectabilis*
Stems with cork thorn	*Bombax ceiba*
Thick leathery leaf	*Osyris wightiana*, *Pistacia weinmanniifolia*, *Glochidion daltonii*
Small leaf	*Phyllanthus emblica*, *Osteomeles schwerinae*, *Osyris wightiana*
Hairy leaf	*Verbascum thapsus*, *Sida mysorensis*, *Cipadessa cinerascens*
Essential oil	*Pistacia weinmanniifolia*, *Cymbopogon pendulus*, *Blumea balsamifera*
Latex	*Secamone sinica*, *Euphorbia hirta*
Thorny	*Sophora davidii*, *Zizyphus yunnanensis*, *Acacia farnesiana*
Succulent	*Euphorbia royleana*, *Opuntia monacantha*, *Sedum* spp.
Water-storing stem/leaf	*Bryophyllum pinnatum*, *Calotropis gigantea*
Hard stem grass	*Heteropogon contortus*, *Bothriochloa pertusa*, *Capillipedium assimile*

Partial data from Jin and Ou (2000)

or succulent, with well-developed root systems, and strong sprouting ability. They are exemplified by plants such as *Phyllanthus emblica* and *Euphorbia royleana* of the Euphorbiaceae. The adaptive characteristics of representative plants are summarized in Table 6.3.

Among the four major river valleys (Nujiang, Lancangjiang, Jinshajiang and Yuanjiang), Yuanjiang harbors the most typical semi-savanna. Semi-savannas including the sparse woody grassland and the thorny succulent shrubland are at 320–1,000 m in the Yuanjiang area (Fig. 6.6). Above 600 m, other vegetation types are present, including semi-evergreen seasonal rain forest (600–800 m), sclerophyllous evergreen scrubland (600–1,100 m), deciduous broad-leaved forest (800–1,500 m), mountain rain forest (1,000–1,400 m), monsoon evergreen broad-leaved forest (1,500–2,000 m), warm temperate coniferous forest (1,000–2,000 m), semi-humid evergreen broad-leaved forest (2,000–2,200 m) and mid-montane moist evergreen broad-leaved forest (2,300–2,500 m).

6.2.1 Sparse Woody Grasslands

In sparse woody grasslands the tree layer is very sparse, with coverage ranging from 5 to 30 % (Fig. 6.7a, b). Most tree species are deciduous, hot- and drought-tolerant. There are only a few evergreen woody species. Tree height is under 10 m. Tree stems are often distorted, barks are rough and thick. Tree canopies often are rounded or umbrella-like. The coverage of the shrub layer is between 10 and 50 %. Most shrubs are deciduous, thorny and dwarfish with a height of 0.5–5 m (sometimes

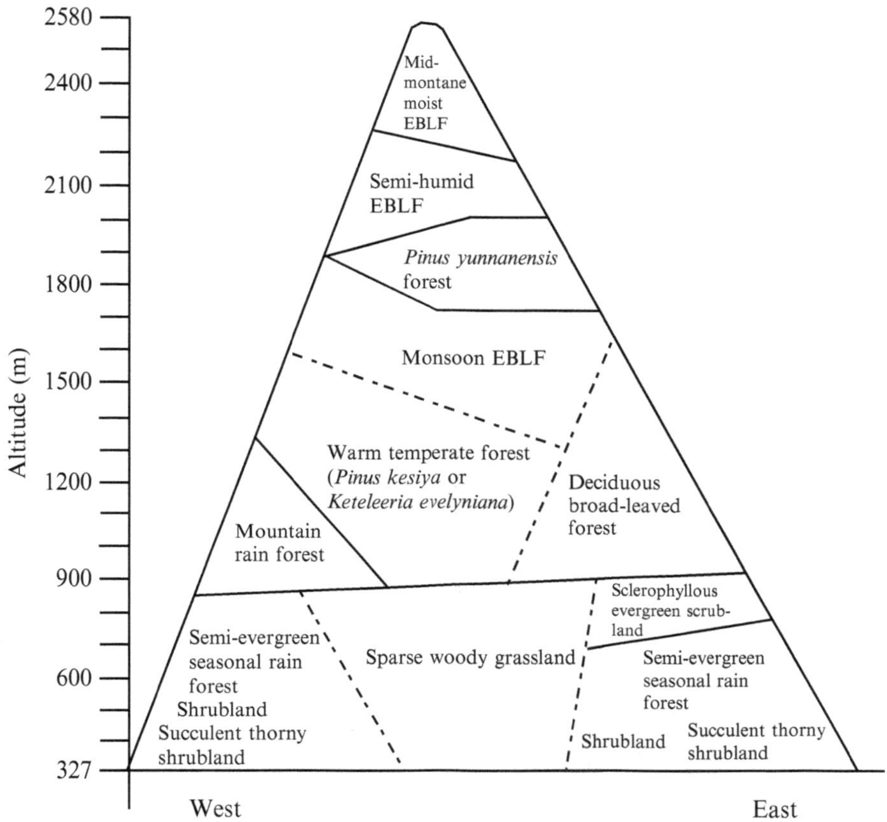

Fig. 6.6 Altitudinal vegetation zones in the Yuanjiang area, Yunnan (After SWFC-KZI-YJCG (2009))

their height is less than that of the grass layer). In the grass layer, the coverage varies from 30 to 90 %, mainly consisting of monocotyledons. Most tussocks are of medium height, 0.5–1 m, and tall tussocks can reach 4 m. During the dry season the tussocks are withered and yellow. They turn green in rainy seasons.

In the Yuanjiang valley, the sparse woody grassland is the most typical semi-savanna and covers the largest area, with diverse communities. *Heteropogon contortus*, *Bothriochloa pertusa*, *Chromolaena odorata*, or *Eulaliopsis binata* are the dominants in the grass layer. *Haldina cordifolia* or *Phyllanthus emblica* or *Lannea coromandelica*, etc. dominate the sparse and stunted tree layer. *Vitex negundo*, *Sida mysorensis* and *Wendlandia uvariifolia*, etc. are common in the shrub layer. Abundant lianas such as *Rhaphidophora decursiva* and *Pueraria alopecuroides* are found in the interlayers. The indicator species *Indigofera cassioides*, *Phyllanthus emblica*, *Woodfordia fruticosa*, *Wendlandia uvariifolia*, *Lannea coromandelica*, *Helicteres isora*, *Polyalthia cerasoides*, *Terminthia paniculata*, *Bombax ceiba*, and *Breynia fruticosa* are of tropical origin (Fig. 6.8a). A number of plants are native to

Fig. 6.7 The sparse woody grassland with major woody species *Lannea coromandelica*, *Garuga forrestii*, *Woodfordia fruticosa*, *Salix myrtillacea*, *Pistacia weinmanniifolia* and herbacious *Heteropogon contortus*, *Solanum aculeatissimum*, and *Tithonia diversifolia* (**a**), and *Garuga forrestii* scatters in the grassland dominated by *Heteropogon contortus* and *Eulalia pallens* along Lüzhi River at 600 asl, Shuangbai, central Yunnan (**b**). The photos were taken during the rainy season (July)

Fig. 6.8 *Phyllanthus emblica* (**a**), *Tamarindus indica* (**b**) in the Yuanjiang valley, Yunnan

Africa, but are widely cultivated in the hot-dry valleys; some may be naturalized, such as *Tamarindus indica* (Fig. 6.8b).

In the Jinshajiang valley, the sparse woody grassland is especially well developed. The grass layer is dominated by *Heteropogon contortus* or *Bothriochloa pertusa* or *Hyparrhenia diplandra* or *Cymbopogon distans*. Woody *Dodonaea angustifolia*, *Dodonaea viscosa*, *Phyllanthus emblica*, *Calotropis gigantea*, *Acacia farnesiana*, *Indigofera lenticellata*, and *Bombax malabaricum* are indicator species, though sparsely distributed in the grassland.

In the Nujiang valley, the shrubs *Phyllanthus emblica*, *Zizyphus calophylla*, *Zizyphus yunnanensis*, *Cipadessa cinerascens*, *Indigofera enneaphylla* and *Engelhardia spicata* are indicator species. *Heteropogon contortus*, *Bothriochloa pertusa*, *Miscanthus floridulus* and *Thysanolaena maxima* are the major grasses.

In the Lancangjiang valley, woody *Eriolaena spectabilis*, *Eriolaena kwangsiensis*, *Woodfordia fruticosa* and *Helicteres angustifolia* are indicator species. *Heteropogon contortus*, *Cymbopogon tortilis* or *Eupatorium odoratum* dominate the grass layer.

Most species mentioned above are distributed throughout tropical Asia, including India, Bhutan, Laos, Myanmar, Nepal, Pakistan, Sri Lanka, Thailand, Vietnam, Cambodia, Indonesia, and Malaysia. Some of them are also found in tropical Africa, such as *Calotropis gigantea*. They are of tropical origin. For instance, *Eriolaena spectabilis*, *Eriolaena kwangsiensis*, *Phyllanthus emblica*, *Calotropis gigantea*, *Helicteres isora*, *Helicteres angustifolia*, *Lannea coromandelica*, *Woodfordia fruticosa* are the closest relatives of the savanna species of India and Africa. *Heteropogon contortus* is also very common in the savannas of India and Africa.

Representative communities of the hot-dry valleys along the four rivers are shown in Table 6.4. The floristic composition of the sparse woody grassland communities is exemplified by the hot-dry valley in Yuanjiang (Table 6.5).

6.2.2 Thorny Succulent Shrublands

Shrubby *Euphorbia royleana* and introduced *Opuntia monacantha* (Fig. 6.9a, b) are indicator species of the thorny succulent shrubland, which is discontinuously distributed from the altitude of 2,000 m in the northern upper stream of the Jinshajiang to the southern Yuanjiang and Nujiang below 800 m. Apart from *Euphorbia royleana* and *Opuntia monacantha* this shrubland contains *Garuga forrestii*, *Lannea coromandelica*, *Haldina cordifolia*, *Grewia celtidifolia*, *Polyalthia cerasoides*, *Terminalia franchetii*, *Pistacia weinmanniifolia*, *Tephrosia purpurea*, *Agave angustifolia*, *Bryophyllum pinnatum*, *Acacia farnesiana*, etc. *Euphorbia royleana* is rather big and tall; it usually is 3–4 m in height, but can reach 7 m, and individuals usually are clumped. *Opuntia* is rather shorter, but may reach 2–2.5 m in height. The coverage of the shrub layer ranges from 20 to 60 %. A grass layer will be 0.25–0.8 m tall, with coverage of 20–80 %. Typical herbaceous species are *Heteropogon contortus*, *Bothriochloa pertusa*, *Arthraxon prionodes* and *Selaginella pulvinata*. The thorny succulent shrubland often occurs scattered in red-earth habitats with thin soils developed from limestone bedrock, with many outcrops. Climbing plants such as *Shuteria involucrata*, *Cardiospermum halicacabum*, *Hiptage minor*, *Porana racemosa* are found in the interlayers.

Euphorbia royleana also grows in Bhutan, N and NE India, Myanmar, Nepal and Pakistan (Ma and Gilbert 2008). *Opuntia monacantha* is native to Argentina, Brazil, Paraguay, and Uruguay; it is widely introduced and naturalized in tropical and subtropical regions (Li and Taylor 2007). *Euphorbia royleana* and *Opuntia monacan-*

Table 6.4 Representative communities of the sparse woody grassland in the hot-dry valleys

Region	Plant community	Distribution
Yuanjiang	*Lannea coromandelica-Vitex negundo-Heteropogon contortus*	450–550 m in Pupiao, Yuanjiang
	Lannea coromandelica–Vitex negundo–Paspalum scrobiculatum	450 m in Xiaohedi, Yuanjiang
	Tamarindus indica-Vitex negundo-Bryophyllum pinnatum	500 m in Pupiao, Yuanjiang
	Albizia julibrissi-Barleria cristata-Oplismenus patens	750 m in Xilahe, Yuanjiang
	Engelhardia colebrookiana-Woodfordia fruticosa-Heteropogon contortus	850 m in Manshahe, Yuanjiang
	Bombax malabarica-Woodfordia fruticosa-Heteropogon contortus	400–1,000 m along Yuanjiang
	Phyllanthus emblica-Hyparrhenia diplandra	1,440–1,470 m in Nanjianhe, Nanjian
	Osteomeles schwerinae-Heteropogon contortus	1,450–1,530 m in Nanjianhe, Nanjian
	Phyllanthus emblica-Heteropogon contortus	1,100–1,300 m in Lüzhijiang, Yimen
	Woodfordia fruticosa-Heteropogon contortus	1,100–1,300 m in Lüzhijiang, Yimen
	Pistacia weinmanniifolia- Heteropogon contortus	1,220–1,300 m in Lüzhijiang, Yimen
	Osyris wightiana-Cymbopogon distans	1,370–1,500 m in Lüzhijiang, Yimen
	Haldina cordifolia-Polyalthia cerasoides-Perotis hordeiformis	795 m in Chongpian, Shidi
	Grewia celtidifolia-Heteropogon contortus	470–600 m in Liuhu
	Buchanania latifolia-Woodfordia fruticosa-Heteropogon contortus	900 m in the river valley, eastern Ailao Mountains
	Phyllanthus emblica-Bridelia tomentosa-Cipadessa cinerascens	900 m in the river valley, eastern Ailao Mountains
Jinshajiang		Below 2,800 m along Jinshajiang
	Dodonaea angustifolia-Heteropogon contortus	1,057–1,630 m in Longjiang, Yuanmou
	Dodonaea angustifolia-Bothriochloa pertusa	1,030–1,290 m in Longjiang, Yuanmou
	Vitex negundo-Terminalia franchetii	1,020–1,600 m in Longjiang, Yuanmou
	Dodonaea angustifolia-Hyparrhenia eberhardii	1,520–1,630 m in Longjiang, Yuanmou
	Eulaliopsis binata-Eriophora comosum	1,190–1,300 m Longjiang, Yuanmou
	Selaginella mairei-Heteropogon contortus	1,520–1,570 m in Pianjiao, Binchuan
	Quercus franchetii-Dodonaea viscosa-Heteropogon contortus	1,540–1,590 m along Puduhe, Luquan
	Barleria cristata-Cymbopogon distans	1,070–1,090 m in Menggu, Qiaojia

(continued)

Table 6.4 (continued)

Region	Plant community	Distribution
	Calotropis gigantea-Bothriochloa pertusa	770–800 m in Menggu, Qiaojia
	Calotropis gigantea-Euphorbia hirta	750–780 m in Menggu, Qiaojia
	Calotropis gigantea-Heteropogon contortus	880–920 m in Yalongjiang, Panzihua
	Phyllanthus emblica-Gossampinus malabarica-Heteropogon contortus	700–1,300 in river valleys, Xichang and Liangshan
	Ficus tikoua-Bothriochloa pertusa	1,150–1,250 m in Anninghe, Miyixian
Nujiang	*Phyllanthus emblica-Pyracantha fortuneana -Heteropogon contortus*	Below 1,100 m in Nujiangba
	Indigofera enneaphylla-Bothriochloa pertusa	690–710 m in Nujiangba
	Cipadessa cinerascens-Heteropogon contortus	690–710 m in Nujiangba
	Bauhinia lecomtei-Bombax ceilba	750–930 m in Liuku
	Pistacia weinmanniifolia-Heteropogon contortus	750–940 m in Lushui
	Terminalia franchetii var. *membranifolia-Bothriochloa pertusa*	720–1,000 m in Nujinagba
	Cryptolepis elegans-Cipadessa cinerascens	790–830 m in Nujiangba
	Zizyphus yunnanensis-Heteropogon contortus	790–830 m in Nujiangba
	Wendlandia scabra-Pinus yunnanensis-Boehmeria platyphylla	830–900 m in Liuku
	Engelhardia colebrookiana-Heteropogon contortus	750–940 m in Lushui
	Wendlandia tinctoria-Terminalia chehula-Heteropogon contortus	770–840 m in Xincheng, Baoshan
	Terminalia chehula-Woodfordia fruticosa-Heteropogon contortus	810–850 m in Mengxing, Longling
Lancangjiang	*Eriolaena spectabilis-Woodfordia fruticosa-Heteropogon contortus*	Below 900 m in Nuozhadu
	Erythrina arborescens-Itoa orientalis-Cymbopogon tortilis	1,010–1,040 m in Nanjian
	Oroxylum indicum-Pistacia weinmanniifolia-Cymbopogon tortilis	1,040–1,170 m in Fengqing, Nanjian
	Symplocos ennestii-Engelhardia spicata-Cymbopogon goeringii	1,270–1,300 in Fengqing

Data sources: SVEG (1980), Wu et al. (1987), Jin (1999), Jin and Ou (2000), SWFC-KZI-YJCG (2009)

Table 6.5 The floristic composition (IV ≥ 3 %) of sparse woody grassland communities in the hot-dry valley in Yuanjiang

Plant community	L-V-H	L-V-P	H-P-P
Site	Pupiao	Xiaohedi	Shidi
Total plot area (m^2)	600	300	550
Coverage of tree layer (%)	23	28	17
Coverage of shrub layer (%)	64	52	41
Coverage of herb layer (%)	85	37	88
Species	IV (%)	IV (%)	IV (%)
Tree layer			
Lannea coromandelica	100	43.5	31.2
Melia toosenden		31.3	
Haldina cordifolia		12.6	44.8
Tamarindus indica			
Bauhinia racemosa			21.9
Shrub layer			
Vitex negundo	74.6	70.3	
Bauhinia racemosa	11.1		
Indigofera hancockii	5.9		
Phyllanthus emblica	5		8.5
Trigonostemon sp.		18.7	
Polyalthia cerasoides		5.5	
Sida mysorensis		3.4	3.8
Desmodium oblatum			16.4
Cipadessa cinerascens			11.2
Sida rhombifolia			9.6
Clausena excavata			7.1
Desmodium gangeticum			5.7
Desmodium rubrum			3.3
Osteomeles schwerinae			3
Sida cordifolia			3
Herb layer			
Heteropogon contortus	48.6		10.5
Boea hygrometrica	14.5		
Aleuritopteris pseudofarinosa	12.2		
Evolvulus alsinoides	10.8		
Boerhavia diffusa	10.3	3.7	
Selaginella moellendorffii	3	3.4	
Paspalum scrobiculatum		46.5	
Gymnopteris vestita		12.4	
Cyanotis arachnoidea		11.3	
Hedyotis auricularia		8.5	
Isoglossa collina		3.2	

(continued)

Table 6.5 (continued)

Plant community	L-V-H	L-V-P	H-P-P
Perotis hordeiformis			58.1
Fimbristylis ovata			9.4
Tephrosia purpurea			8.1
Euphorbia hirta			6.7
Tridax procumbens			3.8

Patial data from SWFC-KZI-YJCG (2009)

For woody species, IV = (relative density + relative basal area)/2; for herbaceous species, IV = (relative density + relative coverage)/2

L-V-H Lannea-Vitex-Heteropogon community, *L-V-P Lannea-Vitex-Paspalum* community, *H-P-P Haldina-Polyalthia-Perotis* community

tha, after being cultivated, spread naturally on the hotter and drier rock banks in the hot-dry valleys (Zhu 1990). The thorny succulent shrubland has the characteristics of similar, dry tropical semi-savanna-like shrublands of India and Africa.

Representative communities of the thorny succulent shrubland are summarized in Table 6.6. The floristic composition of the thorny succulent shrubland communities is exemplified by Yuanjiang (Table 6.7).

The semi-savannas in the river valleys result from the hot and dry climate and from human influence. Within the hot-dry valley areas the ecological environment is particularly fragile. The increasing human population and people's search for a subsistence income have led to deforestation and habitat degradation (e.g. conversion to pasture). In hot-dry environments where vegetation is sparse, soils are likely to be lost in the occasional surface runoff. If the vegetation is further disturbed by human activities, the semi-savannas may degrade to bare ground. Vegetation recovery would then be very unlikely. Better management and effective conservation are needed to ensure the future existence of the semi-savannas.

Fig. 6.9 *Euphorbia royleana* in the Yuanjiang valley (**a**), and *Opuntia monacantha* at Benzilan in the Jinshajiang valley (**b**), Yunnan

Table 6.6 Representative communities of the succulent thorny shrubland in the hot-dry valleys

Region	Plant community	Distribution
Yuanjiang	*Euphorbia royleana-Vitex negundo-Heteropogon contortus*	400–800 m in Pupiao
	Euphorbia royleana-Lannea coromandelica-Heteropogon contortus	450–500 m in Pupiao
	Euphorbia royleana-Sida rhombifolia-Heteropogon contortus	Below 1,000 m in Pupiao
	Osteomeles schwerinae-Euphorbia royleana-Bothriochloa pertusa	695 m in Shidichong
	Securinega virosa-Opuntia monacantha	1,420–1,500 in Nanjianheba
Jinshajiang	*Bauhinia esquirolii-Opuntia monacantha*	920–1,050 m in Longjiang
	Selaginella pulvinata-Opuntia monacantha	790–810 m in Qiaojiaba
	Euphorbia royleana-Jatropha curcas	Below 1,300 m in Daduhe
	Opuntia monacantha-Vitex negundo-Zizyphus jujuba	Below 400 in Chishuihe
Nujiang	*Euphorbia royleana-Opuntia monacantha*	950–990 m in Nujiangba

Data sources: SVEG (1980), Jin and Ou (2000), Chen et al. 2009, SWFC-KZI-YJCG (2009)

Table 6.7 Species (IV ≥ 3 %) composition of the thorny succulent shrubland in the hot-dry valley of Yuanjiang

Plant community	E-V-H	O-E-B	E-S-H
Site	Pupiao	Shidichong	Pupiao
Total plot area (m^2)	500	300	600
Coverage of tree layer (%)	5		
Coverage of shrub layer (%)	12	15	18
Coverage of herb layer (%)	35	30	40
Species	IV (%)	IV (%)	IV (%)
Tree layer			
Euphorbia royleana	26.6		
Albizia bracteata	14.8		
Lannea coromandelica	12		
Pistacia weinmanniifolia	11.7		
Haldina cordifolia	6.3		
Vitex canescens	4.2		
Polyalthia cerasoides	3.5		
Shrub layer			
Euphorbia royleana	15.2	21.2	31.2
Vitex vestita	8.1		
Pistacia weinmanniifolia	7.5		
Barleria cristata	5.6	4.1	4.5
Strophioblachia fimbricalyx	5.5		
Vitex negundo	4.8		

(continued)

Table 6.7 (continued)

Plant community	E-V-H	O-E-B	E-S-H
Selaginella pulvinata	4.3		
Tarenna depauperata	4.2		
Bauhinia brachycarpa	3.9		
Clausena emarginata	3.5		
Breynia fruticosa	3.4		
Opuntia monacantha	3		
Osteomeles schwerinae		24	6.4
Woodfordia fruticosa		5.3	
Terminthia paniculata		4.5	
Polyalthia cerasoides		3.9	
Sida rhombifolia			12.9
Bridelia tomentosa			3.1
Herb layer			
Heteropogon contortus	34.6	5	32.6
Phyllanthus ruinaria	18.7		
Fagopyrum caudatum	11		
Indigofera linnaei	10.4		
Commelina diffusa	9.5		
Paraboea rufesens	7		
Euphorbia hirta	6.8		
Petrocosmea duclouxii	6		
Elsholtzia stachyodes	5.3		
Cymbopogon distans	4.2		
Erianthus rufipilus	3.5		
Tinospora crispa	3.5		
Pellaea nitidula	3.2		
Bothriochloa pertusa		49.5	13.4
Pararuellia delavayana		13	
Phyllanthus virgatus		8.2	5.7
Arthraxon prionodes		5.9	
Scutellaria pingbienensis		5.2	
Alysicarpus monilifer		4.7	
Desmodium triflorum		3.8	
Tupistra grandistigma		3	
Tephrosia purpurea			15.8
Euphorbia hirta			3.9

For woody species, IV = (relative density + relative basal area)/2; for herbaceous species, IV = (relative density + relative coverage)/2

E-V-H Euphorbia-Vitex-Heteropogon, O-E-B Osteomeles-Euphorbia-Bothriochloa, E-S-H Euphorbia-Sida-Heteropogon

References

Chen WW, Wang J, Du F (2009) Characteristics of thorny succulent shrub communities in the Yuanjiang hot-dry valley. Green World Green Sci Technol 4:38–42

He Y-B, Lu P-Z, Zhu T (2000) Causes for the formation of dry-hot valleys in Hengduan Mountain-Yunnan Plateau. Resour Sci 22(5):69–72

Jin ZZ (1999) A phytosociological study on the semi-savanna in the dry-hot valleys of Yuanjiang River, Yunnan. Guihaia 19(4):289–302

Jin ZZ (2002) Floristic features of dry-hot and dry-warm valleys, Yunnan and Sichuan. Yunnan Science and Technology Press, Kunming

Jin ZZ, Ou XK (2000) Yuanjiang, Nujiang, Jinshajiang, Lancangjiang vegetation of dry-hot valleys. Yunnan University Press, Kunming

Li ZY, Taylor NP (2007) Cactaceae. In: Wu ZY, Raven PH, Hong DY (eds) Flora of China, vol 13. Science Press, Beijing, pp 209–211

Ma JS, Gilbert MG (2008) Euphorbia. In: Wu ZY, Raven PH, Hong DY (eds) Flora of China, vol 11. Science Press, Beijing, pp 288–313

Ma H-C, McConchie JA (2001) The dry-hot valleys and forestation in Southwest China. J For Res 12(1):35–39

SVEG (Sichuan Vegetation Editorial Group) (1980) Vegetation of Sichuan. Sichuan People's Press, Chengdu

SWFC-KZI-YJCG (Southwest China Forestry College-Kunming Zoological Institute of the Chinese Academy of Sciences-Yuanjiang County Government) (2009) Yunnan Yuanjiang nature reserve, integrated scientific report. Southwest China Forestry College Press, Kunming

Whittaker RH (1975) Community and ecosystem, 2nd edn. Macmillan, New York

Wu ZY, Zhu YC, Jiang HQ (eds) (1987) The vegetation of Yunnan. Science Press, Beijing

Xu H (1985) Diary on touring in Yunnan Province. Yunnan People's Press, Kunming, pp 214–230

Zhang YZ (1992) Publication series of the scientific survey on Qinghai-Xizang plateau: dry valley in the Hengduan mountainous area. Science Press, Beijing

Zhu H (1990) A study on the thorny succulent shrubs in dry-hot valley of Yuanijang County. Acta Bot Yunnanica 12(3):301–310

Chapter 7
Endemism and Tertiary Relict Forests

Abstract Among China's various regions, the SW harbors the greatest number of endemic plant species. Among them, some taxa from the Tertiary Period, such as the relict maidenhair tree (*Ginkgo biloba*), the dawn redwood (*Metasequoia glyptostroboides*), Sichuan Thuja (*Thuja sutchuenensis*) and Taiwania (*Taiwania cryptomerioides*) have survived as paleoendemics in glacial refuga including the Yangtze River valley and sites in northwestern Yunnan. In addition, over geological time, neoendemics have appeared among some groups of plants.

Within the warm, humid evergreen broad-leaved forest zone and the transitional zone between the evergreen and the deciduous broad-leaved forests, where there is also an admixture of species of gymnosperms, some extant Tertiary relict plant species thrive today, in particular in unstable habitats where natural disturbances occur frequently, such as steep or scree slopes, rocks crevices, cliffs, limestone areas with outcrops, or stream banks. Most of the Tertiary relict plants are light-demanding and appear to be stress-tolerant. In general, their regeneration depends on natural disturbances where there is little competition from other species.

7.1 Endemism

Endemism, that is, confinement to a particular geographic area (Hobohm 2014), is generally attributed to historical processes, such as the persistence of Pleistocene refugia or the occurrence of major geological events. These historical processes are in contrast to present-day ecological and environmental factors, which are normally considered to be the main determinants of patterns of overall species richness (Jetz et al. 2004; Orme et al. 2005). Refugia are geographical locations where natural conditions have remained relatively constant or stable during times of great environmental change, such as glacial advance and retreat. Refugia protect populations of geographically isolated species which may later re-colonize a region when the wider environment returns to levels within the species' tolerance levels. This concept is commonly referred to as the Refugia Theory (Leonard 2013). Locations where species persist during glaciations have generally been described as refugia. In addition, refugia in non-glaciated landscapes are associated with complex landscape topography, such as mountain ranges and deep valleys (Stebbins and Major

C.Q. Tang, *The Subtropical Vegetation of Southwestern China*,
Plant and Vegetation 11, DOI 10.1007/978-94-017-9741-2_7

1966; Taberlet and Cheddadi 2002; Médail and Diadema 2009). These areas have occurred in climatically stable habitats and have allowed plants to persist, differentiate and speciate (Tzedakis et al. 2002; Jetz et al. 2004; Thompson 2005; Médail and Diadema 2009). Quaternary refugia, often located in mountainous areas, probably played a dual role, preserving Tertiary lineages ('species museums') and promoting the evolution of new taxa ('species cradles') (Tzedakis et al. 2002).

Endemic species can be classified into two categories: paleoendemic and neoendemic. A paleoendemic species has typically been isolated for a very long time, and is small in population size, in distribution areas representing the remnants of originally larger ranges. For example, the Ice Ages led to the extinction of many species in North America and Europe, but in southwestern and south central China, a number of species could survive where glaciation was limited. *Davidia involucrata*, for example, is one of the paleoendemics; it grows today as a relict of the Tertiary period, mainly in southwestern and south-central China. A species that has never occurred outside its place of origin is called neoendemic. Various factors could cause populations of a given species to become isolated from one another. For example, as a result of plate tectonics, a land mass on which the species occurs may split apart, forming separate land masses, each with its own population of a given species. Over long periods of time, these populations might evolve differently, and eventually they may be sufficiently different from one another to be classified as separate species. These are known as neoendemic species. Examples in SW China are the liliaceous *Diuranthera inarticulata* (it occurs only in Guizhou, Sichuan and Yunnan), the composite *Nannoglottis yunnanensis* (only in western Sichuan and northwestern Yunnan), and the cucurbit *Belostemma yunnanense* (only in southeastern Yunnan).

Biologists distinguish between taxonomic and biogeographic relicts. Taxonomic relicts are a few or sole survivors from a once diverse taxonomic assemblage, whereas biogeographic relicts descend of once widespread taxa (or populations) that now have a narrow geographic distribution (Lomolino et al. 2006). Taxonomically, paleoendemics are mostly isolated (i.e. 'phylogenetically') owing to the extinction of their sister lineages and/or very long divergence (Stebbins and Major 1965). The two categories (biogeographic and taxonomic relicts) sometimes coincide, as for example in the case of "living fossils" (such as *Ginkgo biloba* in China) that closely resemble their ancient ancestors in overall phenotype (Futuyma 2005; Lomolino et al. 2006). In general, many paleoendemics show disjunct distributions (Kruckeberg 2002), because of their persistence in distinct isolated refugia during the late Cenozoic cooling or earlier geological events. For instance, *Liriodendron chinense* grows in southern China and *Liriodendron tulipifera* in eastern North America.

Endemism is an integral component of plant diversity in SW China. The high level of endemism suggests *in situ* long-term survival, differentiation and speciation. The region was never covered by extensive and unified ice-sheets, owing in part to its complex topography (Shi et al. 1986; Liu 1988). In Fig. 7.1, we can see scattered traces of paleo-glaciers in the mountainous areas. The mountains of the region have had a relatively stable long-term environment, now referred to as Pleistocene glacial refugia, harboring a number of Tertiary relict plants (Ying et al. 1993; Axelrod et al. 1998; López-Pujol et al. 2011a; Qiu et al. 2011). The collision

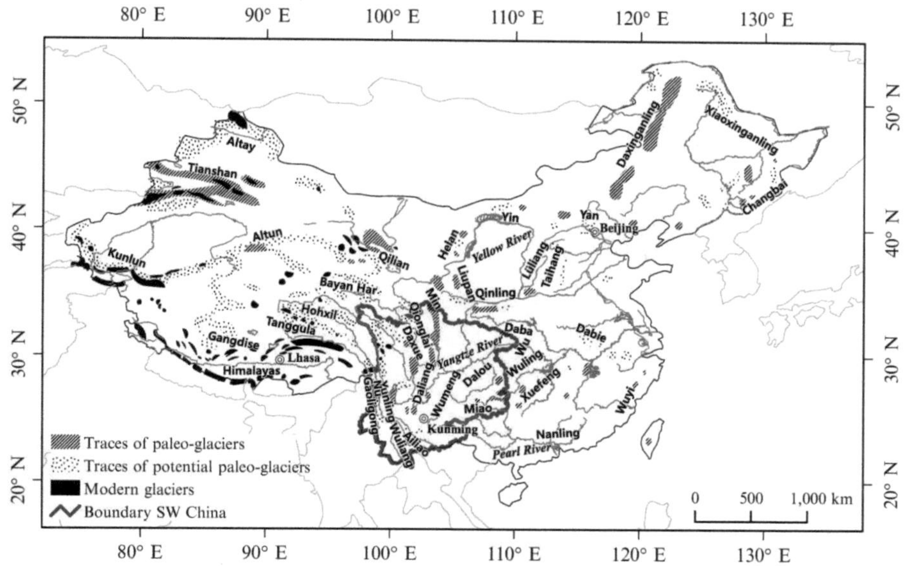

Fig. 7.1 Traces of paleo-glaciers (Data source: CPGC (1980))

of the Indian subcontinent with the Asian continent commencing about 50 Ma created highly dissected and elevated landscapes in SW China, which made the region not only an important center of survival (i.e. a 'living museum' or 'refugium'), but also an important center of speciation and evolution (i.e. a 'cradle') for vascular plants (Qian et al. 2006; López-Pujol et al. 2011b).

In SW China, the extensive and complex mountain ranges and deep valleys certainly yield distinctive species richness and a wide range of taxa, vegetation types, and habitats. Northwestern Yunnan and the Yangtze River valley include two of the 25 "biodiversity hotspots" identified worldwide (Myers et al. 2000). Two plant diversity hot-spots with high levels of endemism are located in SW China (Ying 2001) (Fig. 7.2). The rich flora contains 4,573 endemic seed plant species, accounting for 25 % of the total native seed plant species (18,309) of the region, showing the highest endemism in northwestern Yunnan and western Sichuan (Editorial Committee of Flora Reipublicae Popularis Sinicae 1959–2004; Yang et al. 2013). The vascular species of genera endemic to China occurring in SW China are represented in Table 7.1. Nowhere is the endemic plant species richness more evident than in Yunnan, with 4,008 endemic higher plant species (Chen et al. 2013). Northwestern Yunnan has the highest species richness and endemism, followed by Mongna and Jinghong in southern Yunnan (Fig. 7.3a, b). These two areas have the highest number of narrow-ranging species (Fig. 7.3c).

High species diversity has evolved with the uplift of the Himalayas in northwestern Yunnan (the Hengduan Mountain Range – an extension of the eastern Himalayas), which evolved with the formation of the vast Himalayan ranges that began about 50 Ma, when the Indian and Eurasian plates were driven together. As the two plates are of similar density, composed mainly of geologically light continental materials

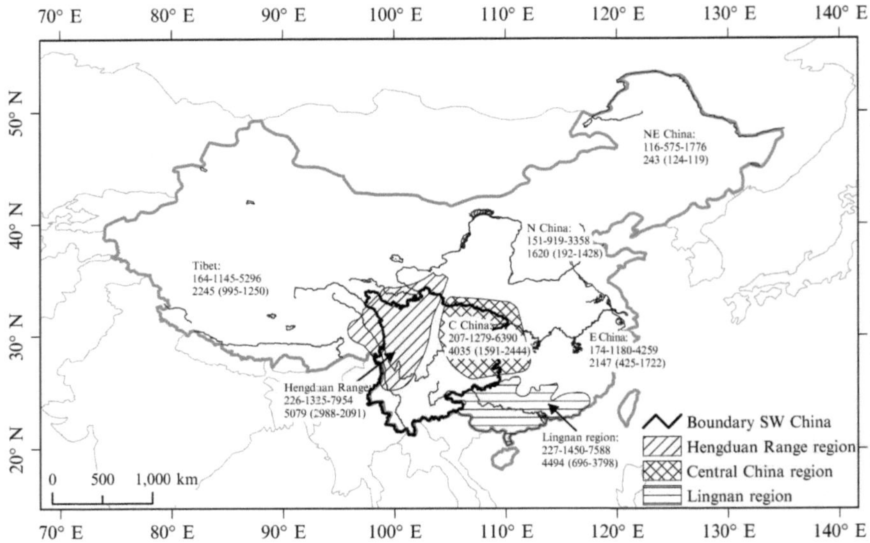

Fig. 7.2 Two plant diversity hot-spots with high levels of endemism in SW China. Key to numbers: *Upper line* = number of families–number of genera–number of species in the region; *Lower line* = total endemic species (number of the region's endemic species–number of the country's endemic species). *NE* Northeast, *N* North, *E* East, *C* Central (Revised from Ying (2001))

Table 7.1 Representative vascular species occurring in SW China, belonging to genera endemic to China

	Distribution		
Species	Yunnan	Sichuan	Guizhou
Pteridophytes			
Aleuritopteris grevilleoides	Yunnan	Sichuan	
Arachniodes tsiangiana			Guizhou
Cyrtogonellum fraxinellum	Yunnan	Sichuan	Guizhou
Cystoathyrium chinense		Sichuan	
Lastreopsis microlepioides	Yunnan		
Sorolepidium glaciale	Yunnan	Sichuan	
Gymnospermae			
Cathaya argyrophylla		Sichuan	Guizhou
Cunninghamia lanceolata	Yunnan	Sichuan	Guizhou
Ginkgo biloba			Guizhou
Metasequoia glyptostroboides		Sichuan	
Pseudolarix amabilis		Sichuan	
Taiwania cryptomerioides	Yunnan	Sichuan	Guizhou
Angiospermae			
Acanthochlamys bracteata		Sichuan	

(continued)

Table 7.1 (continued)

Species	Distribution		
	Yunnan	Sichuan	Guizhou
Allocheilos cortusiflorus	Yunnan		Guizhou
Ancylostemon convexus	Yunnan		
Anemarrhena asphodeloides		Sichuan	Guizhou
Anemoclema glaucifolium	Yunnan	Sichuan	
Anisachne gracilis	Yunnan		Guizhou
Arcuatopterus filipedicellus	Yunnan	Sichuan	
Arundinaria fargesii		Sichuan	
Asteropyrum peltatum	Yunnan	Sichuan	
Beesia calthifolia	Yunnan	Sichuan	Guizhou
Belostemma hirsutum	Yunnan	Sichuan	
Berneuxia thibetica	Yunnan	Sichuan	Guizhou
Bolbostemma paniculatum		Sichuan	
Bostrychanthera deflexa		Sichuan	Guizhou
Bretschneidera sinensis	Yunnan	Sichuan	Guizhou
Briggsiopsis delavayi	Yunnan	Sichuan	
Bulleyia yunnanensis	Yunnan		
Calcareoboea coccinea	Yunnan		
Camptotheca acuminata	Yunnan	Sichuan	Guizhou
Cardioteucris cordifolia	Yunnan	Sichuan	
Centantheropsis rigida			Guizhou
Chamaesium paradoxum	Yunnan	Sichuan	
Changnienia amoena		Sichuan	
Chimonanthus praecox	Yunnan	Sichuan	Guizhou
Chingiacanthus patulus	Yunnan		
Chuanminshen violaceum		Sichuan	
Clematoclethra scandens	Yunnan	Sichuan	Guizhou
Craigia yunnanensis	Yunnan		Guizhou
Cyclocarya paliurus	Yunnan	Sichuan	Guizhou
Cyclorhiza waltonii	Yunnan	Sichuan	
Cyphotheca montana	Yunnan		
Davidia involucrata	Yunnan	Sichuan	Guizhou
Deinocheilos sichuanense		Sichuan	
Dicercoclados triplinervis			Guizhou
Dickinsia hydrocotyloides	Yunnan	Sichuan	
Dipelta yunnanensis	Yunnan	Sichuan	Guizhou
Diplazoptilon picridifolium	Yunnan		
Dipoma iberideum	Yunnan	Sichuan	
Dipteronia sinensis	Yunnan	Sichuan	Guizhou
Diptopanax stachyanthus	Yunnan		
Diuranthera major	Yunnan	Sichuan	Guizhou
Dolichopetalum kwangsiense	Yunnan		Guizhou

(continued)

Table 7.1 (continued)

Species	Distribution		
	Yunnan	Sichuan	Guizhou
Dysosma pleiantha		Sichuan	
Emmenopterys henryi	Yunnan	Sichuan	Guizhou
Eomecon chionantha	Yunnan	Sichuan	Guizhou
Eucommia ulmoides	Yunnan	Sichuan	Guizhou
Eurycorymbus cavaleriei	Yunnan	Sichuan	Guizhou
Ferrocalamus strictus	Yunnan		
Fordiophyton faberi	Yunnan	Sichuan	Guizhou
Formania mekongensis	Yunnan	Sichuan	
Hanceola sinensis	Yunnan	Sichuan	Guizhou
Handeliodendron bodinieri			Guizhou
Haplanthoides yunnanensis	Yunnan		
Harrysmithia heterophylla		Sichuan	
Hemilophia pulchella	Yunnan		
Heterolamium debile	Yunnan	Sichuan	
Holocheila longipedunculata	Yunnan		
Homocodon pedicellatus		Sichuan	
Ischnogyne mandarinorum		Sichuan	Guizhou
Isometrum farreri		Sichuan	
Kingdonia uniflora	Yunnan	Sichuan	
Kinostemon ornatum	Yunnan	Sichuan	Guizhou
Latouchea fokienensis	Yunnan	Sichuan	Guizhou
Loxocalyx urticifolius	Yunnan	Sichuan	Guizhou
Malania oleifera	Yunnan		
Manglietiastrum sinicum	Yunnan		
Melanosciadium pimpinelloideum		Sichuan	Guizhou
Melliodendron xylocarpum	Yunnan	Sichuan	Guizhou
Metanemone ranunculoides	Yunnan		
Mimulicalyx paludigenus	Yunnan	Sichuan	
Musella lasiocarpa	Yunnan		Guizhou
Nannoglottis carpesioides	Yunnan		
Neomartinella violifolia	Yunnan	Sichuan	Guizhou
Notopterygium forrestii	Yunnan	Sichuan	
Notoseris porphyrolepis		Sichuan	Guizhou
Nouelia insignis	Yunnan	Sichuan	
Ostryopsis davidiana		Sichuan	
Parakmeria omeiensis		Sichuan	
Parepigyrum funingense	Yunnan		
Petrocodon dealbatus			Guizhou
Platycraspedum tibeticum		Sichuan	
Poliothyrsis sinensis	Yunnan	Sichuan	Guizhou
Pomatosace filicula	Yunnan	Sichuan	

(continued)

Table 7.1 (continued)

Species	Distribution		
	Yunnan	Sichuan	Guizhou
Przewalskia tangutica		Sichuan	
Psammosilene tunicoides	Yunnan	Sichuan	Guizhou
Psilopeganum sinense		Sichuan	Guizhou
Pteroceltis tatarinowii		Sichuan	Guizhou
Pteroxygonum giraldii		Sichuan	
Pterygiella nigrescens	Yunnan		
Pyrgophyllum yunnanense	Yunnan	Sichuan	
Rhabdothamnopsis sinensis	Yunnan	Sichuan	Guizhou
Rostrinucula dependens	Yunnan	Sichuan	Guizhou
Salweenia wardii	Yunnan	Sichuan	
Saruma henryi		Sichuan	Guizhou
Schnabelia oligophylla		Sichuan	Guizhou
Scrofella chinensis		Sichuan	
Semiliquidambar cathayensis			Guizhou
Sheareria nana	Yunnan	Sichuan	Guizhou
Sinacalia tangutica		Sichuan	
Sinodielsia yunnanensis	Yunnan	Sichuan	
Sinofranchetia chinensis	Yunnan	Sichuan	
Sinojackia xylocarpua		Sichuan	
Sinoliprichtia alpina	Yunnan	Sichuan	
Sinosassafras flavinervium	Yunnan		
Sinowilsonia henryi		Sichuan	
Smithorchis calceoliformis	Yunnan		
Spenceria ramalana	Yunnan	Sichuan	
Styrophyton caudatum	Yunnan		
Syncalathium disciforme		Sichuan	
Tangtsinia nanchuanica		Sichuan	Guizhou
Tapiscia sinense	Yunnan	Sichuan	Guizhou
Tengia scopulorum	Yunnan		Guizhou
Tetradoxa omeiensis		Sichuan	
Tetrapanax papyrifer	Yunnan	Sichuan	Guizhou
Thamnocharis esquirolii			Guizhou
Thyrocarpus sampsonii	Yunnan	Sichuan	Guizhou
Trailliaedoxa gracilis	Yunnan	Sichuan	
Tremacron forrestii	Yunnan	Sichuan	
Tsaiorchis neottianthoides	Yunnan		
Tsoongiodendron odorum	Yunnan		Guizhou
Urophysa henryi		Sichuan	Guizhou
Whytockia chiritiflora	Yunnan		
Xanthopappus subacaulis	Yunnan	Sichuan	
Yinshania acutangula		Sichuan	
Yinshania henryi	Yunnan	Sichuan	Guizhou

Data sources: Ying and Zhang (1984), Wang (1992), Chen (1993), eFloras (2014)

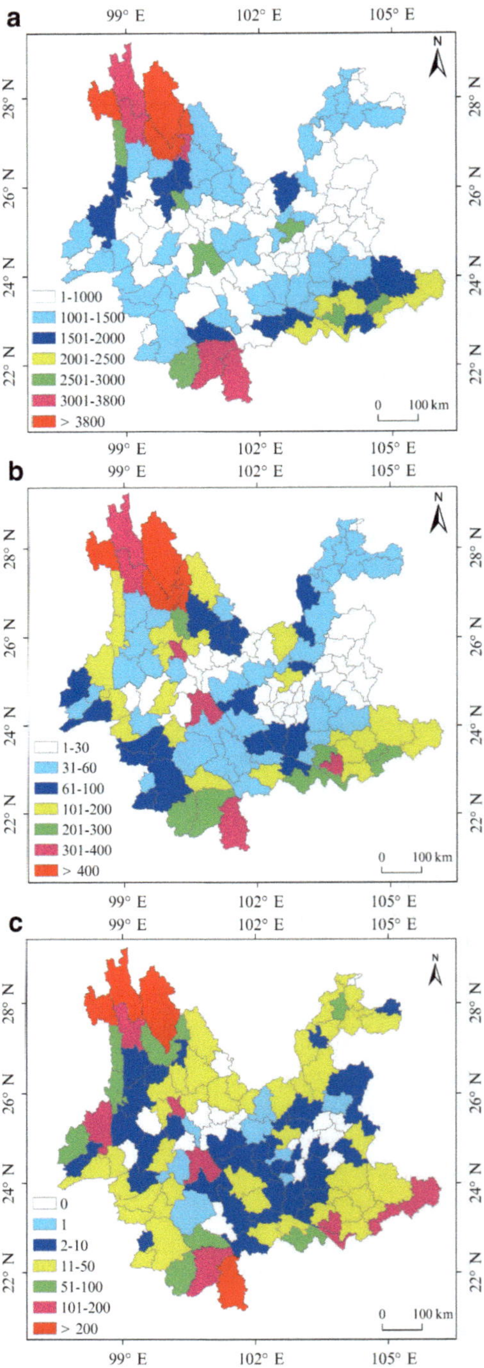

Fig. 7.3 (**a**) Distributional pattern of species diversity for higher plants at the county level (in numbers of species per county) in Yunnan. (**b**) Distributional pattern of endemic species at the county level (in numbers of endemic species per county) in Yunnan. (**c**) Distributional pattern of narrow-ranging species at the county level (in numbers per county) in Yunnan. Narrow-ranging species here are defined as species that occur in one or a few small populations confined to a small area or a few localities (Data source for (a–c): Chen et al. (2013))

(ancient granites, other sediments and metamorphic rock) over-lying heavy oceanic basalts, each resisted being forced down beneath the other (subduction). Instead, they stayed almost level with each other (the Eurasian Plate just over-riding the Indian Plate), thrusting upwards to relieve the pressure, buckling and crumpling in a series of mountain-building events (RBGE 2014).

In general, northwestern and southern Yunnan, with western Sichuan, harbor more endemic species than the areas of eastern China (Yang et al. 2013; Chen et al. 2013). Peculiar to SW China is the paleoendemism of various Tertiary relict plant species surviving in small population sizes in isolated areas, such as *Ginkgo biloba*, *Metasequoia glyptostroboides*, *Cathaya argyrophylla*, and *Thuja sutchuenensis* (in the Yangtze River valley), *Davidia involucrata* and *Liriodendron chinense* (in the Sichuan Basin, Yunnan and Guizhou), *Taiwania cryptomerioides* in the Gaoligong Mountains of northwestern Yunnan, and some primitive evergreen broad-leaved species (e.g. *Pachylarnax sinica* of the Magnoliaceae) in extreme southeastern Yunnan.

The distribution and ecology of the Tertiary relict plants are key to understanding the plant geography and vegetation patterns of SW China. The following sections will be focused on the population persistence of important Tertiary relict plants in the region, except for *Cathaya argyrophylla* and *Taiwania cryptomerioides* (Fig. 7.4a–d) on which our field investigation is not yet completed.

7.2 Tertiary Relict Forests

7.2.1 The Ginkgo biloba *Forest*

7.2.1.1 *Ginkgo biloba*

Ginkgo biloba L. is also known as maidenhair tree, yin xing (silver apricot), baiguo (white nuts) or gongsunshu (grandfather/grandson tree) in Chinese, ichou (duck foot) in Japanese. The name *Ginkgo* derived from “Ginkyo”, a Japanese pronunciation of the Chinese characters “Yinxing” (Hori and Hori 1997). It originates from 1712 when the German botanist Engelbert Kaempfer proposed it. Linnaeus adopted this designation, adding the species name “*biloba*” to denote the bisection of the wedge-shaped lamina of the leaf into two divergent segments (Seward 1911).

In 1691 Kaempfer had discovered the *Ginkgo* in Japan. The species had survived in China where, in historic time, they were mainly found at temples in the mountains and in palace and temple gardens. Buddhist monks cultivated the tree beginning about 1100 AD for its many good qualities. From there it spread (by seed) to Japan (around 1192 AD with some relation to Buddhism) and Korea. *Ginkgo* seeds were brought to Europe from Japan by Kaempfer in the early 1700s and to America later that century (1784 to Philadelphia, USA). The first and oldest *Gingko* tree in the western hemisphere is the one that still grows in the Old Botanical Garden of Utrecht University, the Netherlands. It most likely was planted about 1737. Most of the earlier trees raised in Europe, also that in Utrecht, were males. The first recorded

Fig. 7.4 (**a**) Seed corns of *Cathaya argyrophylla* (Photograph courtesy of Yongchuan Yang). (**b**) A *C. argyrophylla* forest at 1,906 m asl in Jinfoshan, Nanchuan, Chongqing. (**c**) *Taiwania cryptomerioides* forest at 2,300 m asl in Bingzhongluo, the Gaoligong Mountains, northwestern Yunnan. (**d**) A tree trunk of *T. cryptomerioides*

female tree was found near Geneva in 1814, of which scions were grafted onto male trees, e.g. on the tree at Utrecht in circa 1838, and on the tree of 1778 in the Botanic Garden of Montpellier, France, around 1832 (Pulle 1949; Kwant 2014).

Ginkgo biloba has been cultivated for its ornamental value, also for its edible nuts, and for its great commercial value as an alternative medicine. The trees resist dust, disease, and insects, making them a favorite in urban landscaping. *Ginkgo* has been cultivated for about a 1,000 years in China, Japan, and Korea. It reflects the culture of East Asia, is truly as Crane (2013) described: "*Ginkgo biloba* – a symbol of the East." It has also been commonly cultivated in North America for over 200 years and in Europe for close to 300 years, but during that time, it has never become significantly naturalized (Whetstone 2006).

Fig. 7.5 *Ginkgo biloba* forest reaching 35 m in height in the canopy layer, in an unstable limestone habitat in Yangjiagou, Nanchuan County (**a**). *G. biloba* forest as a fengshui forest reaching 38 m in height in the canopy layer in Dazhuyuan, Wuchuan County (**b**)

Ginkgo biloba has a conifer-like architecture: a straight trunk with branches diverging at nearly right angles (Fig. 7.5a, b). The branches are constructed of short and long shoots growing at right angles (Fig. 7.6a). In short shoots, internodes are

Fig. 7.6 (**a**) A *Ginkgo* branch, with long and short shoots. (**b**) A cluster of leaves on a short shoot of *Ginkgo*. (**c**) *Ginkgo* foliage in November

very short, in contrast with long-internodes in long shoots. Long shoots are produced by the terminal meristem of existing long shoots but can also form laterally through the conversion of short shoots. A short shoot may become a long shoot and the tip of a long shoot may change into a short shoot; so, older trees may have a more irregular form. On the ends of short, lateral shoots they grow very slowly, and produce a long shoot with scattered leaves after a number of years. The leaves grow alternate on the long shoots. Long shoots are associated with branch extension, whereas reproductive organs – male and female strobili on different plants – are borne on short shoots. The leaves of short shoots are larger and undivided or slightly bilobed, and more abundant and emerge in early spring from overwintering buds. By contrast, long-shoot leaves are smaller and often deeply grooved in the middle, producing two (sometimes more than two) distinct lobes (Fig. 7.6a); they are not preformed, but instead differentiate as the shoot elongates. Most seedling leaves also have deeply incised blades. *Ginkgo biloba* is deciduous, and has uniquely fan-shaped, leathery and smooth leaves. The leaves have a venation pattern that is open dichotomous; the veins fork in pairs from the base of the leaf, and are not cross-connected (Chamberlain 1966; Critchfield 1970; Leigh et al. 2011; 101GB 2014). They are bright green during the summer, turning bright yellow before dropping in the fall (Fig. 7.6b, c). Leaves generally have a quick fall time (within 7 days). *Ginkgo biloba* is a gymnosperm. It is dioecious. Fruits have a very disagreeable smell, due to the butyric acid contained within the fruit flesh (Fig 7.7a, b). The seeds are tasty and are commonly eaten.

Fig. 7.7 (**a**) *Ginkgo* fruits (Photograph © Ginkgo biloba on Wikimedia Commons). (**b**) Withered fruits; the inside seed germinated after falling on the ground

Today, *Ginkgo biloba* is the sole living representative of the Ginkgoales, one of the most ancient relict species on earth. The oldest probable ginkgophyte is from the Permian, dating back some 270 Ma. The genus *Ginkgo* was widely distributed in the Northern Hemisphere in the Middle Jurassic (c.170 Ma) (Tralau 1968). *Ginkgo biloba* survived to the present in China, and into the Pliocene and Pleistocene epochs elsewhere in East Asia, as attested by fossils found in southwestern Japan (Uemura 1997). Its evolutionary history has recently been the object of intensive study (Zhou and Zheng 2003).

A possible natural or semi-natural population of *Ginkgo* was reported in small areas of Tianmushan Reserve, Zhejiang Province in eastern China (Del Tredici 1992; Ling and Zhang 2004). However, research indicates high genetic uniformity among *Ginkgo* trees from these areas, arguing against a natural origin of these populations and suggesting that these *Ginkgo* trees may have been planted and preserved by Chinese monks over a period of about 1,000 years (Shen et al. 2005). The study of Shen et al. (2005) demonstrates a greater genetic diversity in SW China populations, supporting the probability of glacial refugia in mountains surrounding the eastern Tibetan Plateau. Until 2012, there was no compelling evidence that *Ginkgo* existed in the wild. Tang et al. (2012) provided that evidence, arguing that the current species composition of *Ginkgo* forests in the Dalou Mountains, located between Guizhou Province and Chongqing Municipality, agrees closely with floristic assemblages from *Ginkgo biloba*-bearing fossil records. Current populations are found in habitats similar to those of fossil *Ginkgo*, which, as today, favored rock crevices. Estimated ages for many of the trees show that *Ginkgo* was present in this area prior to human settlement, and indigenous peoples of this area are unlikely to have planted *Ginkgo* because of traditional beliefs. Our results as presented match well with existing genetic studies which have shown that the Dalou Mountains are glacial refugia for *Ginkgo biloba*. The results from Tang et al. (2012) will be summarized and further detail regarding the forests' structure and floristic composition will be provided below.

7.2.1.2 Forest Community Structure and Floristic Composition

In the Dalou Mountains, intensive agricultural development began in the 1950s, resulting in widespread fragmentation of forests and leaving only patches smaller than 1 ha in the vicinity of farmhouses in valley bottoms and close to creeks (EBANC 1991; Fu and Jin 1992; EBAFC 1994; EBAWGMAC 2001). *Ginkgo biloba* is mainly distributed at 840–1,200 m asl, within the subtropical moist evergreen broad-leaved forest zone. The habitats of *G. biloba* in this area are all characterized by limestone outcrops (Fig. 7.8a).

Figure 7.8b, c shows three *Ginkgo* forest community types in the Dalou Mountains: (1) *Ginkgo* forest singly dominated by *G. biloba*, (2) *Ginkgo-Cupressus* forest co-dominated by *G. biloba* and *Cupressus funebris*, (3) *Ginkgo-Liquidambar-Cyclobalanopsis-Cunninghamia-Taxus* forest co-dominated by

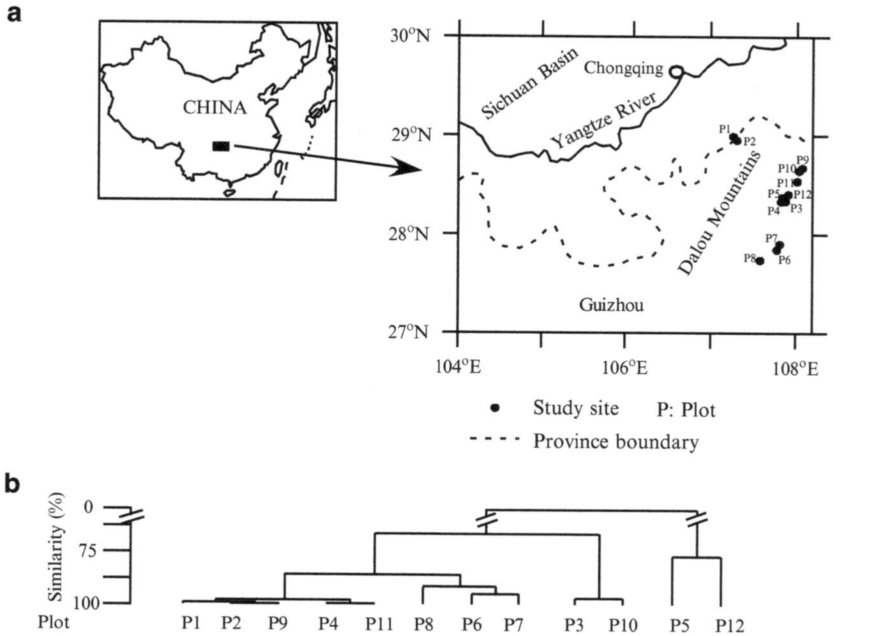

c

Habitat	Limestone outcrops	Limestone outcrops	Limestone outcrops
Coverage of outcrops	75-90%	60-75%	45-60%
Plot area (m^2)	10000	3000	11200
Dominant tree species	*Ginkgo biloba*	*Ginkgo biloba* *Cupressus funebris*	*Ginkgo biloba* *Liquidambar formosan a* *Cyclobalanopsis glauca* *Cunninghamia lanceolata* *Taxus wallichiana* var.*chinensis*
No. of established saplings of *Ginkgo*	53	29	20

Fig. 7.8 (**a**) Study sites of *Ginkgo* communities. (**b**) Floristic similarity dendrogram using Euclidean and group average clustering for the sampling plots of *Ginkgo biloba*. (**c**) Characteristics of *Ginkgo* communities (Data source: Tang et al. (2012))

G. biloba, *Liquidambar formosana*, *Cyclobalanopsis glauca*, *Cunninghamia lanceolata* and *Taxus wallichiana* var. *chinensis*. As the coverage of outcrops increases in these forests, the number of established saplings of *Ginkgo* increases (Fig. 7.8c). The canopy layer reaches a height of 30–38 m in each of the three communities. The floristic composition of each forest type is detailed in Table 7.2. Tang et al. (2012) reported that among the total of 174 woody species, 45 out of 114 genera in the *Ginkgo* forests of the Dalou Mountains are also found in the *G. biloba*-bearing

Table 7.2 Species composition (only species which are taller than 1.3 m in height) for each *Ginkgo* forest

Plant community	*Ginkgo*	*Ginkgo-Cupressus*	*Liquidambar-Ginkgo-Cyclobalananopsis-Cunninghamia-Cupressus*
Species	RBA (%)	RBA (%)	RBA (%)
Ginkgo biloba	**82.5**	**68**	**12.8**
Aphananthe aspera	4.7		4.5
Liquidambar formosana	3.5		**13.1**
Juglans regia	1.6		
Lindera megaphylla	1.3	0.2	1.1
Osmanthus fragrans	0.6		
Diospyros cathayensis	0.5		
Juglans cathayensis	0.5	0.2	
Ilex micrococca	0.4		1.5
Choerospondias axillaris var. *pubinervis*	0.3		3.6
Cornus controversa	0.3		1.2
Machilus nanmu	0.3		0.5
Taxus wallichiana var. *chinensis*	0.3	3.7	3.9
Cunninghamia lanceolata	0.2		**7.8**
Lithocarpus litseifolius	0.2		
Osmanthus yunnanensis	0.2	0.09	
Toxicodendron succedaneum	0.2		
Castanea mollissima	0.1		
Celtis sinensis	0.1		
Eleutherococcus trifoliatus	0.1		
Kalopanax septemlobus	0.1		
Koelreuteria paniculata	0.1		
Platycarya strobilacea	0.1		
Tapiscia sinensis	0.1		
Tilia tuan	0.1		
Toxicodendron vernicifluum	0.1		
Vernicia fordii	0.1		
Myrica rubra	0.09		
Padus wilsonii	0.09		1.2
Eucommia ulmoides	0.08		
Toona sinensis	0.08		
Cupressus funebris	0.07	19.9	**7.4**
Picrasma quassioides	0.07		
Rhus chinensis	0.07		
Sloanea sinensis	0.07		3.1
Castanopsis tibetana	0.06		4.5

(continued)

Table 7.2 (continued)

Plant community	*Ginkgo*	*Ginkgo-Cupressus*	*Liquidambar-Ginkgo-Cyclobalananopsis-Cunninghamia-Cupressus*
Species	RBA (%)	RBA (%)	RBA (%)
Corylus chinensis	0.06		1.5
Phoebe zhennan	0.06		
Lindera communis	0.05		
Hovenia acerba		6.9	
Laurocerasus phaeosticta		0.3	
Ligustrum lucidum		0.2	
Celtis biondii		0.1	3.5
Firmiana simplex		0.1	
Phoebe neurantha		0.07	
Cinnamomum foveolatum		0.05	
Cyclobalanopsis glauca			**9.9**
Emmenopterys henryi			3.9
Castanopsis carlesii var. *spinulosa*			2.1
Acer laevigatum			1.2
Michelia maudiae			1.1
Turpinia pomifera			1
Camellia pitardii			0.9
Acer maximowiczii			0.8
Diospyros kaki			0.8
Pterocarya stenoptera			0.6
Viburnum brachybotryum			0.5

The dominants are indicated by boldface

fossil records of southwestern Japan dating from the Pliocene to the Pleistocene. The point is pertinent because before the development of the East China Sea in the middle Pliocene, southwestern Japan was connected to mainland China (Kitamura and Kimoto 2004). Many Arcto-Tertiary genera (e.g. *Metasequoia, Glyptostrobus, Pseudolarix, Keteleeria, Taiwania, Cunninghamia, Palaeotsuga, Sequoia, Nyssa* and *Carya*) became extinct in Japan during the Pliocene and the Pleistocene (Miki 1938, 1950, 1956; Tanai 1967; Tanai and Huzioka 1967; Momohara 1992). In China most of those species are now confined chiefly to the Yangtze River valley (Chaney 1948; Wang 1961; Wu 1980). These facts would justify a comparison between fossil assemblages in Japan and current *Ginkgo* forests in the Dalou Mountains, which form part of the Yangtze River valley. Among 31 genera of canopy tree species, 25 (e.g. *Ginkgo, Liquidambar, Cunninghamia, Taxus, Lindera, Cyclobalanopsis, Cornus*) are present in *G. biloba*-bearing fossil sites in southwestern Japan (Table 7.3). Fossil records of *Liquidambar*, *Juglans*, *Prunus*, *Corylus*, *Ilex*, *Acer*,

Table 7.3 Canopy tree species of the *Ginkgo biloba* forests in the Dalou Mountains, China

	Canopy tree species
*	*Ginkgo biloba*
*	*Liquidambar formosana*
*	*Cunninghamia lanceolata*
*	*Taxus wallichiana* var. *chinensis*
*	*Lindera megaphylla*
*	*Cyclobalanopsis glauca*
*	*Cornus controversa*
*	*Choerospondias axillaris* var. *pubinervis*
*	*Juglans cathayensis*
*	*Celtis biondii*
*	*Quercus aliena*
*	*Cinnamomum wilsonii*
*	*Machilus nanmu*
*	*Sapium sebiferum*
*	*Acer laevigatum*
*	*Prunus dielsiana*
*	*Tilia tuan*
*	*Ilex micrococca*
*	*Diospyros cathayensis*
*	*Corylus chinensis*
*	*Osmanthus yunnanensis*
*	*Michelia martini*
*	*Salix hypoleuca*
*	*Mallotus apelta*
*	*Ficus henryi*
	Cupressus funebris
	Castanopsis tibetana
	Hovenia acerba
	Aphananthe aspera
	Tapiscia sinensis
	Emmenopterys henryi

Data source: Tang et al. (2012)
An asterisk indicates a species representing a genus that is found in *G. biloba*-bearing fossil sites in southwestern Japan from the Pliocene to the Pleistocene

Quercus, *Tilia*, *Salix*, *Diospyros* and *Celtis* have also been reported co-occurring with *Ginkgo* in other parts of the Northern Hemisphere, as in some Pliocene localities in Germany and Romania (Mädler 1939; Givulescu and Ghiurca 1969). Other remaining genera of canopy trees are also elements of the Tertiary flora, including *Cupressus, Castanopsis, Hovenia, Aphananthe*, *Tapiscia* and *Emmenopterys*. Given that this region has never been covered by extensive and unified ice-sheets (Shi et al.

1986; Liu 1988), the exceptionally high similarity of the present floristic composition of canopy trees in the *Ginkgo* forests of the Dalou Mountains to the *Ginkgo* fossil assemblages provides site evidence that the forests studied are remnants of natural *Ginkgo* forests.

7.2.1.3 Population Structure and Ecological Characteristics

Among the investigated *G. biloba* individuals taller than 1.3 m in height, those younger than 150 years are numerous (Fig. 7.9a, b). Thirty-three *Ginkgo* trees of 160–369 cm DBH are found, corresponding to projected ages of 366–878 years. The population structure of *G. biloba* indicates good regeneration. *G. biloba* trees have survived by developing extensive root systems in thin soils on the outcrops of limestone. They can grow sprouts, suckers and air roots after being injured (Fig. 7.10a, b). In the sites we studied, numerous seedlings and 102 established saplings (1–8 m high) grew in rock crevices in sun-exposed sites. In contrast, seedlings and saplings were absent in the Tianmushan Reserve of eastern China (Del Tredici et al. 1992), which as noted earlier, has been plausibly suggested by some specialists to sustain a natural or semi-natural *Ginkgo* population. The natural habitats in our study sites in the Dalou Mountains are limestone outcrops near creeks and are subject to frequent natural disturbances (Fig. 7.11a–g). *G. biloba* trees also grow in exposed substrates in Dahongshan (Jiang et al. 1990) and Tianmushan (Del Tredici 1992) though their actual ages are unclear. In the fossil record, as well as in the Dalou region, *Ginkgo* has been found in unstable riparian habitats (Royer et al. 2003), confirming the evidence for survival of wild *G. biloba* forests in these mountains.

Tang et al. (2012) further pointed out that the various taxa present today in the *Ginkgo* forests, and also found in the *G. biloba*–bearing fossil sites, are widely distributed in the mixed mesophytic forests of East Asia, with some admixture of

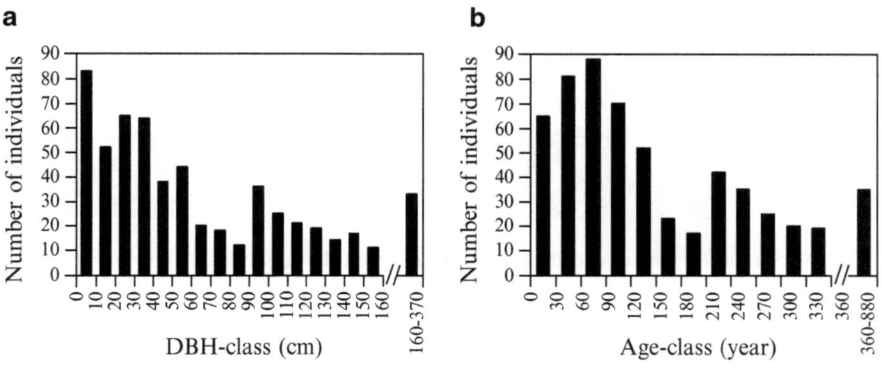

Fig. 7.9 Population structure of *Ginkgo*. The frequency distribution of *Ginkgo* in DBH-class (**a**) and age-class (**b**) shows a continuous pattern with high numbers of individuals in the small classes, indicating good regeneration (Data source: Tang et al. (2012))

Fig. 7.10 Sprouts (**a**), and so-called air roots (not scientifically confirmed) (**b**) of *Ginkgo*

coniferous species. A single exception is the *Ginkgo,* which has nearly disappeared. In other words, other genera found in the *Ginkgo*-bearing fossil sites have various species that apparently evolve and adapt to modern, diverse habitats. In contrast, *Ginkgo* keeps its primitive characteristics and only survives in habitats similar to those it occupied prehistorically.

Fig. 7.11 Part 1 (**a**, **b**) Seedlings of *Ginkgo biloba* thriving in sun-exposed limestone crevices in Dazhuyuan, Wuchuan County. (**c**) A sapling and an adult tree *G. biloba* on a limestone cliff in Xiangshuiyan, Fenggang County. (**d**) A sapling *G. biloba* growing in rock crevices in Yangjiagou, Nanchuan County

7.2.2 *The* Metasequoia glyptostroboides *Forest*

7.2.2.1 *Metasequoia glyptostroboides*

Metasequoia glyptostroboides Hu & Cheng (formerly Taxodiaceae, and now merged with Cupressaceae) known as dawn redwood (shuisha in Chinese, literally water fir), endemic to the Yangtze River valley, is a conifer that sheds its leaves (needles) every year, unlike its cousins *Sequoia sempervirens* (D. Don) Endl. (coast redwood) and *Sequoiadendron giganteum* (Lindl.) J. Buchholz (giant sequoia) of western North America. The tree reaches heights of 30–45 m. The crown has a narrow, conical shape. The species is monoecious. The leaves are opposite, 1–3 cm long, flat, generally

Fig. 7.11 (continued) Part 2 (**e**) Forest of *G. biloba* trees reaching 38 m height in the canopy layer in an unstable limestone habitat in Xiangshuiyan, Fenggang County (Photograph courtesy of Yongchuan Yang). (**f**) Tombs under a big *G. biloba* tree. (**g**) An 878 year-old tree of *G. biloba* with many sprouts as a village fengshui tree in Maopo, Nanchuan County. *Left*: plant ecologist Yongchuan Yang; *Right*: botanist Si-Rong Yi

appearing two-ranked in a flattened display (Fig. 7.12a, b). In spring the color of its foliage is bright green which turns to a beautiful foxy yellow and reddish brown in autumn. Its remarkable colors, and the fact that it is very easy to reproduce by cuttings, have made it a very popular ornamental tree.

Closely related species were widely distributed in the Northern Hemisphere throughout the Tertiary (Florin 1963; Yu 1995; Yang and Jin 2000). The earliest fossils date back to the late Cretaceous, while the latest, from Japan, date from the early Pleistocene (2 Ma) (Itihara et al. 1975). The genus *Metasequoia* Miki was

Fig. 7.12 (**a**) Leaves, a seed cone and male inflorescence of *Metasequoia glyptostroboides*. (**b**) Male inflorescence of *M. glyptostroboides*

described by Miki (1941) based on fossils collected from the Japanese Neogene prior to the discovery of the living plant. In 1941 Shigeru Miki (1903–1974), a Japanese paleobotanist, was reclassifying fossil remains that had been incorrectly classified as *Sequoia* and *Taxodium*. In light of its resemblance to the North

American coast redwood (*Sequoia sempervirens*), he established a new genus *Metasequoia*, derived from the Greek *meta*, meaning alike or akin, and *Sequoia*, the generic name of the coast redwood, which the tree resembles. A living specimen was found in China in 1944. As World War II was in progress, these were not studied further until 1946 (Hu 1946). It was recognized that the species belonged to the already described fossil genus *Metasequoia* and it was only finally recorded as a new, living species of *Metasequoia* in 1948 (Hu and Cheng 1948; Hu 1948). The tree was given the name *Metasequoia glyptostroboides* Hu & Cheng. The specific epithet, *glyptostroboides*, is a reference to the genus *Glyptostrobus*, the Chinese swamp cypress with which the tree was initially confused. The popular common name dawn redwood was a suggestion of the well-known paleobotanist Ralph W. Chaney. The use of "dawn" in the name was an attempt to emphasize the tree's early fossil record (HDOSU 2014). Until that time, the tree was only known as a fossil tree species, extinct for millions of years.

Much excellent research has been conducted from various perspectives since the living fossil was discovered in 1944, including its paleontology (e.g. Chaney 1948, 1951; Florin 1963; Momohara 1994, 2005; Yu 1995; Yang 1999; LePage et al. 2005), its contemporary natural distribution (Hu and Cheng 1948; Cheng and Chu 1949; Fu and Jin 1992; Wang et al. 2005; Leng et al. 2007), its molecular biology and genetics (Schlarbaum and Tsuchiyu 1984; Kuser et al. 1997; Chen et al. 2003; Li et al. 2005; Yang 2005), as well as SSRs having been used to compare the genetic diversity in the wild and the planted populations of *Metasequoia* in China (Li 2006), the finding of buried *Metasequoia* wood in Lichuan (Yang et al. 2004), and its ecological characteristics (Cheng and Chu 1949; Chu and Cooper 1950; Gressitt 1953; Williams 2005; Li and Ban 1989; You et al. 2008). Tang et al. (2011) have analyzed the population structure in the fragmented landscapes to determine whether recruitments have occurred in various habitats, and to evaluate how stand dynamics have changed in the past 60 years by comparison with the results of Chu and Cooper (1950).

7.2.2.2 Distribution and Habitats

The relict dawn redwood *Metasequoia glyptostroboides* currently survives only on wet lower slopes and in montane stream valleys in a border region shared by Chongqing municipality, Hubei and Hunan provinces in the Yangtze River valley. In recent decades the extensive removal of native vegetation in these areas has left most of the remaining vegetation fragmented in patches across the landscape. Complex mosaics of stands of varying age, composition and structure intermingle with cultivated fields. *M. glyptostroboides* is listed by the IUCN (the International Union for Conservation of Nature) as critically endangered (IUCN 2006).

The altitudinal distribution range of wild *M. glyptostroboides* is 900–1,500 m asl in Lichuan, Hubei Province, 1,200–1,508 m in Shizhu, Chongqing Municipality, and 800–850 m in Longshan, Hunnan Province (Fig. 7.13a). On a large geographic scale, this altitudinal range belongs both to the subtropical moist evergreen

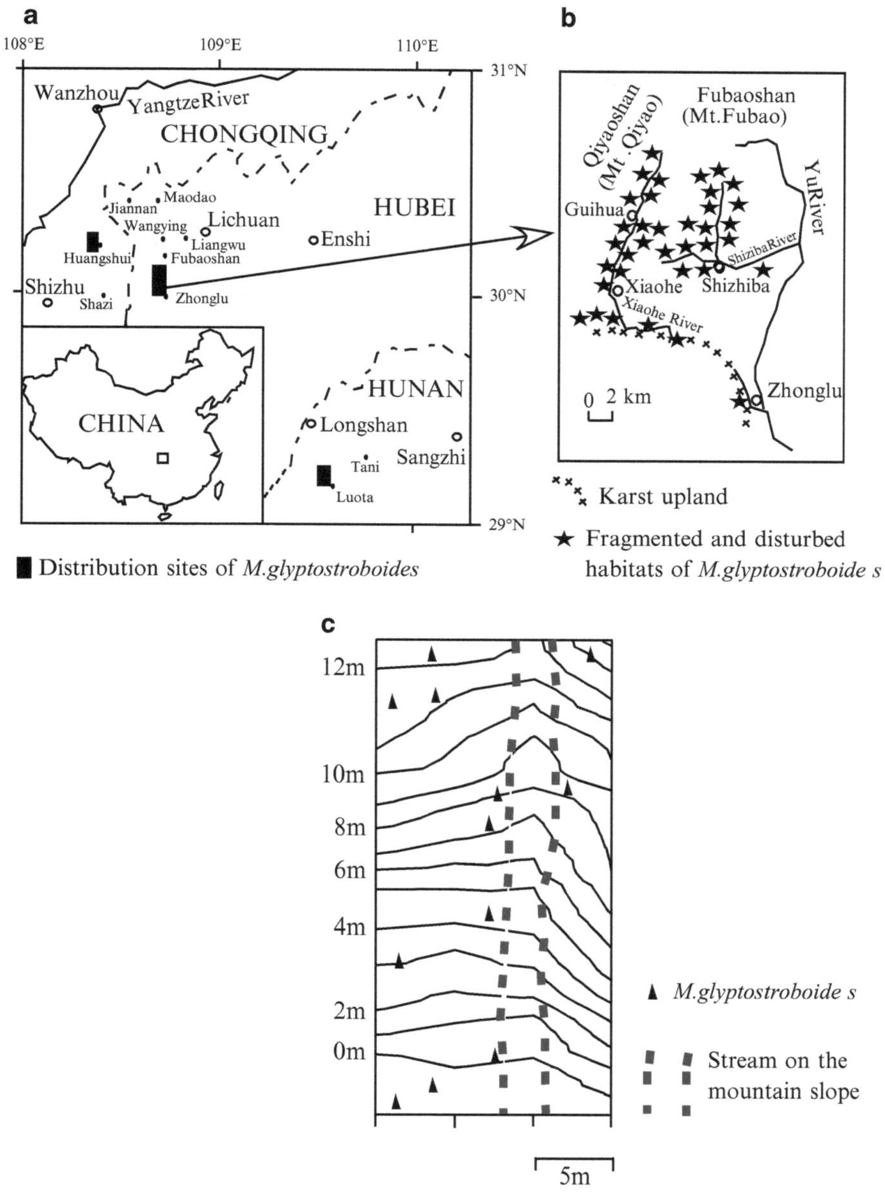

Fig. 7.13 (**a**) The distribution sites of *Metasequoia glyptostroboides* in the Huangshui area in Shizu, the Zhonglu area in Lichuan and the Luota area in Longshan, south-central China. (**b**) Distribution of *M. glyptostroboides* in the Zhonglu area, Lichuan. (**c**) A representative micro-topography (contours at 1 m intervals) of a mountain slope in Gangchangwan, Xiaohe, Zhonglu. It shows that *Metasequoia* trees are distributed along stream banks, since they favor a wet habitat (Data source: Tang et al. (2011))

broad-leaved forest zone and the evergreen, deciduous broad-leaved mixed forest zone, with a humid, warm climate. Lichuan, Hubei is the distribution center of the wild living tree, while only a few individuals were found in Shizhu and in Longshan. The three areas have similar histories of agricultural development and forest management. All the habitats in these areas have been disturbed, primarily by increasing numbers of village houses, rice paddies, cornfields, and the cultivation of *Coptis chinensis*, a cash medicinal herb that is commonly grown in the understory of natural or secondary forests. In Lichuan, about 90 % of the wild *M. glyptostroboides* are found in the Zhonglu area, c. 60 km southwest of the Yangtze River. Zhonglu is located at the confluence of the Xiaohe river valley (Shuisha valley) and Shiziba river valley (Fig. 7.13b). The valleys are bordered on the west by Chiyaoshan, a range of Permian limestone which reaches an altitude of 1,500 m, and the rivers' headwaters are largely in that range. On the east is a roughly parallel range of Jurassic sandstone, Fubaoshan, which attains an altitude of 1,400 m (Chu and Cooper 1950). Wild *M. glyptostroboides* are often found in fragmented forests by streams. The soil depth is from 35 to 80 cm in the fragmented and disturbed forest habitat. pH values range between 6.01 and 6.19 for the surface soil (0–15 cm). Organic matter in the surface soil ranges from 2.9 to 7.4 %. Outcrops are common and the forest floor is very moist. The micro-topography is typified by Gangchangwan, located in Xiaohe (Fig. 7.13c).

All the wild *M. glyptostroboides* trees for the three areas are known (Table 7.4). For Lichuan, Hubei: 5,338 plants excluding unhealthy ones, among them the first specimen recorded in Maodaoqi in 1948, of 248 cm DBH, excluded as unhealthy. 2,107 trees are distributed in fragmented and disturbed forest in

Table 7.4 Number of individuals of wild *Metasequoia glyptostroboides* in the three areas of south-central China

Location	Habitat type	No. of adult plants[a]	No. of saplings/ seedlings
Lichuan of Hubei	HA (Habitat A): fragmented and disturbed habitats nearby streams	2,107	0
	HB (Habitat B): habitats by rice paddies or roadsides or yards of village houses	3,231	0
Shizhu of Chongqing	HB (Habitat B): habitats by rice paddies or roadsides or yards of village houses	28	0
Longshan of Hunan	HB (Habitat B): habitats by rice paddies or roadsides or yards of village houses	5	0
	Total	5,371	0

Data source: Tang et al. (2011)

[a]The number of adult plants only contain healthy individuals of wild *M. glyptostroboides*

Fig. 7.14 (**a**) *Metasequoia glyptostroboides* by a stream in Zhonglu, Lichuan (HA). (**b**) *M. glyptostroboides* by a cornfield in Longshan, Hunan (HB). (**c**) *M. glyptostroboides* in a farmhouse yard (HB)

valley bottoms and by streams. This type of habitat we call habitat A (HA). 3,231 trees are scattered along rice paddies, roadsides, or yards of village houses. This type of habitat we call habitat B (HB). In Shizhu, Chongqing: 28 plants are found, scattered along rice paddies or roads or in yards of village houses (HB). In Longshan, Hunan: 5 individuals are found, near rice paddies or roads or yards of village houses (HB). No natural seedlings or saplings are found in any of the habitats. Figure 7.14a–c shows wild *M. glyptostroboides* trees in HA (habitat A) and HB (habitat B).

Tang et al. (2011) evaluated the degree of disturbance by scores assigned as follows to indicate the susceptibility to planting of other profitable species in the understory, selective cutting, harvesting for fuel wood and/or seed collection: 3 for one of these treatments; 5 for two treatments; 11 for three treatments; 15 for more than three treatments. Therefore, the fragmented and disturbed habitat (HA) in Zhonglu of Lichuan can be further classified into three categories: Habitat A1 (with a score of 5 as the degree of disturbance), Habitat A2 (with a score of 11 as the degree of disturbance), and Habitat A3 (with a score of 15 as the degree of disturbance). The representative habitats are shown in Fig. 7.15a–c.

Fig. 7.15 (**a**) A *Metasequoia* community in valley bottoms and by mountain streams, where it is less frequently disturbed (HA1). (**b**) In the *Metasequoia* community, *Cryptomeria japonica* var. *sinensis* has been planted since the 1960s (HA2). (**c**) The cash crop medicinal herb *Coptis chinensis* has been planted in the understory of the *Metasequoia* community (HA3)

7.2.2.3 Community Structure

Figure 7.16a shows three communities in which *Metasequoia glyptostroboides* forms the canopy. Habitat A1 (HA1; 13 plots) is relatively natural with a score of 5 as regards the degree of disturbance. Here the subcanopy is dominated by evergreen and deciduous broad-leaved trees including *Phoebe*, *Cyclobalanopsis*, *Machilus*, *Ficus*,

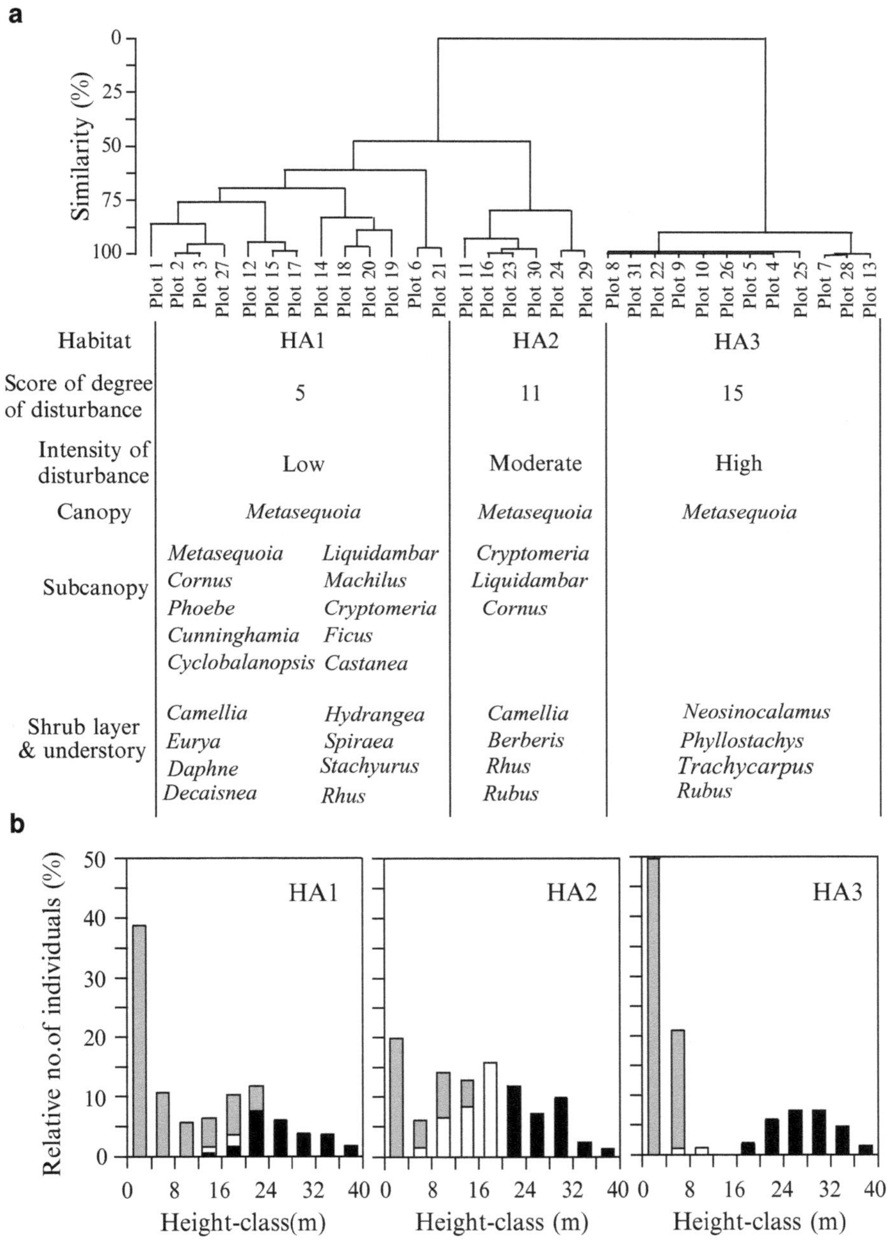

Fig. 7.16 Similarity dendrogram using relative Søerensen's similarity index and group average clustering (**a**), and the frequency distribution over height-class for all the plants over 1.3 m tall in habitat A1 (HA1), habitat A2 (HA2) and habitat A3 (HA3) of *Metasequoia glyptostroboides* stands (**b**). *Black bar* indicates *M. glyptostroboides*, *white bar* is for cultivated *Cryptomeria japonica*, *grey bar* is for the other species (Data source: Tang et al. (2011))

Liquidambar, *Cornus*, *Castanea* and coniferous *Cunninghamia*, as well as cultivated coniferous *Cryptomeria*, while the shrub and understory are composed of *Camellia*, *Hydrangea*, *Eurya*, *Spiraea*, etc. Habitat A2 (HA2; 6 plots) has a score of 11 for the degree of disturbance. Here *Cryptomeria* has been planted since the 1960s, while the shrub layer and understory consist of *Camellia*, *Berberis*, *Rhus* and *Rubus*. Habitat A3 (HA3; 12 plots) has a score of 15 for the degree of disturbance. Here bamboos are frequently cultivated, including *Neosinocalamus affinis*, *Phyllostachys heteroclada* and *Phyllostachys sulphurea*, with some medicinal herbs (*Coptis chinensis*) in the understory. In habitat A1 (HA1), the stands are multilayered, containing canopy, subcanopy, shrub and understory layers (Fig. 7.16b). In habitat A2 (HA2), the shrub layer is sparse. In habitat A3 (HA3), subcanopy and shrub layers are absent. The coverage of herbaceous species in HA1 is 70–100 %, including *Impatiens dicentra, Elatostema involucratum, Galium odoratum, Viola vaginata, Polygonum hydropiper, Iris tectorum,* etc., and various ferns, such as *Parathelypteris japonica*, *Selaginella tamariscina*, *Polystichum neolobatum*, *Coniogramme intermedia*. The herbaceous coverage in HA2 and HA3 is only between 10 and 30 %. All the *M. glyptostroboides* are over 16 m tall in the three habitats, except for one just 13 m tall individual in HA1. No seedlings or saplings of this species exist in these habitats.

In the various mosaic plant communities containing *M. glyptostroboides*, a total of 91 woody species were recorded (Table 7.5). Cultivated plants, including *Cryptomeria japonica* var. *sinensis*, are common in these habitats. Quite a number of exotic species are found too. As regards diversity, habitat A1, with a low degree of human interference, has a significantly higher diversity index as measured by the Shannon–Wiener index, than habitats A2 (HA2) and A3 (HA3), with moderate and high degrees of anthropogenic disturbance (Table 7.5), though as such each habitat has a rather low diversity index.

In 1948 the floristic composition in the natural habitat of *M. glyptostroboides* included many Tertiary relicts, among them *Nyssa sinensis*, *Cercidiphyllum japonicum*, *Tapiscia sinensis* and *Fagus longipetiolata* (Chu and Cooper 1950). These species have disappeared at present (Table 7.5). On the other hand, more cultivated species, including *Cryptomeria japonica* var. *sinensis*, *Neosinocalamus affinis, Phyllostachys heteroclada, Phyllostachys sulphurea, Trachycarpus fortunei* and *Coptis chinensis,* are common. The species diversity of the plant communities containing *M. glyptostroboides* has declined as human encroachment increased (Tang et al. 2011).

7.2.2.4 Population Structure

Tang et al. (2011) have reported that stem diameter at a height of 1.3 m (DBH) (x) and age (y) for wild *M. glyptostroboides* are correlated ($y=1.4x+6.03$, $R=0.83$, $n=86$). According to ring widths, their growth rates generally decreased with increasing age. Before the age of 80 years, the widths of annual growth rings varied from 5.79 to 2.87 mm per year; for ages of 80–170 years from 2.36 to 0.711 mm per year. The distribution over DBH-classes for all the *M. glyptostroboides* individuals of the three

Table 7.5 Floristic composition (woody species, bamboo and *Trachycarpus* with RBA ≥ 0.08 %) of the *Metasequoia* communities in the fragmented and disturbed stands in Zhonglu, Lichuan

Habitat	HA1	HA2	HA3
Intensity of disturbance	Low	Moderate	High
Altitude (m)	1,000–1,300	1,000–1,301	900–1,330
Number of plots	13	6	23
Area of plots (m^2)	9,350	3,400	17,600
Shannon–Wiener index	1.12 ± 0.33[a]	0.61 ± 0.15[b]	0.11 ± 0.17[c]
Relative basal area (RBA)	RBA (%)	RBA (%)	RBA (%)
Dominant species			
Metasequoia glyptostroboides	77.9	86.1	98.3
Companion species			
Liquidambar formosana	4.9	0.9	0.1
Cornus controversa	4.3	0.4	0.1
Machilus pingii	2.3		
Phoebe hunanensis	2.1		
Cryptomeria japonica var. *sinensis*	1.7	6	0.4
Cunninghamia lanceolata	1.3	5.2	0.3
Ficus heteromorpha	0.9		
Cyclobalanopsis multinervis	0.7		
Euscaphis japonica	0.7		
Castanea seguinii	0.6		
Diospyros sp.	0.5		
Lindera pulcherrima	0.4		
Betula sp.	0.3		+
Acer sinense	0.3		
Prunus padus	0.2		
Corylus sp.	0.1		
Rhus potaninii	0.1		+
Vernicia fordii	0.1	+	
Toxicodendron succedaneum	+	0.8	
Broussonetia papyrifera	+	+	
Xylosma racemosum	+		
Styrax japonicus	+		
Eurya hebeclados	+		
Hydrangea villosa	+		
Quercus serrata var. *brevipetiolata*	+		
Acer davidii	+		
Cinnamomum wilsonii	+		
Camellia oleifera	+		
Camellia caudata	+		
Abelia chinensis	+		
Lindera glauca	+		
Camellia japonica	+		

(continued)

Table 7.5 (continued)

Habitat	HA1	HA2	HA3
Intensity of disturbance	Low	Moderate	High
Altitude (m)	1,000–1,300	1,000–1,301	900–1,330
Number of plots	13	6	23
Area of plots (m^2)	9,350	3,400	17,600
Shannon–Wiener index	1.12 ± 0.33^a	0.61 ± 0.15^b	0.11 ± 0.17^c
Relative basal area (RBA)	RBA (%)	RBA (%)	RBA (%)
Viburnum erosum	+		
Lindera aggregata	+		
Dendrobenthamia angustata	+		
Taxus chinensis var. *mairei*	+		
Magnolia officinalis	+		
Decaisnea insignis	+		
Dichroa febrifuga	+	+	
Cephalotaxus fortunei	+		
Rhus chinensis	+		+
Stachyurus chinensis	+		
Toricellia angulata	+	+	
Morus australis	+		
Pittosporum glabratum var. *neriifolium*	+		
Euonymus alatus	+		
Daphne odora	+		
Hydrangea aspera	+	+	
Trachycarpus fortunei	+	+	+
Dalbergia dyeriana	+		
Robinia pseudoacacia		+	
Rhamnus utilis	+		
Quercus sp.	+		
Idesia polycarpa		+	
Glochidion wilsonii	+	+	
Rhododendron sp.	+		
Cotoneaster divaricatus	+		
Spiraea japonica var. *fortunei*	+		
Coriaria nepalensis	+	+	
Berberis bergmanniae	+	+	
Debregeasia orientalis	+	+	
Phellodendron chinense	+	+	
Nothopanax davidii	+		
Desmodium racemosum	+		
Rubus flagelliflorus	+	+	+
Phyllostachys bambusoides		+	+
Phyllostachys heteroclada		+	+

(continued)

Table 7.5 (continued)

Habitat	HA1	HA2	HA3
Intensity of disturbance	Low	Moderate	High
Altitude (m)	1,000–1,300	1,000–1,301	900–1,330
Number of plots	13	6	23
Area of plots (m^2)	9,350	3,400	17,600
Shannon–Wiener index	1.12 ± 0.33^a	0.61 ± 0.15^b	0.11 ± 0.17^c
Relative basal area (RBA)	RBA (%)	RBA (%)	RBA (%)
Phyllostachys sulphurea			+
Neosinocalamus affinis			+
Bambusa surrecta	+	+	

Data source: Tang et al. (2011)
Significantly different values for the Shannon-Wiener index (±SE) are indicated by different superscript letters ($p < 0.001$) (the non-parametric Kruskal–Wallis all-pairwise comparisons test)

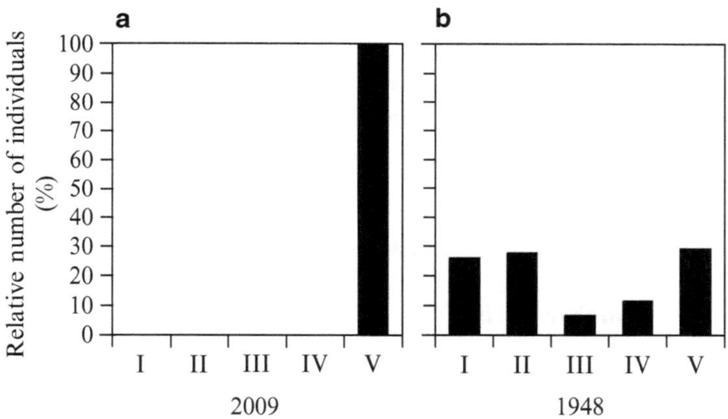

Fig. 7.17 Frequency distribution per DBH-class for *M. glyptostroboides* in the fragmented and disturbed habitat (HA) in 2009 (**a**) and in the natural habitat in 1948 (**b**). *Class I* height less than 30.5 cm, *Class II* height greater than 30.5 cm but diameters at breast height (DBH) less than 2.5 cm, *Class III* DBH 2.5–7.6 cm, *IV* DBH 10.2–25.4 cm, *V* DBH greater than 25.4 cm (Data for 2009 are from Tang et al. (2011). Data for 1948 are from ten 100 m^2 plots of Chu and Cooper (1950))

areas shows a total lack of young trees. Most of the *M. glyptostroboides* trees are between 45 and 85 cm DBH, and around 69–125 years old. For comparison with the data taken in 1948, Tang et al. (2011) adopted the categories of Chu and Cooper (1950) for Fig. 7.17a, b. In 2009 all individuals of *M. glyptostroboides* have been found greater than 25 cm DBH, which corresponds to an age of c. 41 years, in habitats fragmented and disturbed during the past 60 years (Fig. 7.17a), whereas in the natural habitat in 1948 individuals were found in all classes from I to V (Fig. 7.17b).

7.2.2.5 Regeneration

In this region land clearance and conversion to agriculture and plantation forestry have led to a loss of large areas of actual and potential habitats. The habitats of *M. glyptostroboides* have been exposed to disturbance, principally by cultivation of other plants and by seed collecting. Seedlings are often at an especially sensitive stage in the plant life cycle, requiring considerably more specific conditions than adults to support their growth. Seedlings and saplings of *M. glyptostroboides* were found in the dense and shady understory in the ecological survey of 1948 (Chu and Cooper 1950). But neither seedlings nor saplings are found in the current environment. The present disturbed mosaic habitats of *M. glyptostrobioides* have been created within the past 60 years. These habitat changes have negatively influenced seedling recruitment, community structure, species richness and diversity. So far, *M. glyptostroboides* has shown no ability to recover from the human disturbances in these areas.

Chinese official policy allows seeds to be collected yearly from every adult tree, for sale by local villagers. If seedlings are found in the stands, they are collected and moved to non-native habitats as landscape plants. Cultivation of other, profitable plant species in the stands containing *M. glyptostroboides* has been common. And our findings (Tang et al. 2011) emphasize how plant populations in fragmented and disturbed habitats may fail to persist in the long term.

7.2.3 The Thuja sutchuenensis *Forest*

7.2.3.1 *Thuja sutchuenensis*

The genus *Thuja* (Cupressaceae) is represented in the fossil record dating back to the late Cretaceous. Fossil records reveal a wide distribution in the Northern Hemisphere over geological time, including *Thuja smieya* in the late Cretaceous in Alaska (LePage 2003), *T. ehrenswaerdi* in the Paleocene of Greenland (Schweitzer 1974), *T. polaris* in the Paleocene of Ellesmere Island, Canada (McIver and Basinger 1989), *T. nipponica* in the Miocene of the Sikhote Alin Mountains, Russia, and in the late Miocene of Akita Prefecture, Japan (Akhmetiev 1973; Huzioka and Uemura 1973), *T. standishii* and *T. protojaponica* in the Plio-Pleistocene in Japan (Miki 1958), and *T. occidentalis* in the Plio-Pleistocene of Peary Land, North Greenland (Bennike 1990). *Thuja* derives from the Greek *thyia* or *thya,* an ancient name used by Theophrastus (370–288 BC) for some resin-bearing evergreen, possibly a juniper. It may be related to *thuon,* a sacrifice; the resin used as an incense in Eastern sacrifices (Jacobson 2014). *Thuja* currently comprises five extant species occurring disjunctly in East Asia and North America. *T. sutchuenensis* Franchet (Yabai or Sichuan *Thuja*) is endemic to the Daba Mountains at the northern fringe of the Sichuan Basin in the Yangtze River valley. *T. koraiensis* Nakai (Korean *Thuja*) is found in Korea and on the boundary between Korea and northeastern China (Changbaishsan). *T. standishii* (Gordon) Carrière (Japanese *Thuja*) is native to

Japan. *T. occidentalis* L. (northern white cedar) and *T. plicata* Donn *ex* D. Don (western red cedar) grow in eastern and western North America, respectively (Farjon 2001). *Thuja* disappeared from Europe and the high latitudes during the ice ages of the Pleistocene (Farjon 2008). The ancestral area of *Thuja* had been presumed to be eastern Asia. Two dispersal events would account for the modern distribution of *Thuja* in North America. Both the North Atlantic land bridge and the Bering land bridge have been considered possible routes for the migration of ancestral populations to North America (Li and Xiang 2005). By contrast, Peng and Wang (2008), using multiple gene sequences, have inferred that *Thuja* could have originated in the high-latitude areas of North America in the Paleocene or earlier, with subsequent expansion into eastern Asia through the Bering land bridge. The two eastern Asia species *T. standishii* and *T. sutchuenensis* have a sister relationship, and could have differentiated in the Oligocene or early Miocene.

Plants of *Thuja* are small to large evergreen trees, monoecious; branchlets arranged in a plane, flattened. Leaves are 4-ranked, scalelike, base decurrent only on leading branchlets, dimorphic along branchlets: facial pairs obovate-rhombic; lateral pairs boat-shaped, ridged abaxially, overlapping facial pairs, without conspicuous, white stomatal bands abaxially. Seed cones are terminal, solitary, with four to six pairs of imbricate scales, dehiscent when mature in 1st year. Seeds are flat, with two lateral, narrow wings (eFloras 2014). Both North American *Thuja* species have strongly aromatic foliage, so that when they are being pruned, or merely if a twig is scratched or the foliage bruised, a volatile odor is released. *Thuja plicata* emits a strong fruity smell, and *Thuja occidentalis* is more acrid (Jacobson 2014). As for the Japanese *Thuja*, its shoots are often curved at the tip and have a lemony smell when crushed (Phillips 1978). In contrast with the two American and the Japanese species, the Sichuan *Thuja* does not have distinctively scented foliage, and its leaves and corns are often smaller.

The first specimens of *T. sutchuenensis* were collected from its only known location at 1,400 m asl in Chengkou County (formerly Sichuan Province, now part of Chongqing Municipality) by the French missionary P. G. Farges in the period from 1892 to 1900 (Xiang et al. 2002) and were named *T. sutchuenensis* Franchet in 1899 (Franchet 1899). Botanical collecting trips to Chengkou by Chinese botanists later in the twentieth century failed to find *T. sutchuenensis,* thus it was considered extinct in the wild (Fu and Jin 1992) and known only by Farges' specimens in various European herbaria (Xiang et al. 2002). It was the only conifer listed as extinct in the wild on the global red list of conifers (Farjon and Page 1999). In October 1999 it was rediscovered in Chengkou by a regional team of Chinese botanists near the location where P. G. Farges made his last collections almost 100 years earlier (Liu et al. 2000; Xiang et al. 2002). It had been subject to extremely intensive logging, as local people used the soft, easily-worked wood for home construction, shingles, and other purposes requiring resistance to rot, such that the remaining specimens were limited to those left in inaccessible areas. Furthermore, Liu et al. (2013), using ISSR markers show that *Thuja* populations have a relatively high level of within-population genetic diversity, though this species have small number of extant individuals. It is now ranked as an endangered species on the IUCN red list (IUCN 2013).

T. sutchuenensis is confined to the Dabashan Nature Reserve of Chengkou County and the Xuebaoshan Nature Reserve of Kaixian County in the Daba Mountains on the borders of Chongqing Municipality and Sichuan, Hubei and Shaanxi provinces (31°35′52″–31°41′50″ N, 108°34′52″–108°50′34″ E) (Fig. 7.18). During the past century, logging and agricultural development greatly altered the

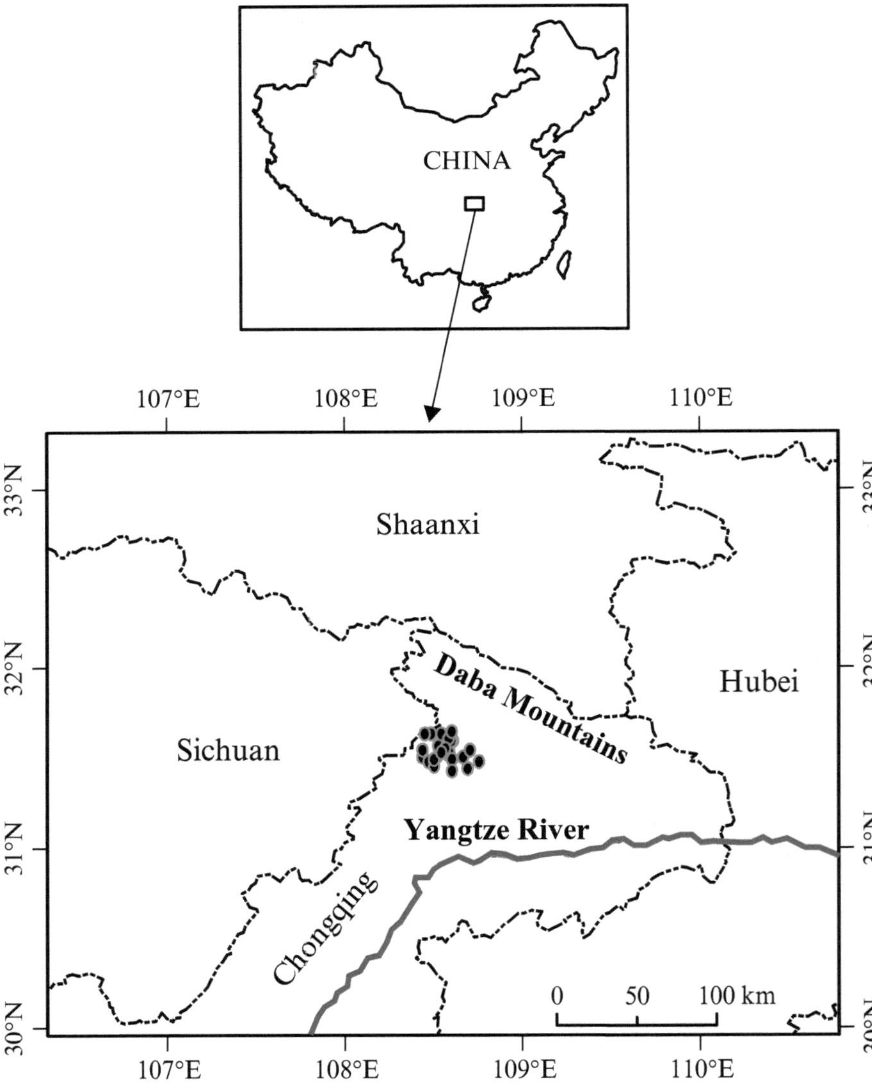

Fig. 7.18 Distribution of *Thuja sutchuenensis* in Kaixian and Chengkou counties in the Daba Mountains, SW China. The study plots representing the distribution range of *T. sutchuenensis* are indicated by *black dots*. The *broken line* shows the provincial boundary

composition and structure of the region's vegetation and landscape. A wide variety of forest types once intermingled at varying spatial scales in the Daba Mountains. Today, plantations (e.g. *Cunninghamia lanceolata*), early successional plant communities (dominated by *Populus adenopoda* or *Betula utilis*), and particularly coppices for firewood and shrublands that regenerated after the abandonment of agricultural fields, comprise the most common vegetation. Embedded within these secondary communities are some remnant forests on the limestone cliffs (inclination: 75–90°), steep slopes (inclination: 30–75°), and crest ridges (rounded tops of cliffs and slopes), some including stands of *T. sutchuenensis*. The steep slopes and crest ridges are partially covered by soils composed of moist and fertile mountain yellow-brown earth. The soil layer is relatively thin, and rich in organic matter (2.3–3 % of total dry weight), with a thick humus layer (c. 20 cm), and with pH values of 6.2–7.0 (Xiang et al. 2002). The limestone landscape with its steep precipices and cliffs is one of the most pronounced topographic features of the Sichuan Basin (Fig. 7.19a–c). The cliffs are especially landslide-prone, being composed of limestone, clay shale, and volcanoclastic and siliceous rock. The rock crevices on the cliffs result from tectonic instability and landslide-prone formations of various

Fig. 7.19 Topographic futures of the limestone cliffs (**a**–**c**) in Chengkou, Chongqing, which form the natural habitat of *Thuja sutchuenensis*

rock strata. The main triggering factors of landslides are active faults and rainstorms, especially during May to October (Luo 1990; Shi et al. 2009). On average, in the range of our study sites there are four landslides per year of various intensities, including cliff rockslides and debris flows (Luo 1990). This is one of the most geologically active regions in China. A subtropical monsoon climate (humid and warm) prevails, with a mean annual precipitation of 1,100–1,380 mm, of which 78 % occurs in May to October. The mean annual relative humidity is 73 %. Mean annual temperature ranges from 13.2 to 14.7 °C among years, with mean January (coldest month) temperatures ranging from 1.4 to 3.6 °C and mean July (hottest month) temperatures ranging from 23.4 to 25.2 °C.

Tang et al. (2015) conducted a comprehensive study on the spatial distribution patterns and population structure of *T. sutchuenensis* and examined how its population has been maintained in the Daba Mountains. The results from their work will be summarized here.

7.2.3.2 Forest Community Structure

Three forest communities (Fig. 7.20a) containing *T. sutchuenensis* have been distinguished: (1) the *Thuja* community dominated by *T. sutchuenensis*, on cliffs (inclination of 75–90°); (2) the *Thuja-Quercus-Cyclobalanopsis* community dominated by *T. sutchuenensis*, *Quercus engleriana* and *Cyclobalanopsis oxyodon*, along with *Fagus engleriana* and *Carpinus fargesiana*, on steep slopes (inclination of 30–75°); (3) the *Thuja-Tsuga-Quercus* community dominated by *T. sutchuenensis*, *Tsuga chinensis*, and *Quercus spinosa*, on crest ridges (round tops on ridges). The relative basal area (RBA) of *T. sutchuenensis* decreases with decreasing intensity of natural disturbance (rockslide), while *Tsuga* increases with decreasing intensity of the slide (Fig. 7.20b). The major woody species composition of each community is shown in Table 7.6. The community on the steep slopes has significantly higher Gleason richness values than those on the crest ridges and the cliffs. A few other less abundant conifers, including *Torreya fargesii, Taxus wallichiana* var. *chinensis*, *Pinus tabuliformis* var. *henryi*, and *Cephalotaxus fortunei*, grow in all three habitats. Evergreen sclerophyllous broad-leaved *Quercus engleriana* is absent from the crest ridges. Deciduous *Fagus engleriana* exists only on the steep slopes. The tallest *T. sutchuenensis* tree reached 16 m in height, but few exceeded 10 m in all three habitats. The cliffs support the greatest number of *Thuja* seedlings/saplings (15 cm < height < 1.3 m) in the study area.

7.2.3.3 Regeneration Dynamics

DBH-class distributions of the five major species (Fig. 7.21) can be categorized by inverse-J, sporadic, or unimodal types of DBH-size frequency distributions. On the cliffs (*Thuja* community), the DBH of *T. sutchuenensis* ranged from 0.5 to 40 cm

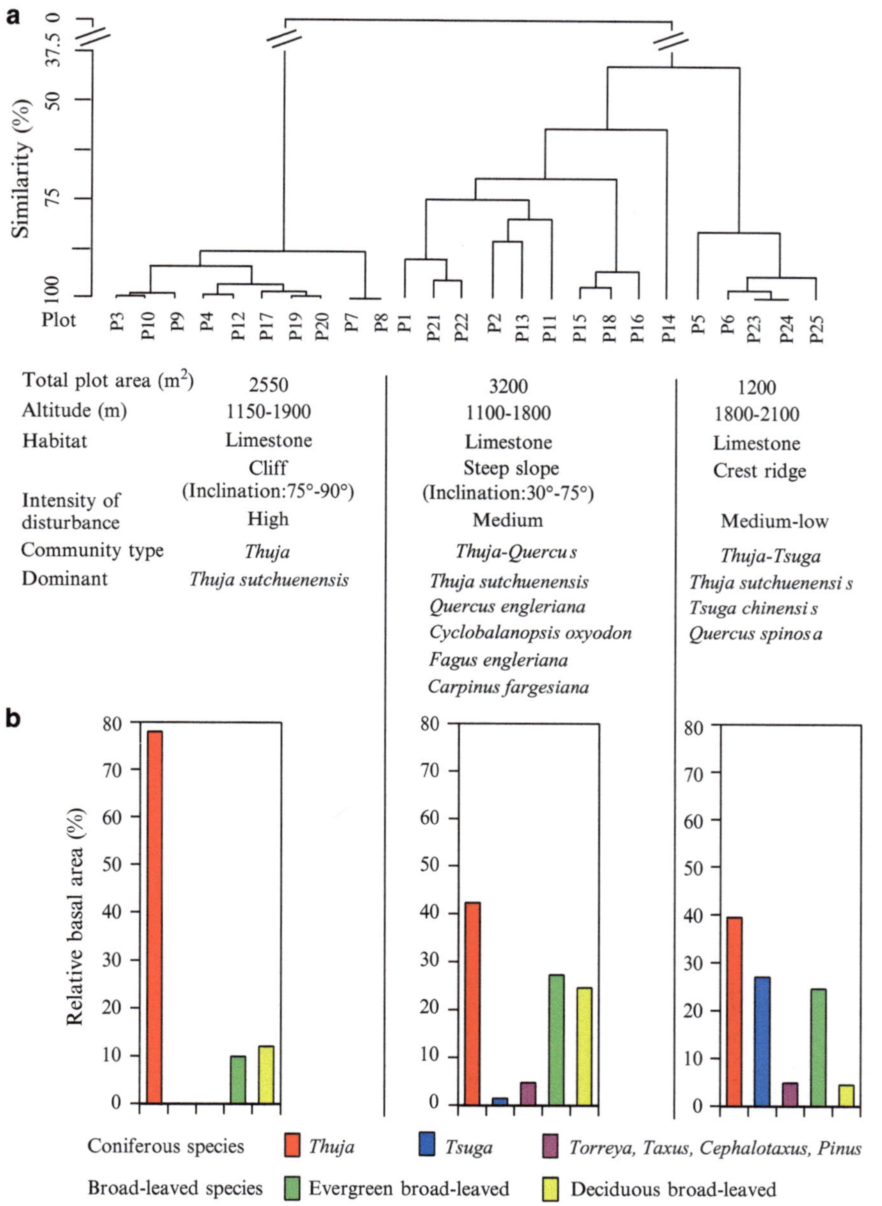

Fig. 7.20 (**a**) Floristic similarity dendrogram and characteristics of plant communities containing *Thuja sutchuenensis*. (**b**) Relative basal area (%) distribution of species (height ≥1.3 m). The evergreen broad-leaved species include the subtropical evergreen broad-leaved species and the evergreen sclerophyllous species of *Quercus* (Data source: Tang et al. (2015))

Table 7.6 Floristic composition for woody species (RBA ≥ 0.5 % in at least one plant community) of the *Thuja* communities

Community type		*Thuja*	*Thuja-Quercus-Cyclobalanopsis*	*Thuja-Tsuga-Quercus*
Topographic habitat		Cliff	Steep slope	Crest ridge
No. of plots		10	10	5
Plot area (m^2)		2,550	3,200	1,200
Gleason richness (d_{GI}) ± SD		2.78 ± 1.4[a]	3.73 ± 1.4[b]	2.27 ± 0.18[a]
Species	Life form	RBA (%)	RBA (%)	RBA (%)
Thuja sutchuenensis	C	**78**	**41.8**	**36**
Quercus engleriana	ESBL	3.5	**10.2**	
Quercus dolicholepis	ESBL	2.9	0.6	
Juglans cathayensis	DBL	2.5		
Carpinus fargesiana	DBL	2.3	**3.7**	2.3
Cyclobalanopsis oxyodon	EBL	2.3	**4.3**	2.2
Quercus phillyreoides	ESBL	1.7	0.7	
Platycarya strobilacea	DBL	1.3	3.1	
Daphniphyllum macropodum	EBL	1.2	1.7	
Buxus sinica	EBL	0.8		1.4
Symplocos multipes	EBL	0.8	0.6	
Zanthoxylum ovalifolium	EBL	0.6	0.3	
Betula austrosinensis	DBL	0.4		1.1
Cornus oblonga	DBL	0.3	3	
Cyclobalanopsis myrsinifolia	EBL	0.2	3.3	
Emmenopterys henryi	DBL	0.2	0.5	
Euptelea pleiosperma	DBL	0.1	0.5	
Pinus kwangtungensis	C	0.2	0.02	0.5
Viburnum atrocyaneum	DBL	0.1	0.5	
Tsuga chinensis	C	0.08	1.4	**31**
Pinus tabuliformis var. *henryi*	C	0.07	0.7	
Osmanthus armatus	EBL	0.07	0.2	0.5
Viburnum betulifolium	DBL	0.06	0.5	0.07
Acer davidii	DBL	0.05	0.5	0.02
Torreya fargesii	C	0.03	3.6	4.8
Taxus wallichiana var. *chinensis*	C	0.02	1	
Fagus engleriana	DBL		**3.7**	
Quercus spinosa	ESBL		1.3	**18.1**
Illicium micranthum	EBL		0.9	
Sorbus wilsoniana	DBL		0.9	
Ilex sp.	EBL		0.6	0.009
Machilus sp.	EBL		0.5	
Rhododendron sp.	EBL		0.2	1.3
Sorbus glomerulata	DBL			0.6

Data source: Tang et al. (2015)

Dominant species are indicated by boldface. Significantly different values for Gleason richness are indicated by different letters ($p < 0.05$) (Tukey–Kramer test)

RBA relative basal area, *C* coniferous, *EBL* evergreen broad-leaved, *ESBL* evergreen sclerophyllous broad-leaved, *DBL* deciduous broad-leaved

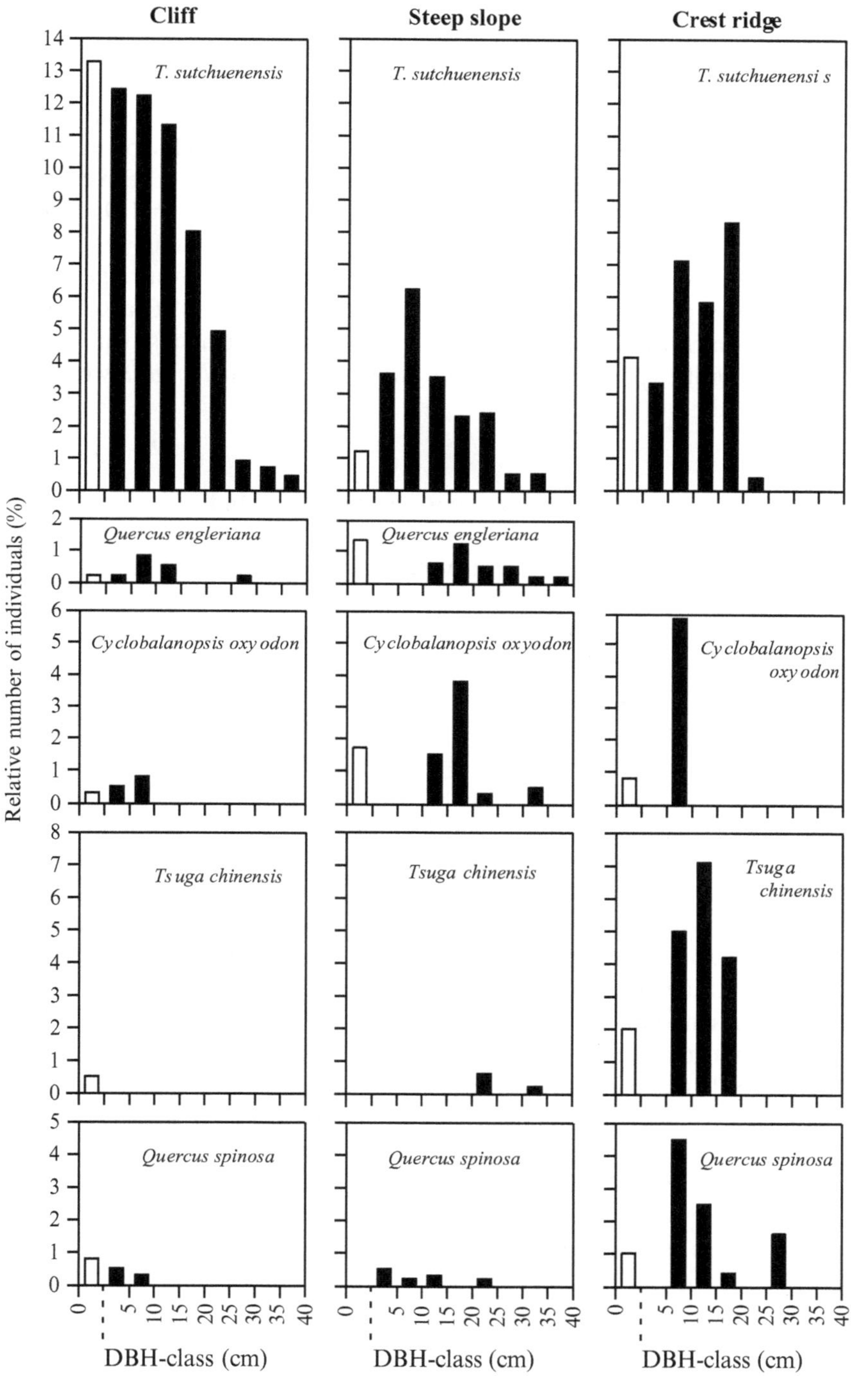

Fig. 7.21 DBH-class frequency distribution for five major species of plant communities containing *Thuja sutchuenensis* in various topographic habitats. *Empty bars* indicate the seedlings/saplings (After Tang et al. (2015))

and showed the inverse-J type distribution, suggesting that regeneration is active with continual recruitment of seedlings/saplings. Only a few individuals of *Quercus engleriana, Cyclobalanopsis oxyodon, Quercus spinosa* and *Tsuga chinensis* were present, almost all in small size classes. On the steep slopes (*Thuja-Quercus-Cyclobalanopsis* community), *Thuja* reached 35 cm DBH and showed the unimodal distribution type, suggesting that recent conditions have not been conducive to regeneration. *Quercus engleriana, Cyclobalanopsis oxyodon* and *Quercus spinosa* displayed the sporadic distribution type, suggesting periodic regeneration, and *Tsuga* had very few relatively large individuals. On the crest ridges (*Thuja-Tsuga-Quercus* community), *Thuja* ranged from 3 to 25 cm DBH and showed a distribution intermediate between the cliff and slope populations, with *Quercus spinosa* fitting the sporadic type, while *Tsuga* and *Cyclobalanopsis* were of the unimodal type.

7.2.3.4 Population Age Structure and Regeneration Modes

The growth rate of *T. sutchuenensis* is very low. The average radial growth rate per year is 1.1 mm below the age of 20 years, 0.75 mm from 20 to 40 years old, 0.5 mm from 40 to 90 years old, and varying between 0.8 and 0.5 mm above 90 years (Tang et al. 2015). The age structure of *T. sutchuenensis* with populations combined is shown in Fig. 7.22. The peak in number of individuals is between the ages of 30–60 years with DBH ranging from 3 to 7 cm. The largest tree, at 40 cm DBH, is c. 293 years old, located on the cliffs. Many trees having a greater DBH were logged during the past century.

Figure 7.23a–e show growth forms of *T. sutchuenensis*. Its substantial banks of seedlings/saplings were found on the cliffs. Most seedlings/saplings grew in rock crevices, cliff-faces scarred from rockslides, and cliff-edges where the micro-sites are dry. Some seedlings/saplings were found on fallen logs, in canopy gaps, and at forest margins near slopes and crest ridges. The growth forms varied; some stems grew straight, but most stems were twisted. Roots were well developed, and widely spread on the outcrops, sometimes reaching over 10 m. The stems showed evidence of varying intensity of disturbances. Some showed signs of cutting, and several adult trees had limbs cut off. There was no evidence of fire disturbance. After natural disturbances (i.e. rockslides or landslides) or cutting, *T. sutchuenensis* proved able to resprout. The ratio of resprouts to main stems was roughly 15:1. For example, many individuals on the cliffs had 2–18 resprouts after damage by rockfall. On a main stem of *Thuja* of 140 cm DBH cut in the 1970s, there were 20 resprouts ranging from 2 to 35 cm DBH in 2013.

7.2.3.5 Persistence

Tang et al. (2015) reported that the spatial distribution patterns of *T. sutchuenensis* as found in the study result from the interactions among natural disturbances, life history traits and stress-tolerance. Marginal habitats formed under conditions of

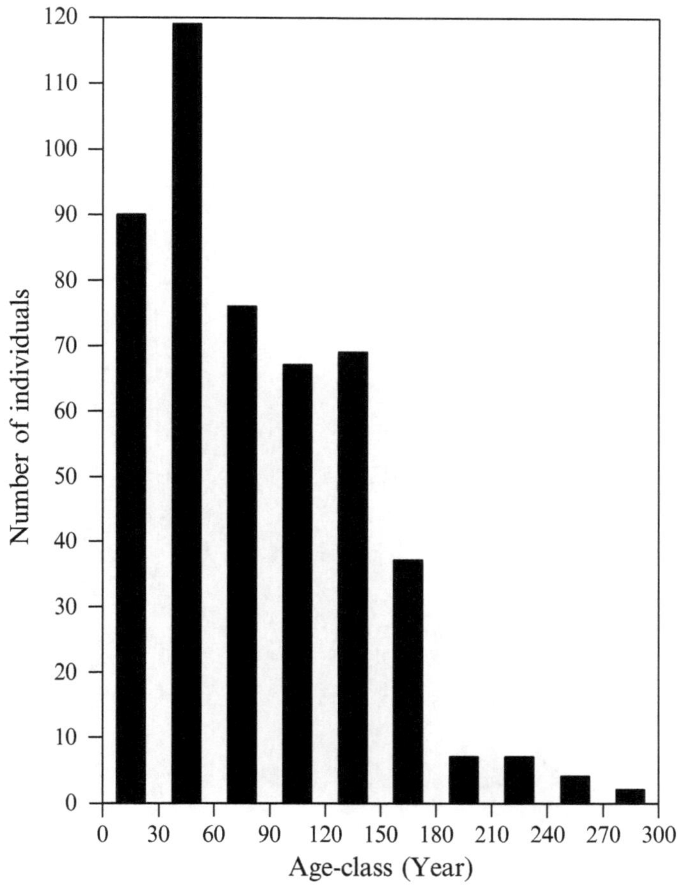

Fig. 7.22 Age structure of the *Thuja sutchuenensis* population

disturbance support distinctive flora and rare species because of differing competitive abilities and tolerance limits among species (Wisheu and Keddy 1992). The cliffs with high intensity of rockslides are dominated solely by *T. sutchuenensis.* In contrast, the slopes and crest ridges are dominated by more than two species and capable of forming dense canopies. The size-structures of *T. sutchuenensis* and its major associated species, including *Quercus engleriana*, *Cyclobalanopsis oxyodon*, *Tsuga chinensis* and *Quercus spinosa* in the three topographic communities, suggest a trade-off among those species (Fig. 7.21). Species with narrow regeneration niches often require special sites (Grubb 1977). Seedlings/saplings of *T. sutchuenensis* growing in crevices, on scarred cliff-faces and cliff-edges, on fallen logs, in canopy gaps and forest margins at steep slopes and crest ridges, show that the species' regeneration depends on disturbances. Its seeds are minute (1.13 g per 1,000 seeds) (Wang et al. 2007) and can take advantage of wind-dispersal to fall into rock crevices on the cliffs, unlike the acorns of *Quercus* and *Cyclobalanopsis. Tsuga*

Fig. 7.23 Part 1 (**a**) Foliage and seed corns of *Thuja sutchuenensis* (Photograph courtesy of Arthur Lee Jacobson). (**b**) A sapling *T. sutchuenensis* in a crevice on the cliff. (**c**) An adult tree *T. sutchuenensis* on a cliff

chinensis is shade-tolerant but susceptible to drought, and so cannot establish on the exposed cliffs. Resprouting can also be a strategy in maintaining population stability and continuity, especially in habitats where disturbances frequently do not kill the whole plant. *T. sutchuenensis* resprouts vigorously after disturbances that damage stems and branches. *Quercus engleriana, Quercus spinosa* and *Cyclobalanopsis oxyodon* also showed resprouting abilities (the mean ratio of resprouts to main stems is 9:1) after disturbances on the cliffs, but we did not find many resprouts for these four species on the slopes and crest ridges, where natural disturbances are less frequent. *T. sutchuenensis* shows relatively strong recruitment on the cliffs as compared to the slopes and the crest ridges. This also suggests its strong tolerance of xeric sites. The species appears to be of the stress-tolerant type as classified by Grime (1974).

Fig. 7.23 (continued) Part 2 (**d**) An adult tree *Thuja sutchuenensis* on a cliff. (**e**) *T. sutchuenensis* resprouting after cutting

Clearly, the *T. sutchuenensis* population thrives on cliffs where there is little competition from other species because of harsh conditions and rockslide disturbances. It is shade-intolerant but stress-tolerant. Its regeneration has been maintained by natural disturbances.

7.2.3.6 *Thuja sutchuenensis* as compared to the Other *Thuja* Species

Tang et al. (2015) compared general characteristics of *T. sutchuenensis* and other four *Thuja* species (Table 7.7). We pointed out that *T. sutchuenensis* is unique from other *Thuja* species in that it falls within the subtropical evergreen broad-leaved forest zone and mixed evergreen, deciduous broad-leaved and coniferous forest zone (altitudinal range (800) 1,100–2,100 m; Table 7.6). In contrast, the other four species of *Thuja* grow in the cool temperate and cold temperate forest zones. *T. koraiensis* is mainly found at 400–1,900 m in Korea and on the boundary (Changbaishan) between Korea and northeastern China. It grows in humid sites in valleys, mountain slopes and ridges, sometimes in rocky crevices (Fu and Jin 1992; Kong and Watts 1993). *T. standishii* is distributed from (250) 1,000–2,000 (2,150) m in Honshu and Shikoku, Japan. It is shade tolerant and occurs on moist, mesic and xeric sites and mountains (Nakamura and Krestove 2005). *T. occidentalis* has a wide distribution range throughout southeastern Canada and the northeastern United States; its primary range extends from southeastern Manitoba to Nova Scotia, and from James Bay to southern New York State, Michigan, and Ohio (Hofmeyer et al. 2009). Its altitudinal range is (10) 150–600 (670) m (Farjon 2013). It grows in various habitats including wet forests, coniferous swamps, and limestone cliffs. It is also found in abandoned pasture sites. It appears to be a highly variable species that can adapt to a wide range (both wet and dry) of environmental stresses. It is found at all successional stages (Kelly et al. 1992; de Blois and Bouchard 1995; Hannah 2004; Larouche et al. 2010). *T. plicata* grows at 0–1,500 (2,000) m along the Pacific Coast of North America from southeastern Alaska to northern California, and in the interior from east-central British Columbia into the Panhandle of Idaho and western Montana. The coastal and interior parts of the range are essentially isolated from each other but a few stands occur between the Coast Range and the Selkirk Mountains near the southern border of British Columbia (Minore 1990). *T. plicata* is a giant tree. It is shade-tolerant (Franklin and Debell 1988). It occurs in all successional stages (Minore 1983).

As in the three communities studied, some associated genera such as *Tsuga*, *Pinus* and *Fagus* are also found in other *Thuja* communities in Korea, Japan, eastern and western North America, though there they are accompanied by different species. Only in southwestern China are *T. sutchuenensis* communities accompanied by evergreen *Quercus* and *Cyclobalanopsis*. Other *Quercus* species associated with *Thuja*, such as *Quercus mongolica* in the *T. koraiensis* community in Korea and northeastern China, are deciduous. The genera composing the *Thuja* communities described in this study are also representative of glacial refugia communities that do not necessarily include *Thuja*, such as the Apalachicola bluffs and ravines in subtropical

Table 7.7 General characteristics of *Thuja sutchuenensis* as compared with the other four species of *Thuja* existing in the world

Region	Species	Height (m)	DBH (cm)	Age (year)	Radial growth (mm/year)	Habitat and altitude (m)	Shade tolerance	Major associated species as examples	Source
a	*T. sutchuenensis*	2–8 (16)	3–7 (40)	30–60 (293)	0.5–1.1	Limestone cliffs, steep slopes, and crest ridges (800) 1,100–2,100	Intolerant	*Quercus engleriana*	1
								Cyclobalanopsis oxyodon	
								Quercus spinosa	
								Carpinus fargesiana	
								Fagus engleriana	
								Tsuga chinensis	
								Pinus tabuliformis var. *henryi*	
								Torreya fargesii	
b	*T. koraiensis*	3–10	10–30 (80)	Not available	Not available	Humid sites in valleys, mountain slopes and ridges	Not available	*Abies nephrolepis*	2, 3
						Sometimes in crevices of rocks 400–1,900		*Taxus cuspidata*	
								Pinus pumila	
								Quercus mongolica	
								Betula ermanii	
								Acer ukurunduense	
c	*T. standishii*	15–30 (35)	40–60 (100)	Not available	c. 1.4–2.3	Mountains	Tolerant	*Abies firma*	4, 5, 6
						Rocky or steep slopes		*Abies homolepis*	
						Moist, mesic, and xeric sites (250) 1,000–2,000 (2,150)		*Tsuga sieboldii*	
								Tsuga diversifolia	
								Pinus parviflora	
								Thujopsis dolabrata	
								Fagus crenata	

(continued)

Table 7.7 (continued)

Region	Species	Height (m)	DBH (cm)	Age (year)	Radial growth (mm/year)	Habitat and altitude (m)	Shade tolerance	Major associated species as examples	Source
d	*T. occidentialis*	12–15 (38)	30–600 (175)	Cliff: 200–500 (1032)	Limestone cliffs: 0.05–0.17	Dry, mesic and wet sites	Tolerant, and intermediate	*Abies balsamea*	7, 8, 9, 10, 11, 12, 13
		Stunted or prostrate in harsh limestone cliffs		Noncliff: 80 (400)	Acadian forest: 0.43–0.82	Limestone cliffs	Stress-tolerant	*Larix laricina*	
					Bogs: 1.55	Swamps and abandoned pasture sites (10) 150–600 (670)	Occuring at all stages of forest succession	*Picea mariana*	
								Tsuga canadensis	
								Pinus banksiana	
								Pinus strobus	
								Acer rubrum	
								Betula papyrifera	
								Populus balsamifera	
e	*T. plicata*	40–50 (60)	100–300 (584)	(1460)	10–20 on the best moist sites, but 200 years can be 7.6 cm DBH	Gentle and steep slopes with moist soils, rocky slopes and river valleys 0–1,500 (2,290)	Tolerant	*Tsuga heterophylla*	14, 15, 16, 17, 18, 19
							Occurring at all stages of forest succession	*Pseudotsuga menziesii*	
								Abies amabilis	
								Taxus brevifolia	
								Abies grandis	
								Pinus monticola	

After Tang et al. (2015)

a. Daba Mountains, southwestern China; b. Korea and Changbei Mountains at the boundary between Korea and northeastern China; c. Honshu and Shikoku, Japan; d. Eastern North America; e. Western North America

1. Tang et al. (2005); 2. Fu and Jin (1992); 3. Kong and Watts (1993); 4. Nakane et al. (1988); 5. Nakamura and Krestov (2005); 6. Sawada et al. (1995); 7. Larson and Kelly (1991); 8. Kelly et al. (1992); 9. de Blois and Bouchard (1995); 10. Hannah (2004); 11. Hofmeyer et al. (2009); 12. Larouche et al. (2010); 13. Farjon (2013); 14. Franklin and Debell (1988); 15. Minore (1983); 16. Minore (1990); 17. Chamber (1993); 18. Daniels et al. (1995); 19. Christopher (2015)

northern Florida, which similarly harbor species of *Acer, Betula, Carpinus, Cornus, Fagus, Ilex, Illicium, Juglans, Osmanthus, Pinus, Quercus, Rhododendron, Symplocos, Taxus, Torreya,* and *Viburnum* (Harper 1914) (Table 7.6).

T. sutchuenensis is of small stature, with a slow radial growth rate (0.5–1.1 mm/year) on limestone formations such as the three topographic habitats in this study. It can reach very old ages. Although the largest tree found (40 cm DBH) was 293 years old, many stems with a larger DBH were logged in the 20th century. Some stems reached 140 cm DBH, based on our interviews with local villagers and our observations of logged stumps. The tree architecture varied greatly. The distinctive traits parallel *T. occidentalis* on exposed cliff edges and faces of the Niagara Escarpment of southern Ontario, where the species is stunted and morphologically deformed, slow-growing, and exceptionally long-lived (over 1,000 years) (Larson and Kelly 1991; Kelly et al. 1992). On the cliffs of the Niagara Escarpment, a high probability of survival may be linked to slow use of limited resources compensating for slow growth (Kelly et al. 1992). Cliff-dwelling trees of *T. occidentalis* almost without exception have lower growth rates than trees growing in more favorable environments (Kelly et al. 1992). Biotic interactions on cliffs are low because individuals are widely spaced, competition for light is low or nonexistent, and the root systems of neighboring plants are often completely separated by rock formations (Matthes-Sears et al. 1995). Survival of trees on cliffs may therefore be unrelated to growth rate and may instead be primarily controlled by the accident of finding safe sites with adequate rooting volume (Matthes-Sears and Larson 1999).

The extant *Thuja* species and their associated taxa in East Asia and eastern and western North America serve as some of the most remarkable examples of the survival of the Arcto-Tertiary relict flora. Knowing the distribution patterns, ecological characteristics and their survival mechanisms should lead to our better understanding of the East Asia and North America disjunct paleoendemics.

7.2.4 The **Davidia involucrata** *Forest*

7.2.4.1 *Davidia involucrata*

The genus *Davidia* is named for Father Armand David (1826–1900), "Père David", a French Lazarist missionary and keen naturalist who lived and traveled in China from 1862 to 1874. He first described the tree in 1869 after finding a single one at over 2,000 m asl in Muping, Sichuan, and sent dried specimens to Paris; in 1871, Henri Baillon classified it as a new genus and species (Gardener 1972). The Irish plant hunter Augustine Henry again found a single tree, this time in the Yangtze Ichang gorges, Hubei, and sent both flowering and fruiting specimens to the herbarium at Kew. This plant material was illustrated and discussed by Dr. Daniel Oliver, keeper of the herbarium and library at Kew. Oliver commented at the end of his description, "*Davidia* is a tree almost deserving a special mission to Western China with a view to its introduction to European gardens…" In 1899 plant collector Ernest Henry Wilson was employed by Sir Harry Veitch to find Henry's tree, but arrived to find that it had been felled for building purposes; however, in 1900 he found a grove of the trees overhanging a sheer drop, 600 km away in Ichang. During his return, Wilson's boat was wrecked, but he managed to save his *Davidia* specimens (Gardener 1972; Wikipedia 2014a, b). Contrary to the original speculation concerning the rarity of *Davidia* in the wild, Wilson commented as follows: "This remarkable tree is fairly common in moist woods throughout western Szechuan between altitudes of 1,600 and 2,500 m… The tree grows to the height of 20 m (±60 ft.) with a girth of trunk of 2 m (±6 ft.); the branches are ascending-spreading and the tree is of a loose pyramidal habit similar to that of the common wild Pear. On old trees the lower branches are wide-spreading. The bark is dark gray, rough, with corky lenticels and exfoliates in thin, small, flat irregularly oblong flakes. The wood is white, tough and heavy…When in full flower the tree is more conspicuous on dull days and in the early morning and evening than when the sun is shining…" (Sargent 1913).

Davidia is a monotypic genus constituting the *Davidiaceae* (formerly included in the family *Nyssaceae*). In the species *Davidia involucrata* Baillon there are two varieties differing slightly in their leaves: *Davidia involucrata* var. *involucrata* (Dode) Wangerin has the leaves thinly pubescent (short-haired) on the underside, and *Davidia involucrata* var. *vilmoriniana* Hemsl. has glabrous (hairless) leaves (Qin and Phengklai 2013). It is known as the dove tree or handkerchief tree in English, Gongtong in Chinese. It was first introduced to England in 1903. Since then it has become a popular garden tree in many parts of the world. Wilson once referred to the dove tree as "the most interesting and beautiful of all trees of the north-temperate flora…The flowers and their attendant bracts are pendulous on fairly long stalks, and when stirred by the slightest breeze they resemble huge butterflies hovering amongst the trees" as cited by Jiang (2011). It is a deciduous broad-leaved tree up to 20 m tall, endemic to China. It bears simple leaves with long petioles and ovate, cordate-based laminae which are serrate with nearly right-angled teeth (Fig. 7.24a). *Davidia involucrata* is best known for its flowers. Inflorescences consist of globose terminal heads positioned some distance below on long peduncle. The Latin specific epithet *involucrata* literally means "wrapped up": in *Davidia* 2 (or 3) pure white bracts surround flowers. The flowers are at their best in late May or early June. On a breezy day, the bracts flutter in the wind like white doves or gathered handkerchiefs (Fig. 7.24b), hence the English names for this tree. Nearly all flowers in the head are staminate, but a single bisexual flower, upon fertilization, develops into the fruit. The resulting elliptical fruit is borne on an odd-looking structure representing the remnant of the flowering head with scars of the shed staminate flowers, terminating in a substantial peduncle (Fig. 7.24c, d). The fruit stone is woody, composed of fibers, and has several radially arranged locules, each with a tongue-shaped, elongate germination valve that hinges open from the base (Manchester 2014a, b). The nut contains 3–6 seeds.

Davidia fossils are found dating from the Upper Cretaceous (c. 58 Ma) in Canada (Serbet et al. 2004), the early Eocene in Washington State (Wolfe and Wehr 1987) and the late Eocene in Oregon (Manchester and McIntosh 2007), the early Paleocene

Fig. 7.24 Leaves (**a**), flowers (**b**), fresh fruits (**c**) and a dry fruit (**d**) of *Davidia involucrata*

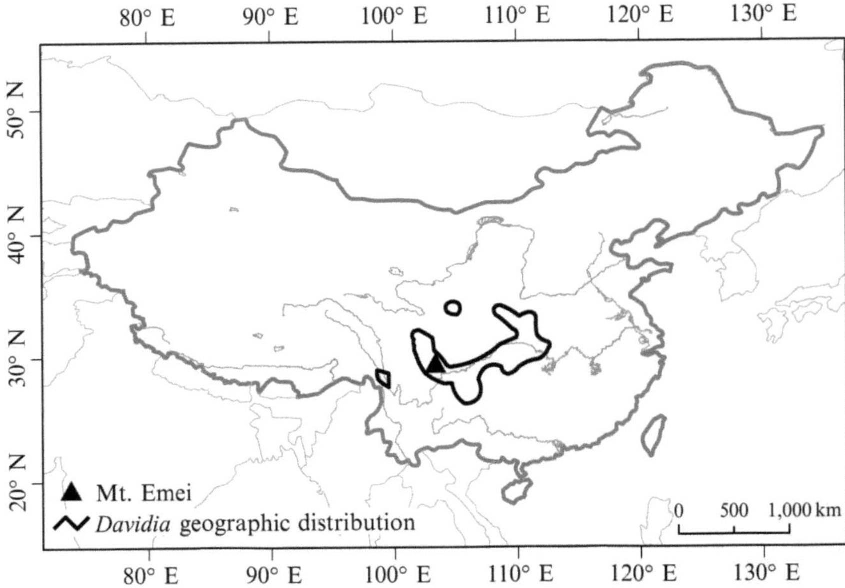

Fig. 7.25 Geographic distribution range of *Davidia involucrata* (Data source: Tang and Ohsawa (2002))

in Russia (Manchester 2002), and the Pliocene to Pleistocene in Honshu, Japan (Kokawa 1965; Ozaki 1984; Tsukagoshi et al. 1997). The fossil records suggest that *Davidia* traversed the Bering Land Bridge during the late Cretaceous or early Tertiary, allowing it to become established both in North America and East Asia by the Paleocene. This biogeographic pattern is similar to that of the related extinct genus *Amersinia*/*Beringiaphyllum* (Manchester et al. 1999), and is distinct from that of other cornalean genera *Alangium*, *Cornus*, *Mastixia* and *Nyssa*, which successfully populated Europe as well as North America during the early Tertiary and appear likely to have traversed the North Atlantic land bridge. The restricted modern native range of *Davidia* in China is the result of extinctions that obliterated the North American species, apparently at the end of the Eocene, and the reduction of a once broader Asian range (Manchester 2014a, b).

The distribution of extant *Davidia* is confined to subtropical mountains in southwest and south-central China from Hubei to southern Gansu, and south to Guizhou, Sichuan and Yunnan (Fig. 7.25). Its altitude ranges from 550 to 3,400 m (Fig. 7.26).

7.2.4.2 Age Structure and Regeneration Modes

Many *Davidia involucrata* trees, along with the other Tertiary relict deciduous trees *Tetracentron sinense*, *Cercidiphyllum japonicum* and *Euptelea pleiospermum*, found at middle altitudes on Mt. Emei, Sichuan, appear to favor a particular topography – the scree slope (Tang and Ohsawa 1997). Under more favorable conditions,

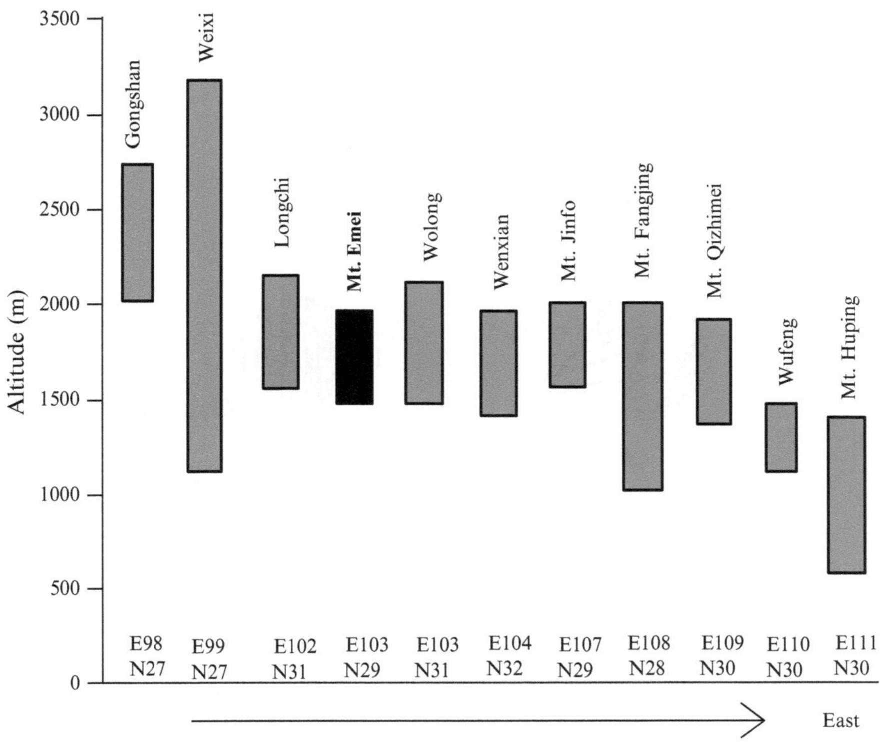

Fig. 7.26 Altitudinal distribution range of *Davidia involucrata* (Data source: Tang and Ohsawa (2002))

they cannot successfully compete with associated evergreen broad-leaved trees (Tang and Ohsawa 2002).

Near the altitude of 1,600 m on Mt. Emei, there are three typical topographies: (1) a steep scree slope at 1,620 m, (2) a slope with moderately rocky soils at 1,640 m, (3) a gentle slope with less rocky soils at 1,616 m. The age structure of the *Davidia* populations in the three topographic habitats is shown in Fig. 7.27. On the scree slope, *Davidia* occurred in all age classes, with most individuals being 5–10 years old. The oldest *Davidia* tree was 130 years old when our field survey was conducted. *Davidia* seedlings were found mostly in relatively well-lit sites, such as tree-fall gaps or on irregular topography. On the slope with moderately rocky soils, *Davidia* did not occur in all age classes, though again there was a single 130-year-old specimen in the plots. On the gentle slope with less rocky soils, *Davidia* only reached the age of 70 years, with very few individuals. The age structure indicates that the regeneration of *Davidia* is vigorous on the unstable scree slope; on the other hand, it is poor in relatively stable habitats, where evergreen trees monopolize the canopy layer. The 130-year-old *Davidia* trees, now c. 145 years old, may be the oldest ones now living in natural Chinese forests. On Mt. Emei, as elsewhere, their life span has been limited by human activity (Yang and Li 1989). Moreover, at around 1,600 m altitude, especially on the scree slope, landslides are common after heavy rains or storms.

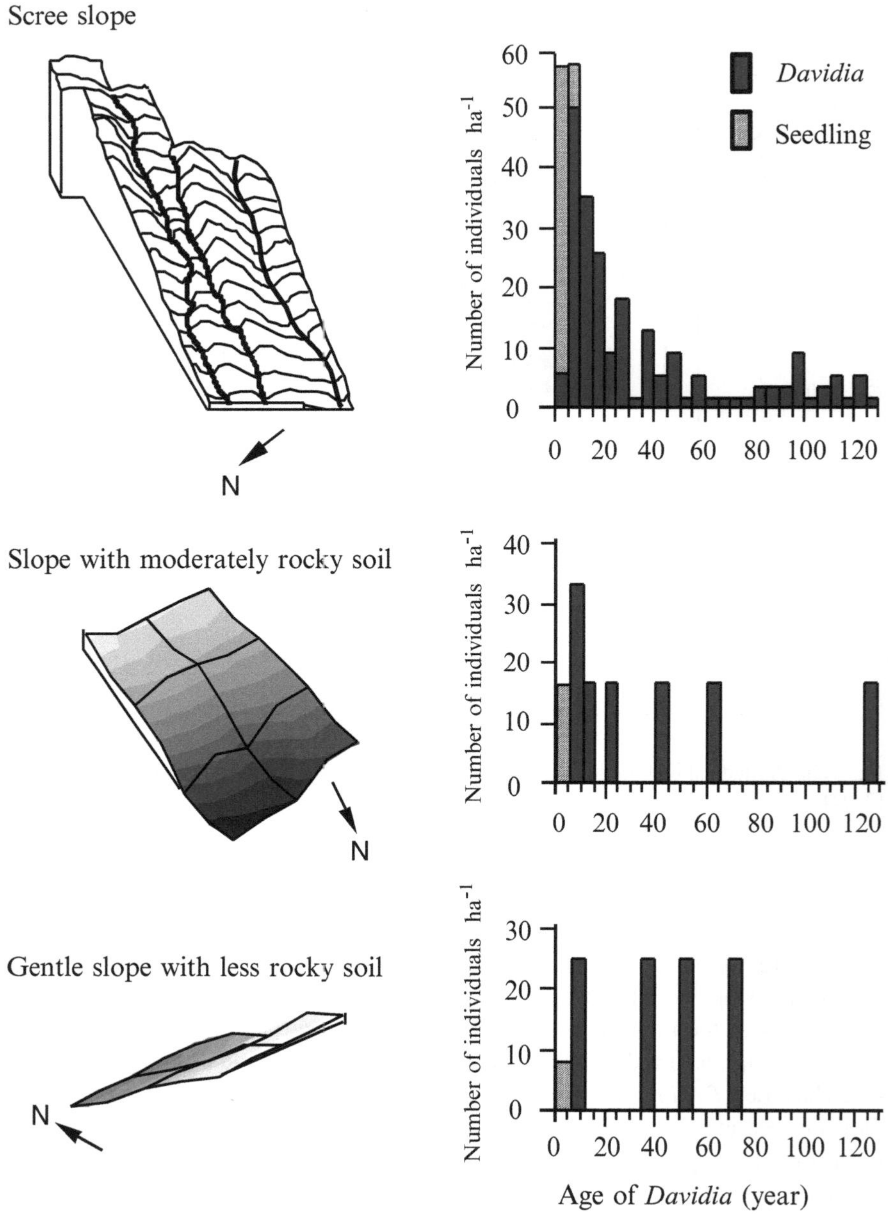

Fig. 7.27 Age structure of *Davidia involucrata* in three topographically different habitats on Mt. Emei. On the scree slope, regeneration is continuous with abundant saplings younger than 20 years; seedlings are also found, indicating good regeneration. By contrast, on the slope with moderately rocky soil and the gentle slope with less rocky soil, the distribution of individuals in the age classes is discontinuous, and few seedlings are found, indicating weak regeneration (Data source: Tang and Ohsawa (2002))

The regeneration modes of *Davidia* are by means of seedlings and sprouts. *Davidia* is light-demanding. With its large seeds dispersed by monkeys, rodents or other animals, it has better chances of survival and growth of seedlings than the other three species, *Cercidiphyllum*, *Tetracentron* and *Euptelea*, which depend on the wind for seed dispersal. This may explain why *Davidia involucrata* is the most dominant species in the relict deciduous forest. All these four relict deciduous tree species show a high ratio (from 1.3 to 2.4) of sprouts to main steams. These are much higher than those of *Castanopsis* at 0.5 and *Machilus* at 0.1. The abundant sprouting of the relict trees is a key factor in their survival in an environment where damage to tree stems is frequent.

7.2.4.3 Tree Dispersion Patterns and Forest Composition

On the scree slope there are three dispersion patterns (Fig. 7.28): a concave pattern, with Tertiary relict deciduous trees, including *Davidia, Tetracentron, Cercidiphyllum* and *Euptelea*, on concave sites; a random pattern, with non-relict deciduous (hereafter called other deciduous) trees, such as *Styrax hemsleyana, Pterocarya hupehensis,* and *Prunus brachypoda*, randomly occurring on both concave and convex sites; and a convex pattern, with evergreen broad-leaved (hereafter called evergreen)

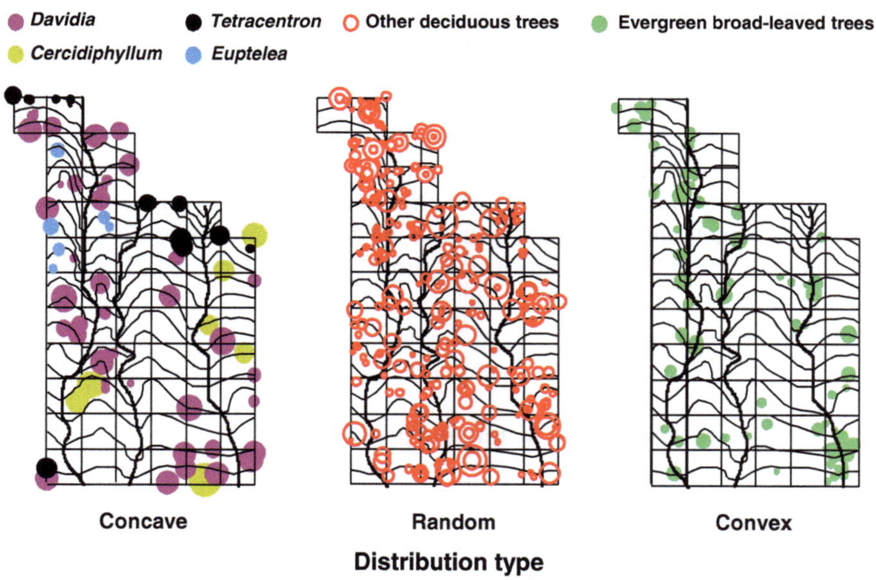

Fig. 7.28 Dispersion patterns of the Tertiary relict tree species on Mt. Emei as compared with non-relict tree species. On the scree slope, Tertiary relict deciduous *Davida*, *Tetracentron, Cercidiphyllum* and *Euptelea* are concentrated in the concave sites, while other non-relict deciduous trees are randomly distributed on the slope, and most of the evergreen broad-leaved trees are found at convex sites (Data source: Tang and Ohsawa (2002))

trees, such as *Lithocarpus cleistocarpus, Cyclobalanopsis oxyodon, Castanopsis platyacantha* and *Machilus pingii*, on convex sites. *Davidia* is the most dominant on the scree slope, probably because of frequent erosion or landslides there which keep the stands rather open. As the habitat becomes more stable, even on the scree slope, such as in the convex sites, the broad-leaved deciduous relict species give way to non-relict, evergreen broad-leaved trees (Tang and Ohsawa 2002; Tang et al. 2014). On the unstable scree slope the relict species are most abundant and there they have their greatest relative basal area.

On the other two non-scree slope types, the evergreen tree species prevail over the deciduous relict trees. The slope with moderately rocky soils is co-dominated by seven species, with evergreen *Machilus* and *Castanopsis* as the first and second dominants, while *Davidia* is the fourth dominant. The gentle slope with less rocky soils is co-dominated by evergreen *Machilus* and *Castanopsis*, along with deciduous *Styrax hemsleyana. Davidia* is not a dominant in this relatively stable habitat. *Cercidiphyllum, Tetracentron* and *Euptelea* are only found on the scree slopes. The woody species make-up on the three topographic habitats is shown in Table 7.8.

7.2.4.4 Ecological Performance

Figure 7.29 shows the relation between ecological performances (ability of plants to thrive) and habitat for the relict deciduous and non-relict evergreen trees on Mt. Emei. Tertiary-relict deciduous trees grow well on the frequently-disturbed unstable scree slopes where new soil resources are regularly available and competition from other species is limited, and where they also benefit from their high sprouting ability as an adaption to the frequent damage they are subjected to at ground level. In contrast, evergreen trees such as *Castanopsis* and *Machilus* manage to dominate on the stable habitats with deeper soils in the convex parts of the landscape, which allows them to increase their height and canopy density, and thus where, consequently, competition is more rigorous (Tang and Ohsawa 2002; Tang et al. 2014).

7.2.5 The Tetracentron sinense *Forest*

We have seen that *Tetracentron sinense* is a companion species in the *Davidia* forest on Mt. Emei in Sichuan. In contrast, it dominates some forest patches along stream banks at altitudes of 2,000–2,800 m on the eastern slopes of the Ailao Mountains of Yunnan (Fig. 7.30). Here *Tetracentron sinense* and its forest communities in the Ailao Mountains will be described in detail.

Table 7.8 Woody species composition in the three topographic habitats on Mt. Emei, Sichuan

		Plot number	1	2	3
		Altitude (m)	1,620	1,640	1,616
		Physical stability	Unstable	Moderately unstable	Relatively stable
		Exposure (degrees)	N58W	N21E	N79W
		Inclination (degrees)	30	18	9
		Maximum height (m)	25	23	23
		Maximum dbh (cm)	107	84	80
		Plot size (m^2)	5,400	600	400
		Diversity index (Shannon-Wiener)	3.551	3.172	2.535
		Total BA (cm^2/100 m)	3,410	5,500	5,170
		Total stem number	652	135	81
		Total species number	42	21	15
			RBA (%)	RBA (%)	RBA (%)
Family	Distribution of genus	**Relic deciduous trees**			
Davidiaceae	China	*Davidia involucrata*	*23.8	*11.2	6.1
Cercidiphyllaceae	EAs	*Cercidiphyllum japonicum*	*9.5		
Tetracentraceae	EAs	*Tetracentron sinense*	*4.4		
Eupteleaceae	EAs	*Euptelea pleiospermum*	0.001		
		Subtotal	37.701	11.2	6.1
		Other deciduous trees			
Styracaceae	Tr, SubTr	*Styrax hemsleyana*	*17.4	2.7	*19
Juglandaceae	NTem	*Pterocarya hupehensis*	*9.5	*6.6	
Rosaceae	NTem	*Prunus brachypoda*	*8.5	*13.4	5.3
Rosaceae	NTem	*Prunus padus*	*6.2		
Staphyleaceae	NTem	*Staphylea holocarpa*	*4.3		
Aceraceae	NTem	*Acer oliverianum*	3.0		
Sabiaceae	TemAs, Am	*Meliosma beaniana*	3.0		
Hippocastanaceae	NTem	*Aesculus wilsonii*	1.2		
Rosaceae	NTem	*Acer flabellatum*	1.1	1.8	0.5
Cornaceae	Eu, Am, As	*Macrocarpium chinense*	1.1	1.3	
Styracaceae	SEAs	*Rehderodendron macrocarpum*	0.8	0.5	
Lardizabalaceae	SEAs	*Decaisnea fargesii*	0.6	0.01	
Betulaceae	NTem	*Carpinus fangiana*	0.3		
Aceraceae	NTem	*Acer maximowiczii*	0.2		0.0003

(continued)

Table 7.8 (continued)

		Plot number	1	2	3
		Altitude (m)	1,620	1,640	1,616
		Physical stability	Unstable	Moderately unstable	Relatively stable
		Exposure (degrees)	N58W	N21E	N79W
		Inclination (degrees)	30	18	9
		Maximum height (m)	25	23	23
		Maximum dbh (cm)	107	84	80
		Plot size (m^2)	5,400	600	400
		Diversity index (Shannon-Wiener)	3.551	3.172	2.535
		Total BA (cm^2/100 m)	3,410	5,500	5,170
		Total stem number	652	135	81
		Total species number	42	21	15
			RBA (%)	RBA (%)	RBA (%)
Rosaceae	SEAs	*Stranvaesia davidiana*	0.2		
Alangiaceae	Af, As, Fiji	*Alangium chinense*	0.1		
Cornaceae	NAm	*Cornus controversa*	0.1	1.1	0.9
Celastraceae	NTem	*Euonymus porphyreus*	0.07	0.006	0.004
Caprifoliaceae	NTem, SubTr	*Viburnum betulifolium*	0.04		8.0
Caprifoliaceae	NTem, SubTr	*Viburnum erubescens*	0.03		
Lauraceae	TrAs, SubTrAs, Am	*Litsea veitchiana*	0.02		
Elaeagnaceae	EEu, As, NAm	*Elaeagnus pungens*	0.01		
Aceraceae	NTem	*Acer stachyophyllum*	0.008		
Sabiaceae	TemAs, Am	*Meliosma cuneifolia*	0.005	0.02	
Hydrangeaceae	NTem	*Hydrangea macrophylla*	0.003		
Grossulariaceae	NTem, SAm	*Ribes tenue*	0.001		
Aceraceae	NTem	*Acer sinense*		*14	
Aquifoliaceae	PanTr	*Ilex micrococca*		0.07	
Rosaceae	NTem	*Prunus grayana*		0.006	
Aceraceae	NTem	*Acer laxiflorum*			0.03
		Subtotal	57.787	41.512	33.734
		Evergreen broad-leaved trees			
Ericaceae	NTem	*Rhododendron argyrophyllum*	1.9	1.4	
Fagaceae	SEAs	*Lithocarpus cleistocarpus*	0.9		1.3
Fagaceae	NTem,Tr, SubTr	*Cyclobalanopsis oxyodon*	0.6		
Theaceae	SEAs	*Camellia oleifera*	0.5	0.6	1.2

(continued)

Table 7.8 (continued)

		Plot number	1	2	3
		Altitude (m)	1,620	1,640	1,616
		Physical stability	Unstable	Moderately unstable	Relatively stable
		Exposure (degrees)	N58W	N21E	N79W
		Inclination (degrees)	30	18	9
		Maximum height (m)	25	23	23
		Maximum dbh (cm)	107	84	80
		Plot size (m^2)	5,400	600	400
		Diversity index (Shannon-Wiener)	3.551	3.172	2.535
		Total BA (cm^2/100 m)	3,410	5,500	5,170
		Total stem number	652	135	81
		Total species number	42	21	15
			RBA (%)	RBA (%)	RBA (%)
Fagaceae	As	*Castanopsis platyacantha*	0.1	*17.2	*24.6
Caprifoliaceae	NTem, SubTr	*Viburnum oliganthum*	0.09	0.04	
Oleaceae	As, Am	*Osmanthus delavayi*	0.08		
Lauraceae	EAs	*Machilus pingii*	0.05	*20.3	*30.3
Aquifoliaceae	PanTr	*Ilex yunnanensis*	0.02		
Symplocaceae	Tr, SubTr	*Symplocos caudata*	0.02	*7.7	2.6
Stachyuraceae	EAs	*Stachyurus yunnanensis*	0.005		
Cornaceae	EAs	*Aucuba omeiensis*	0.003		
Cornaceae	EAs	*Aucuba himalaica* var. *oblanceolata*		0.002	
Aquifoliaceae	PanTr	*Ilex rockii*			0.3
		Subtotal	4.268	47.242	60.300
		Total	100	100	100

Data source: Tang and Ohsawa (2002)
The general distribution of genera is also given. Dominant species of each plot are indicated by an asterisk
RBA relative basal area, *E* East, *SE* Southeast, *As* Asia, *N* North, *Tem* Temperate, *Tr* Tropical, *Am* America, *Af* Africa, *Eu* Europe. Coding of exposure values: N58W = 58° W of N

7.2.5.1 *Tetracentron sinense*

Tetracentron Oliver (previously in Tetracentraceae), is a monotypic genus, that together with the monotypic genus *Trochodendron* Sieb. & Zucc., forms the family Trochodendraceae, now endemic to East Asia. *Tetracentron sinense* Oliver (shuiqingshu in Chinese) is a deciduous tree, growing up to 20–40 m, in southwestern and

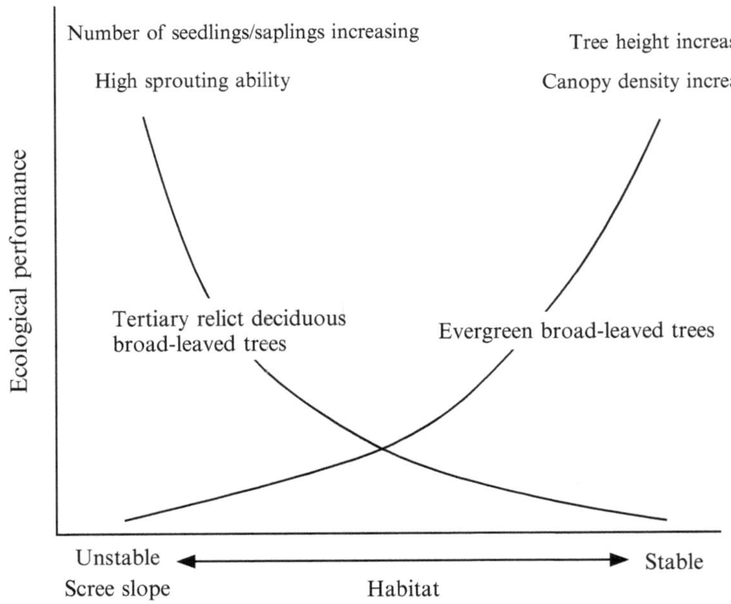

Fig. 7.29 A model showing the ecological performance of Tertiary relict trees as compared to evergreen broad-leaved trees on Mt. Emei. Tertiary relict deciduous broad-leaved trees thrive on the unstable scree slope by increasing their number of seedlings/saplings and by a high sprouting ability, while non-relict evergreen broad-leaved trees dominate the stable habitat by increasing tree height and canopy density. In the moderately unstable habitat, Tertiary relict trees and evergreen broad-leaved trees may co-dominate (Revised from Tang and Ohsawa (2002))

central China, the Himalayas, northern Myanmar and Vietnam (Doweld 1998; Fu and Bartholomew 2001; Qin 2004), whereas *Trochodendron aralioides* Sieb. & Zucc. is a large evergreen shrub or small tree native to the mountains of Japan, South Korea and Taiwan (Smith 1945; Mabberley 1987). Both species are without petals and have loculicidal capsules that dehisce to release winged seeds (Smith 1945; Pigg et al. 2001; Doweld 1998). Although earlier researchers reported that the wood of Trochodendrales lack vessels and accordingly suggested that the species was among the earliest-diverging angiosperms, recent research has documented the presence of vessels in the wood of both genera (Smith 1945; Pigg et al. 2001; Ren et al. 2007; Li et al. 2011). Unlike *Trochodendron*, *Tetracentron* has inconspicuous, wind-pollinated flowers, with a catkin-like inflorescence and broadly ovate leaves (Fig. 7.31a, b). The fruit is a follicle of 2.5–5 mm diameter containing 4–6 seeds (Fu and Bartholomew 2001).

Tetracentron fossils have been found from the Eocene (Pigg et al. 2001, 2007) and the Miocene of western North America (Manchester and Chen 2006), the Miocene of Iceland (Grímsson et al. 2008) and central Japan (Suzuki et al. 1991).

T. sinense yields a fine timber, as listed in CITES (Convention on International Trade in Endangered Species). Its populations in the wild have been reduced by forest destruction and overexploitation, its distribution is scattered, and it is listed as a

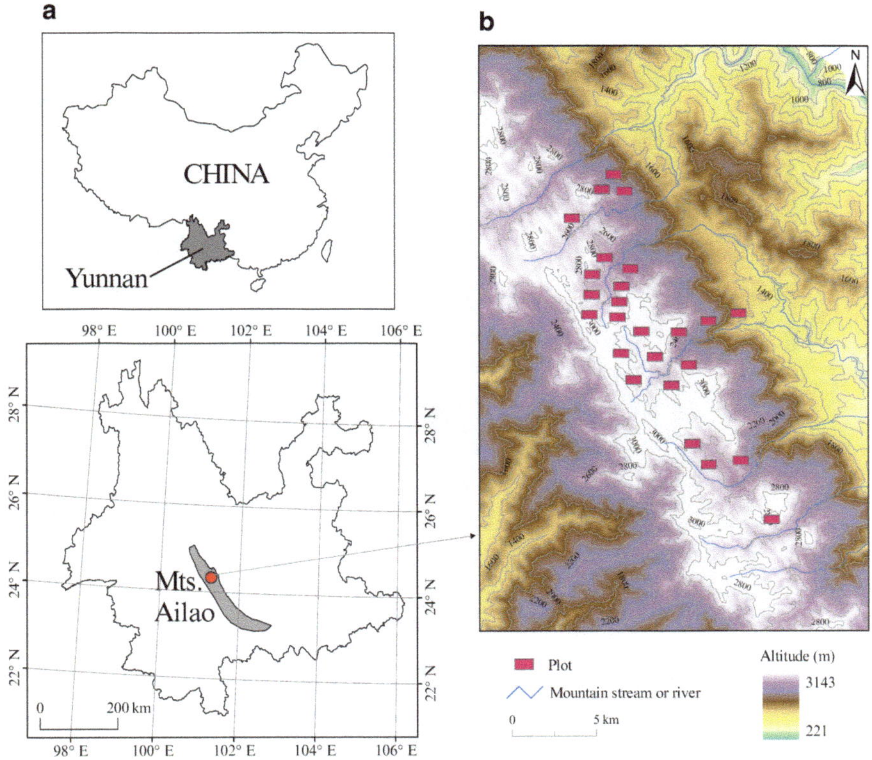

Fig. 7.30 Study area of *Tetracentron sinense* in the eastern Ailao Mountains, Yunnan

rare and threatened species in China (Fu and Jin 1992). Its altitudinal distribution range has been found to be 2,500–2,900 m in ravines or land sliding habitats of the Himalayas (Ohsawa 1987a; Ohsawa 1991), on stream banks at 1,400–1,580 m in the Shennongjia Mountains, Hubei (Jiang et al. 2000), on scree slopes in valleys at 1,450–1,900 m on Mt. Emei, Sichuan (Tang and Ohsawa 2002), and on mountain stream banks at 2,000–2,800 m on the eastern slopes of the Ailao Mountains, Yunnan (Tang et al. 2013a). Tang et al. (2013a) focused the study of *T. sinense* forests in the Ailao Mountains on the population structure (size, age) and regeneration patterns of *T. sinense*. The results of our study reveal ecological characteristics and survival mechanisms of this East Asian paleoendemic tree species.

7.2.5.2 Forest Community Structure

On the eastern slopes of the Ailao Mountains of central Yunnan, there are four forest community types (Groups) (Fig. 7.32a):

Fig. 7.31 (**a**) Leaves and inflorescence of *Tetracentron sinense*. (**b**) A *T. sinense* tree by a mountain stream

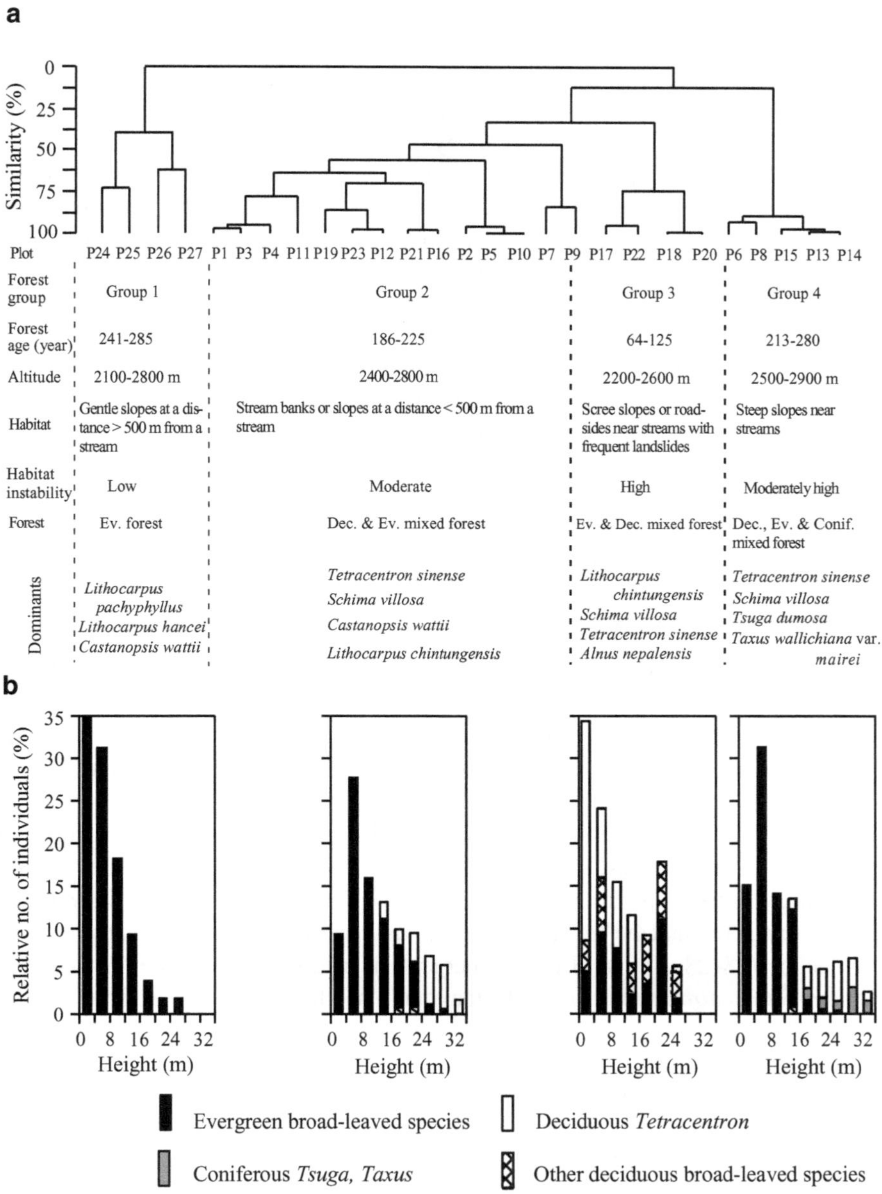

Fig. 7.32 (**a**) Floristic similarity dendrogram using relative Sørensen and group average clustering for the 27 plots in the eastern Ailao Mountains. (**b**) Height-class frequency distribution of species (height ≥1.3 m). *Ev.* evergreen broad-leaved, *Dec.* deciduous broad-leaved, *Conif.* Coniferous (Data source: Tang et al. (2013a))

1. The evergreen broad-leaved *Lithocarpus-Castanopsis* forest with an age of 240–285 years (Group 1), on stable, gentle slopes at a distance of >500 m from any stream, at altitudes of 2,100–2,800 m, where deciduous *T. sinense* is absent, while evergreen *Lithocarpus pachyphyllus*, *Lithocarpus hancei* and *Castanopsis wattii* exclusively dominate the canopy layer.
2. The deciduous and evergreen broad-leaved mixed *Tetracentron-Schima-Castanopsis-Lithocarpus* forest with an age of 186–225 years (Group 2), on unstable sites including stream banks or slopes at a distance of <500 m from a stream, at altitudes of 2,400–2,800 m, where *T. sinense* along with *Schima villosa*, *Castanopsis wattii*, and *Lithocarpus chintungensis* dominate in the canopy layer.
3. The evergreen and deciduous broad-leaved mixed *Lithocarpus-Schima-Tetracentron-Alnus* forest with an age of 64–125 years (Group 3), on very unstable sites including scree slopes, and roadsides near streams, with frequent landslides, at altitudes of 2,100–2,600 m, where the dominant trees are *Lithocarpus chintungensis, Schima villosa, T. sinense* and *Alnus nepalensis*.
4. The deciduous and evergreen broad-leaved and coniferous mixed *Tetracentron-Schima-Tsuga-Taxus* forest with an age of 213–280 years (Group 4), on steep slopes near streams, at the relatively higher altitudes of 2,500–2,900 m, where *T. sinense*, *Schima villosa*, and coniferous *Tsuga dumosa* and *Taxus wallichiana* var. *mairei* co-dominate the forest.

The floristic composition of woody species in the four forest groups is shown in Table 7.9. The species diversity in the mixed forests containing *T. sinense* in the unstable habitats is significantly higher on the Shannon-Wiener index H′ (2.90–3.26) than for the evergreen broad-leaved forest on the stable gentle slope (2.16) (Table 7.9).

Figure 7.32b depicts the vertical structure of the four forest types. In the evergreen broad-leaved *Lithocarpus-Castanopsis* forest, no individuals of *T. sinense* are found in the canopy, subcanopy or shrub layer. In the *Tetracentron-Schima-Castanopsis Lithocarpus* forest and the *Tetracentron-Schima-Tsuga-Taxus* forest, *T. sinense* reaches the canopy (18–28 m) and also the emergent layer (28–34 m). In the *Lithocarpus-Schima-Tetracentron-Alnus* forest, the maximum height of *T. sinense* is 25 m.

7.2.5.3 Population Age Structure

The age structure can provide an indication of the regeneration processes of the population. The age structure and DBH-classes of *T. sinense* (height ≥1.3 m) in various forest types is shown in Fig. 7.33a–f, omitting the evergreen broad-leaved forest where no *T. sinense* grows. In the *Tetracentron-Schima-Castanopsis-Lithocarpus* forest (Fig. 7.33a, b), the age of 164 individuals (≥1.3 m high) of *T. sinense* ranged from 19 to 225 years. Both the frequency of DBH-class and age structure are multimodal, indicating that the regeneration varied by chance,

Table 7.9 Floristic composition of woody species (with RBA ≥ 0.5 %) for four forest groups on eastern slopes on the Ailao Mountains

Forest group		Group 1	Group 2	Group 3	Group 4
Shannon-Wiener H′		2.16±0.3[a]	3.26±0.2[b]	2.9±0.46[b]	3.05±0.17[b]
Species	Life form	RBA (%)	RBA (%)	RBA (%)	RBA (%)
Lithocarpus pachyphyllus	ET	**45.5**	1.3		
Lithocarpus hancei	ET	**14.3**			
Castanopsis wattii	ET	**10.2**	**11.3**	2.1	4.5
Schima villosa	ET	4.3	**17.9**	**23.6**	**22.6**
Ilex forrestii	EST	4.1	1.8		2.8
Rhododendron irroratum	EST	3.6	1.8	6.3	0.6
Manglietia insignis	ET	3.5	3.7	6.1	1.8
Symplocos dryophila	ES	2.7			
Hartia sinensis	ET	2.3	0.5		
Myrsine africana	ES	2.0			
Camellia yunnanensis	EST	1.4			
Neolitsea homilantha	EST	0.9			
Cinnamomum porrectum	ET	0.8			
Lyonia ovalifolia	DT	0.7			
Eurya groffii	ES	0.6			
Eurya cavinervis	ES	0.6			
Eurya tsaii	ES	0.5			
Schefflera shweliensis	ET	0.5	1.3	1.3	1.4
Tetracentron sinense	DT		**31.4**	**14.7**	**30.5**
Lithocarpus chintungensis	ET		**10.8**	**26.2**	2.1
Acer amplum	DT		3.9		0.5
Ilex triflora	EST		1.9		0.6
Ilex macrocarpa	DT		1.6		
Neocinnamomum delavayi	ET		1.4		0.5
Taxus wallichiana var. *mairei*	CT		1.2		**10.5**
Illicium simonsii	ET		1.1		0.7
Machilus viridis	ET		0.9		
Acer elegantulum	DT		0.8		
Magnolia liliiflora	DT		0.6		
Alnus nepalensis	DT			**13.4**	
Betula alnoides	DT			4.8	
Tsuga dumosa	CT				**15.4**
Schima argentea	ET				2.5
Illicium macranthum	DT				0.6
Skimmia arborescens	DT				0.6
Lonicera japonica	DS				0.5

Data source: Tang et al. (2013a)

Dominant species are indicated by boldface. Significant differences in species diversity indices (±SE) are indicated by different superscript letter ($p < 0.05$)

ET evergreen broad-leaved tree, *DT* deciduous broad-leaved tree, *CT* coniferous tree, *EST* evergreen broad-leaved small tree, *ES* evergreen broad-leaved shrub, *DS* deciduous broad-leaved shrub, *RBA* relative basal area

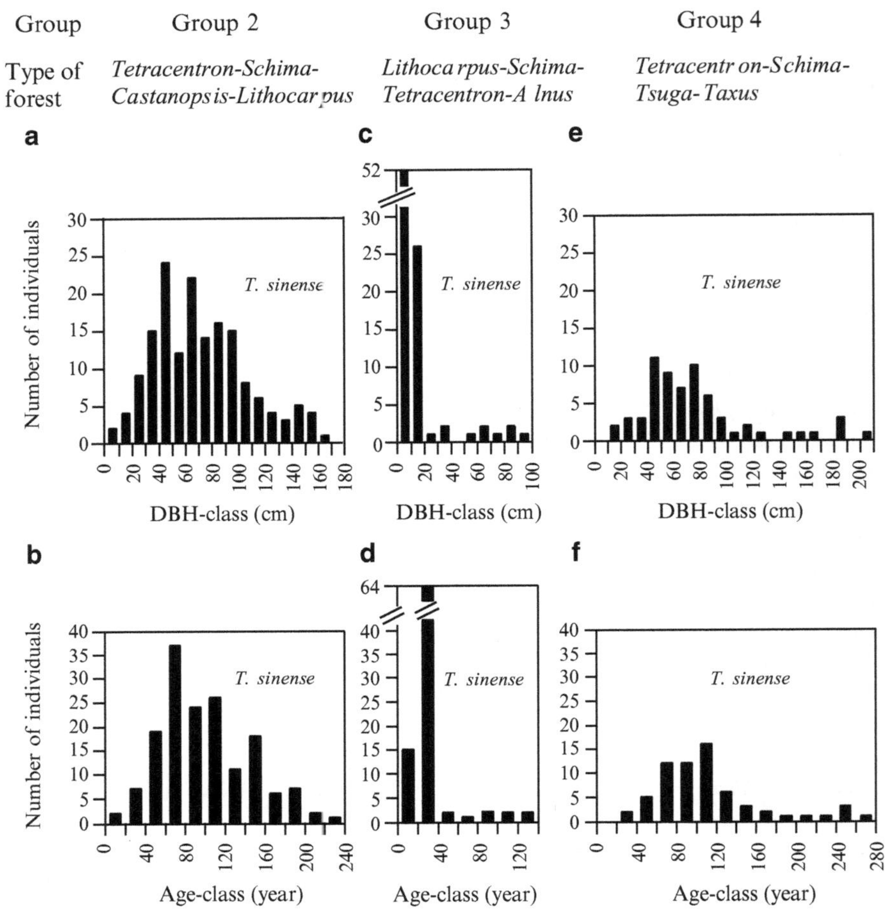

Fig. 7.33 DBH-class frequency distribution (**a**, **c**, **e**) and age-class frequency distribution (**b**, **d**, **f**) of *Tetracentron sinense* (height ≥1.3 m) in three forest types on the eastern slopes of the Ailao Mountains (Data source: Tang et al. (2013a))

depending on natural disturbances that reduce the competition from other plants. In the *Lithocarpus-Schima-Tetracentron-Alnus* forest (Fig. 7.33c, d), 102 individuals of *T. sinense* ranged from 15 to 125 years in age. This forest is younger than the other three. There are more individuals in small age classes than in the medium and large classes, indicating good regeneration. In the *Tetracentron-Schima-Tsuga-Taxus* forest (Fig. 7.33e, f), the age structure is nearly bell-shaped, and, the peak numbers of individuals are 60–120 years old, indicating weak regeneration. Among the 65 *T. sinense* trees, the oldest is 279 years old with a DBH of 200 cm. In all three forest types a number of *T. sinense* sprouts are found.

7.2.5.4 Regeneration Modes

Bamboo (*Yushania crassicollis*) is common on the forest floor at altitudes of 2,200–2,900 m in these mountains. The number of seedlings/saplings of *T. sinense* correlated negatively with the coverage of bamboo in the understory. Among the 134 established seedlings/saplings, 103 are by mountain paths or nearby streams. Ninety out of 134 seedlings/saplings grow in large gaps (>120 m^2), and 92 out of 134 have survived in rock crevices on scree slopes, on roadsides with frequent landslides, and in alluvial soil without litter near streams (Fig. 7.34a–c). A general regeneration pattern along the habitat instability gradient is shown in Fig. 7.35. The size of canopy gap increases and the coverage of bamboo decreases with greater habitat instability. On the unstable scree slopes or roadsides the high light intensity provides favorable conditions for establishment of *Tetracentron* seedlings/saplings. In general, the number of seedlings/saplings increases with greater habitat instability. *T. sinense* is also able to develop extensive roots in very shallow soils or in rock crevices. Its root networks increase resistance to erosion in the stream forest ecosystem. It resprouts after disturbances on the stream banks, or on steep slopes where landslides are frequent. Sprouts supplement the number of seedlings/saplings (Tang et al. 2013a).

7.2.5.5 Natural Disturbance and Persistence of *T. sinense*

The instabilities of the habitats, including stream banks, steep slopes, scree slopes, and roadsides near streams, along with outcrops, bamboo coverage in the understory and canopy gap size, are the principal controlling factors of the distribution and survival of *T. sinense*. The population structure makes clear that the patchy forests containing this species are of different ages. In the younger forest, *Tetracentron* trees are not just found along with late-successional *Lithocarpus chintungensis*, but also with the other two pioneers, *Schima vilosa* and *Alnus nepalensis*. This may indicate that the *Lithocarpus-Schima-Tetracentron-Alnus* forest is at an early middle-successional stage between the pioneer and the late stage that develops after formation of big canopy gaps under varying conditions. In a topo-community, species have specific habitat niches (Ohsawa 1987b). Sprouting allows a species to persist in a site after disturbances (Bond and Midgley 2001). *T. sinense* is a pioneer tree characterized by a long life span, reproducing by means of abundant minute wind-dispersed seeds in unstable habitats and reinforced by resprouting after frequent landslides. It survives well, along with the other long-lived, light-demanding pioneer evergreen broad-leaved *Schima villosa*, into the late-successional stages, coexisting in this area with late-successional species including evergreen broad-leaved *Castanopsis wattii*, *Lithocarpus chintungensis*, coniferous *Tsuga dumosa* and *Taxus wallichiana* var. *mairei*. In the eastern Ailao Mountains, varying topographic conditions allow many tree species with diverse life history traits to coexist.

Fig. 7.34 (**a, b**) Saplings of *Tetracentron sinense* growing in rock crevices. (**c**) *T. sinense* reaching the emergent layer (above the canopy)

Micro-habitats	Gentle slopes	Stream banks	Steep slopes	Scree slopes or roadsides
Habitat instability	**Very low**	**Moderate**	**Moderately high**	**High**
Coverage of bamboo in the understory	High	Medium	Medium	Low
Size of canopy gap	Small	Medium	Medium-large	Large
Light intensity	**Low**	**Moderate**	**Moderately high**	**High**

Seedlings/saplings

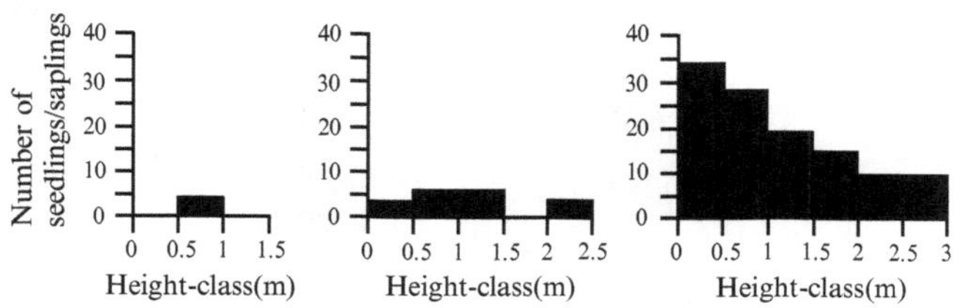

Fig. 7.35 Regeneration patterns of *Tetracentron sinense* in various topographic habitats in the eastern Ailao Mountains. As the habitat instability increases from gentle slopes to stream banks or steep slopes, then to very unstable habitats (e.g. scree slopes, roadsides), the number of seedlings/saplings increases in various height classes, because of the decrease of bamboo coverage and the increase of light intensity

7.2.6 *The* Liriodendron chinense *Forest, with Special References to Village Fengshui Forests*

7.2.6.1 *Liriodendron chinense*

Liriodendron is a genus of two species in the magnolia family (Magnoliaceae): *Liriodendron tulipifera* L. from eastern North America and *Liriodendron chinense* (Hemsl.) Sarg. from southern China. *Liriodendron* actually means "lily or tulip tree", though the resemblance is only superficial. *L. tulipifera* is widely known by the common name tulip tree for its large flowers superficially resembling tulips, though they are not closely related to the true tulips. The tree is sometimes referred to as "tulip poplar" or "yellow poplar", and the wood simply as "poplar", although again, it is not closely related to the true poplar. The tree has characteristic leaves, which are distinctive, having four lobes in most cases and a cross-cut notched or straight apex (Fig. 7.36a–c). Both the American and Chinese species are commonly used horticulturally, and hybrids have been produced between these two allopatrically distributed species (Liu et al. 1991; Wikipedia 2014c). *Liriodendron* trees are characteristically large, sometimes exceeding 32 m in height. The American poet William Cullen Bryant (1794–1878) wrote:

> The tulip-tree, high up,
> Opened in airs of June,
> Her multitude of golden chalices to humming birds,
> And silken-winged insects of the sky.

Fig. 7.36 **(a)** Leaves of *Liriodendron chinense*. **(b)** Seedlings of *L. chinense* in rocky soil. **(c)** A *L. chinense* tree with an age of 150 years

Various species of the genus *Liriodendron* have been found in fossil records from the Oligocene to the Pleistocene in the Northern Hemisphere, from North America to Central Asia and Europe (Berry 1941; Bùžek 1971; Tzedakis et al. 2006), most of which became extinct during the glacial period. It is believed that aridity associated with the last ice age contributed to its extinction in Europe.

Today, the extant two morphologically similar species represent a well-known example of an eastern North American and eastern Asian disjunct distribution also shown by several other temperate forest species (Parks and Wendel 1990). *L. tulipifera* is a fast-growing timber tree that is widespread and common in eastern North American deciduous forests (Braun 1950; Beck and Della-Bianca 1981), while *L. chinense* has been regarded as an endangered species because it occurs in small populations scattered in montane broad-leaved forests of southern China and northern Vietnam (Fu and Jin 1992; Hao et al. 1995).

There is no written record of the distribution of *L. chinense* before 1875, when the first specimen was collected on Lushan, northern Jiangxi Province, in the Yangtze River valley (Sargent 1903). Early botanists documented *L. chinense* in northern Jiangxi, western Hubei and eastern Sichuan, and in various parts of the Yangtze River valley (Forbe and Hemsley 1886; Sargent 1903; Wilson 1905). Wilson (1905) noted that *L. chinense* was common in forests and copses in western Hubei, parts of the same valley, and later photographed a *L. chinense* growing on a steep slope in Badong, Hubei (Wilson 1910). There was no documentation of its population size until 1988. From extensive plant surveys done in the 1970–1980s, Huang et al. (1988), Fu and Jin (1992) and He and Hao (1999) reported that human abuses before the1970s had left the species in seriously fragmented and degraded habitats in southern China. Relatively large populations of *Liriodendron* are scattered in the Dalou and Wuling Mountains (Hao et al. 1995). Some locations include remote mountain slopes where, although logged early in the twentieth century, subsequent reduction of intensive land use has allowed persistence of the species. Also, near certain villages, traditional beliefs of the people have led to preservation and management of patches of forest, called village fengshui (literally "wind and water") forests by inhabitants and described by Guan (2002). Their orientation, species composition, canopy density and other forest community attributes are thought to influence the health and wealth of the villagers (Guan 2002). Over hundreds of years, village fengshui forests have been preserved and managed by local people whose beliefs forbid the overlogging and deforestation that have recently been typical of the broader landscape. Because of their structure, composition, and rarity in the landscape, the village fengshui forests harbor various plant species that are not found elsewhere in the region. Tang et al. (2013b) investigated and compared the various landscape contexts in which *L. chinense* persists, to determine its habitat requirements and the role of natural processes and human intervention in its maintenance. We explored the forest characteristics and the village fengshui forests' roles in ecological processes regarding this Tertiary relict species in the Dalou and Wuling Mountains near the borders of Guizhou, Hunan and Hubei provinces and Chongqing Municipality. We have concluded that cultural practices and traditional sustainable forest resource uses provide regeneration niches for *L. chinense* that complement the low frequency of natural disturbances in mountain valleys, while on uninhabited mountain slopes the species depends on natural disturbances to survive.

7.2.6.2 Habitats and Forest Community Structure

In the Dalou and Wuling mountains, wild *L. chinense* populations are scattered from altitudes of 900 to 1,600 m (Fig. 7.37). On a large geographic scale, this altitudinal range belongs to both the subtropical moist evergreen broad-leaved forest zone and the mixed mesophytic forest (the evergreen and deciduous broad-leaved mixed forest) zone (Huang et al. 1988). These study areas are in a limestone region subject to landslides, but with virtually no fire. They were subjected to intensive agricultural development in the 1930s, including expansion of rice paddies, fields of corn and tobacco, and plantations of *Cunninghamia lanceolata* (China fir). The result has been

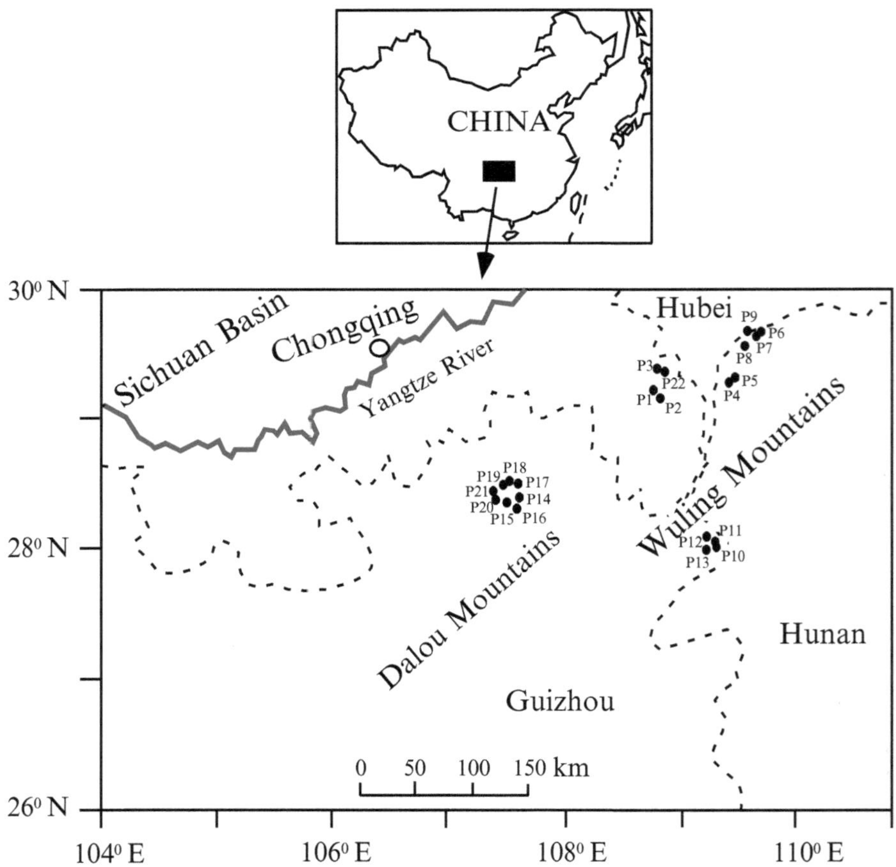

Fig. 7.37 Study sites of *Liriodendron chinense*

Fig. 7.38 (**a**) A village fengshui forest dominated by *Liriodendron chinense* in the North Kangjin Village of Songtao County, Guizhou. (**b**) A village fengshui forest dominated by *L. chinense* in the Panbao Village of Songtao County, Guizhou. (**c**) Tertiary relict species *L. chinense* (indicated by a *red arrow*) and *Emmenopterys henryi* (*yellow arrow*) growing together in a village fengshi forest

extreme fragmentation of forests, leaving only the village fengshui forests (Fig. 7.38a–c), typically smaller than 1 ha, in the vicinity of farmhouses, and remote forest patches on steep slopes close to creeks. As noted above, the village fengshui forests are preserved and managed by local villagers in accordance with traditional culture, beliefs, and sustainable forest resource uses, including limited harvesting of trees for firewood and collection of fruits and mushrooms for food and herbs for Chinese medicine. This mosaic of small patches of remnant vegetation constitutes the only link to forests that once may have been widely distributed throughout the areas.

The *Liriodendron* forest communities are classified into three types at a floristic similarity level of 38 % (Fig. 7.39a, b). The types are (1) *Liriodendron*: young

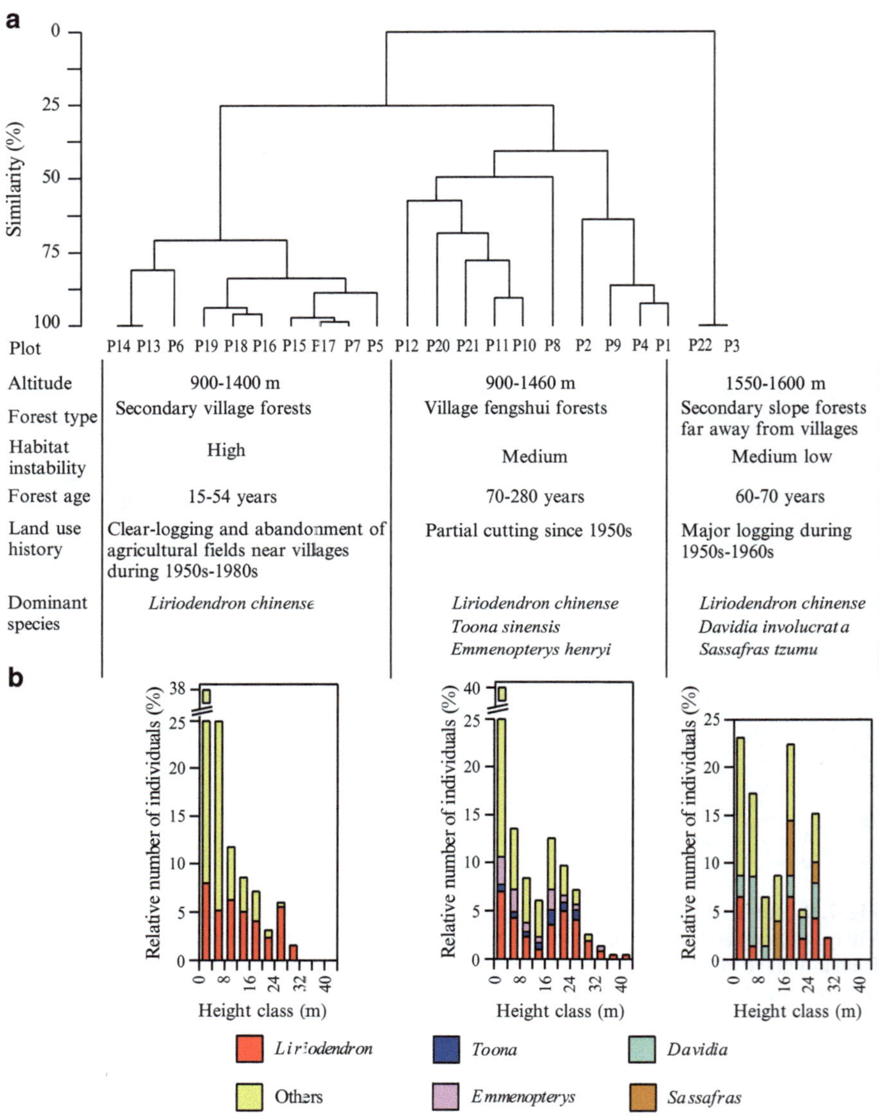

Fig. 7.39 (**a**) Floristic similarity dendrogram and *Liriodendron* forest community characteristics. (**b**) Height-class frequency distribution of species (height ≥1.3 m) (Data source: Tang et al. (2013b))

secondary forests near villages with an age of 15–54 years, after clear-logging in the 1950s and field abandonment during the 1980–1990s, with limited harvesting for firewood since then; (2) *Liriodendron-Toona-Emmenopterys* mature village fengshui forests with an age of 70–280 years, with limited harvesting of trees for firewood since the 1950s; (3) *Liriodendron-Davidia-Sassafras*: old secondary forests on mountain slopes (moderately steep) far from villages, with an age of 60–70 years, where clear-cutting occurred during the 1950–1960s but with no further land uses because of slope steepness and remoteness. On the mountain slopes, a number of *L. chinense* and *Davidia involucrata* have been able to sprout after the clear-cutting. Among the dominants, *L. chinense*, *Emmenopterys henryi* and *Davidia involucrata* are rare and endangered Tertiary relicts. The frequency of human and natural disturbances as a whole was judged to decrease from secondary forests near villages to village fengshui forests to mountain slope secondary forests. The three communities, taken together, contain 152 woody species, representing 97 genera in 53 families. The floristic composition for each community type is shown in Table 7.10. The relative basal area (RBA) of *L. chinense* decreased with decreasing habitat instability. Species diversity (Shannon-Wiener index H′) was highest in the mountain slope secondary forests, followed by the village fengshui forests and the secondary forests near villages (Table 7.10).

In each community type, *Liriodendron* dominates the canopy (18–32 m). In the village fengshui forests, *Liriodendron* reaches the emergent layer (32–42 m). In the subcanopy (8–18 m), shrub layer (3–8 m) and understory (<3 m) of the secondary forests near villages and the village fengshui forests, *Liriodendron* were commonly found, while in the mountain slope secondary forests *Liriodendron* was absent in the subcanopy (8–18 m) and infrequent in the shrub layer and understory. *Toona* and *Emmenopterys* also reached the canopy in the village fengshui forests; *Davidia* and *Sassafras* reached the canopy in the mountain slope secondary forests (Tang et al. 2013b).

7.2.6.3 Population Structure

In both the secondary forests near villages and the village fengshui forests, there are more individuals in small DBH and age classes than in the medium and large classes, indicating continuous regeneration (Fig. 7.40a, b). In contrast, mountain slope secondary forests have a rather low density of individuals overall, and the DBH and age structure are relatively even, resulting from clear-cutting during the 1950–1960s.

7.2.6.4 Regeneration Patterns and Processes

In general, seedlings/saplings of *L. chinense* are only found in disturbed micro-sites with a high light intensity, including canopy gaps, forest margins, roadsides in each of the three forest types, and abandoned fields near villages (Table 7.11; Fig. 7.36b). A few seedlings/saplings of other deciduous canopy tree species, such as *Betula luminifera*, *Castanea sequinii*, and *Emmenopterys henryi,* are found. Seedlings/saplings of evergreen canopy tree species such as *Lindera megaphylla* and *Lithocarpus henryi* are rare. Recruitment is dominated by *Liriodendron.* Very few seedlings/saplings of *Liriodendron* are found in the mountain slope secondary forests, which have been free of human disturbances since clear-cutting in the 1950–1960s (Tang et al. 2013b).

Table 7.10 Floristic composition of woody species (with RBA ≥ 0.5 %) for the three forest types containing *Liriodendron chinense*

Forest type		SVF	VFF	SSF
Shannon-Weiner diversity H′ (±S. E.)		1.6±0.6	2.1±0.2	2.8±0.1
Species	Life form	RBA (%)	RBA (%)	RBA (%)
Liriodendron chinense	DT	**71.7**	**48.2**	**36.6**
Castanea seguinii	DT	9.4	2.4	
Toona sinensis	DT	3.1	**10.7**	
Daphniphyllum macropodum	EST	2.4	0.5	
Liquidambar acalycina	DT	2.3		
Zelkova schneideriana	DT	2	2.4	
Populus davidiana	DT	1.3	3.2	
Morus cathayana	DT	1.3	1.1	
Carpinus pubescens	DT	0.7		
Sassafras tzumu	DT	0.5		**9.3**
Kalopanax septemlobus	DT	0.5		
Emmenopterys henryi	DT		**10.1**	
Cornus controversa	DT		3.8	3.4
Cinnamomum bodinieri	ET		2.6	
Lindera megaphylla	ET		1.9	
Liquidambar formosana	DT		1.5	
Dendrobenthamia japonica var. *chinensis*	DST		1.5	
Ginkgo biloba	CT		1.2	
Paulownia fortunei	DT		1.0	
Pinus fenzeliana	CT		0.9	
Cyclocarya paliurus	DT		0.6	
Sycopsis sinensis	DT		0.5	
Davidia involucrata	DT			**27.7**
Juglans sigillata	DT			4.4
Idesia polycarpa	DT			4
Acer franchetii	DT			3.2
Sorbus caloneura	DST			2.2
Cyclobalanopsis multinervis	ET			1.9
Tilia tuan	DT			1.7
Lithocarpus henryi	ET			1.7
Cinnamomum wilsonii	ET			1
Aphananthe aspera	DT			0.9
Laurocerasus zippeliana	DT			0.7
Cyclobalanopsis oxyodon	ET			0.6
Magnolia sprengeri	DT			0.5

Data source: Tang et al. (2013b)
Dominant species are indicated by boldface
SVF secondary forests near villages, *VFF* village fengshui forest, *SSF* mountain slope secondary forests far from villages, *RBA* relative basal area, *ET* evergreen broad-leaved tree, *DT* deciduous broad-leaved tree, *CT* coniferous tree, *EST* evergreen broad-leaved small tree, *ES* evergreen broad-leaved shrub, *DST* deciduous broad-leaved small tree

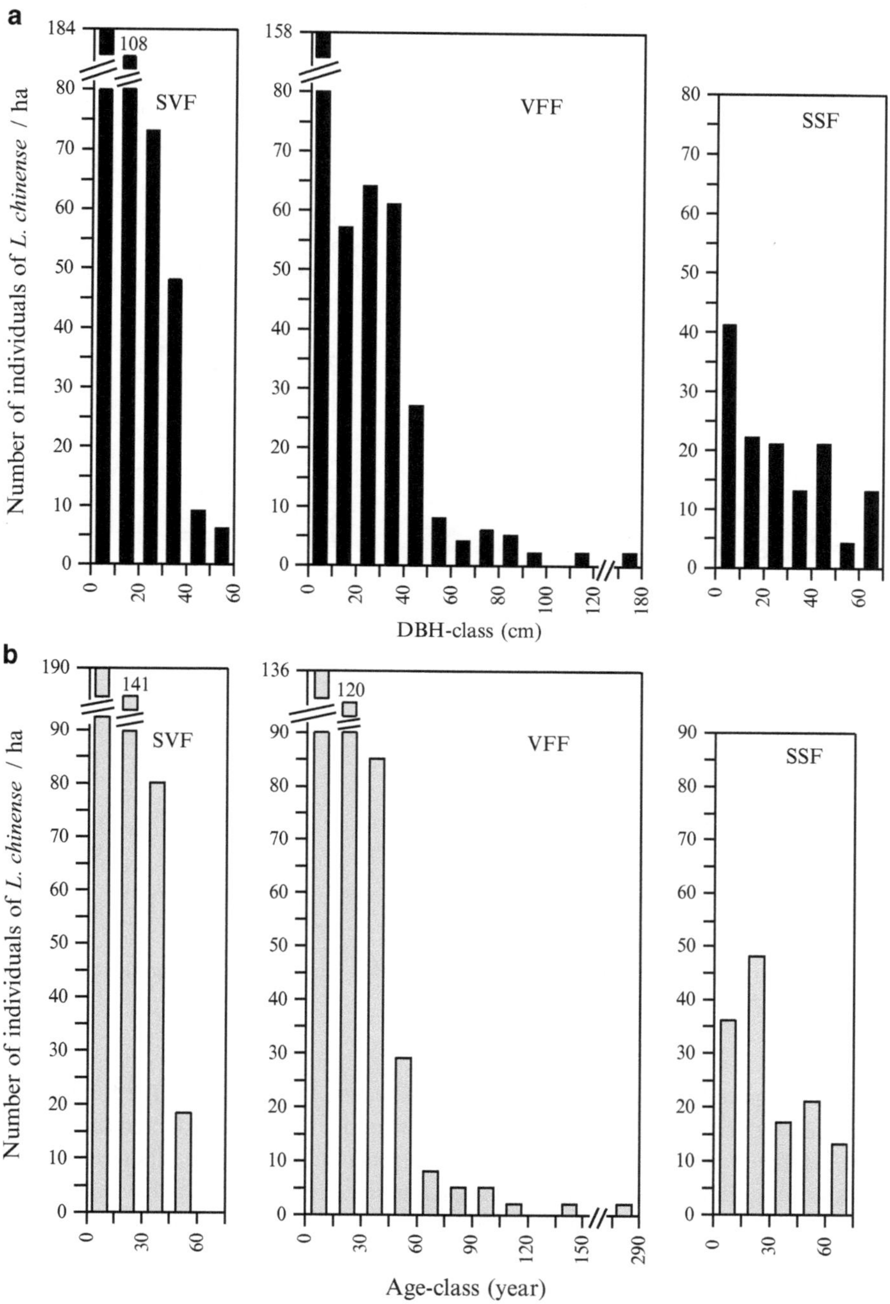

Fig. 7.40 DBH-class (**a**) and age-class (**b**) frequency distribution of *Liriodendron chinense* (height ≥1.3 m). *SVF* secondary forests near villages, *VFF* village fengshui forests, *SSF* mountain slope secondary forests (Data for age-class are from Tang et al. (2013b))

Table 7.11 Number of seedlings/saplings of *Liriodendron chinense* and other woody species under given conditions: beneath the canopy, in the canopy gap, at the forest margins, on roadsides, and in abandoned fields

Forest type	Habitat feature	Condition	GS (m^2) mean ± SD (maximum)	Number of seedlings & saplings of *L. chinense*	Number of seedlings & saplings of other deciduous trees and shrubs	Number of seedlings & saplings of evergreen trees and shrubs
SVF	Valleys and gentle slopes near villages	Closed canopy		0	0	0
		Canopy gaps	192 ± 103 (377)	39	25	0
		Forest margins		41	13	1
		Roadsides		70	19	3
		Abandoned fields		79	65	0
Total				229	122	4
VFF	Valleys and gentle slopes near villages (fengshui treatment)	Closed canopy		0	0	3
		Canopy gaps	120 ± 49 (202)	31	1	0
		Forest margins		84	12	2
		Roadsides		96	13	0
Total				211	26	5
SSF	Moderately steep mountain slopes far from villages	Closed canopy		0	0	1
		Canopy gaps	64 ± 16 (80)	0	0	0
		Forest margins		5	0	0
Total				5	0	1

Data source: Tang et al. (2013b)
SVF secondary forests near villages, *VFF* village fengshui forests, *SSF* mountain slope secondary forests, *GS* gap size, *SD* standard deviation

As compared with other associated tree species, *L. chinense* in the forests near villages appears to have been competitively superior and better adapted to the present, relatively open environments. The occurrence of the Tertiary relict species and its sprouting in the mountain slope forests indicate that the forests have developed after clear-cutting of primeval stands. The primeval forests that once may have prevailed in those areas would have been maintained by natural disturbances. Though true primeval and intact forests are now practically nonexistent in the study sites, those tree species in the village fengshui forests and mountain slope secondary forests are relicts of forest communities that have existed over millions of years. The village fengshui forests have undoubtedly provided seed sources to their surroundings, where secondary forests near villages have regenerated naturally after the abandonment of agricultural fields. Although once much more extensive than today, *Liriodendron* communities stand in contrast with the once regionally extensive forests of evergreen broad-leaved *Castanopsis* spp., *Lithocarpus* spp., and *Cyclobalanopsis* spp., which occupy more stable, mesic sites currently limited to remote areas or nature reserves (Huang et al. 1988). The regeneration patterns and processes show *L. chinense* to be a shade-intolerant pioneer species that has fast growth but is long-lived, dominating the canopy and emergent layers and not being replaced by other species during forest development. While the Tertiary relict *L. chinense* appears in moist sites and early successional stands, it is also co-dominant with other Tertiary relict deciduous broad-leaved species in old secondary and mature forests. A noteworthy parallel is seen in its close relative, *L. tulipifera* of eastern North America, which is usually confined to perennially moist but infrequently flooded sites in ravines and stream heads protected from fire, is shade-intolerant, requires light gaps for regeneration, grows in early-successional stands, but is also long-lived and is often a canopy dominant in old-growth mixed mesophytic forests (Olson 1969; Beck and Della-Bianca 1981; Nowacki and Abrams 1994; Lafon 2004).

7.2.6.5 Associated Village Fengshui Forest

Local villagers consider the fengshui forests to be a natural shelter around the village, creating a physical barrier and providing protection against evil spirits (Zhong and Boris 2007). The sustainable use of forest resources by the villagers have created heterogeneous forest margins, roadsides, light gaps, and associated regeneration niches favorable to *L. chinense*, mimicking and supplementing infrequent natural disturbances. This rare species and its peculiar forest communities near villages have survived through symbiotic relationship with the human population, while on the remote mountain slopes, where there are no humans, the species has depended on natural disturbances to provide conditions for seedlings/saplings establishment (Tang et al. 2013b).

7.2.7 Tertiary Relict Trees and Their Habitats

We have discussed the vegetation and population structures and species diversity of a number of important forests in SW China dominated by Tertiary relict tree species. China harbors many Tertiary relict species, many of which are endemic to China. A number of these are at the same time representatives of the remarkable long-distance disjunct floristic distribution pattern of genera with closely related species in East Asia and North America. Without attempting to be complete,

Tang et al. (2014) listed 40 Tertiary relict tree species found in southern China, including their life forms and habitats. They comprise both coniferous and broad-leaved species, and evergreen as well as deciduous species. 52.5 % of these Tertiary relict tree species are conifers as against 47.5 % broad-leaved. About 86 % of the conifer relict tree species, and about 21 % of broad-leaved relict tree species, are evergreen. Thus, among the Tertiary relict tree species, evergreen coniferous species and deciduous broad-leaved species form by far the largest groups, whereas deciduous coniferous and evergreen broad-leaved species are relatively few.

All the Tertiary relict tree species now occur exclusively in a limited number of rather restricted areas in China, mainly central (Yangtze River valley) and southern China, and partly also in adjacent parts of the Himalayas, Myanmar, Laos and northern Vietnam. However, some relict coniferous tree species are found in the cold temperate forests of China. These regions generally have a rugged topography, and have provided long-term suitable habitats, enabling the survival of the relict-endemics (López-Pujol et al. 2011a). Their long-term survival obviously was possible because most of this region was never covered by ice-sheets during the Quaternary glacial period.

Many Tertiary relict tree species, including those discussed in this chapter, occur in particular, unstable habitats within the warm, humid transitional zone between the evergreen and the deciduous broad-leaved forests of China, with some admixture of coniferous species. The spatial distribution patterns of those species result from the interactions among natural disturbances, life history traits and stress-tolerance. As mentioned before, marginal habitats formed under conditions of disturbance support distinctive flora and rare species because of differing competitive abilities and tolerance limits among species (Wisheu and Keddy 1992), and some of these species appear to be of the stress-tolerant type as classified by Grime (1974). Probably, the relict tree species persisted in the unstable habitats because erosive forces kept the habitats relatively open. As a result, forest stands are not so dense on such sites, and thus light conditions are less limiting (Tang et al. 2014). Open, well-lighted stands may be particularly important for the coniferous species, since their photosynthetic capacity is considerably lower than that of broad-leaved species (Larcher 1976), and as a result they are more likely to be outcompeted in dense mixed forest stands. Furthermore, screes and other unstable substrates, as well as thin soils between rock outcrops, more frequently undergo short episodes of drought in the top soil which can cause embolism in the xylem vessels. Coniferous species possess a particular type of xylem vessel that makes them less susceptible to xylem embolism during drought conditions (Choat et al. 2012). Also many of the deciduous broad-leaved species are more or less confined to screes and unstable slopes and outcrops, or to stream banks. Apart from more favorable light conditions, such sites generally have a rapid turn-over and replenishment of nutrient resources, which is beneficial for deciduous species, as they have a faster nutrient turn-over than evergreen species, and thus a larger nutrient loss rate (Tang et al. 2014). And since moving rocks and gravel slides on unstable screes and steep slopes frequently seriously damage the vegetation, the strong sprouting capacity of the relict broad-leaved tree species is an important advantage for regeneration and persistence in those habitats.

The Tertiary relict species provide a unique opportunity to understand past and recent biogeographical and evolutionary processes because of their taxonomical isolation, rarity and phylogenetic traits. Today they are highly vulnerable and even in danger of extinction, because their populations in the wild have been greatly threatened by overexploitation of the land. Their conservation is of great concern for China and for the world.

References

101GB (101 Ginkgo Biloba) (2014) Ginkgo Biloba: morphology. http://ginkgo.dm.pagesperso-orange.fr/GINKGO/GbMorphology.htm. Retrieved 18 Sept 2014

Akhmetiev MA (1973) Sikhote-Alin Miocene flora (Botchi River). Geol Inst Tr Akad Nauk USSR 247:1–124

Axelrod DI, Al-Shehbaz I, Raven PH (1998) History of the modern flora of China. In: Zhang AL, Wu SG (eds) Floristic characteristics and diversity of East Asia plants. China Higher Education Press, Beijing, pp 43–55

Beck DE, Della-Bianca L (1981) Yellow-poplar: characteristics and management. Agriculture Handbook Number 583. United States Department of Agriculture, Forest Service, Washington, DC

Bennike O (1990) The Kap København formation: stratigraphy and paleobotany of a Plio-Pleistocene sequence in Peary Land, North Greenland. Meddelelser Grønland 23:1–85

Berry EW (1941) *Liriodendron* in the Miocene of America and eastern Asia. Torreya 41:82–84

Bond WJ, Midgley JJ (2001) Ecology of sprouting in woody plants: the persistence niche. Trends Ecol Evol 16:45–51

Braun EL (1950) Deciduous forests of eastern North America. Hafner, New York

Bůžek Č (1971) Tertiary flora from the northern part of the Pětipy area (North-Bohemian Basin). Edice Rozpravy Ustř. Ust. Geologického 36

Chamber KL (1993) *Thuja*. In: Flora of North America Editorial Committee (eds) Flora of North America North of Mexico, vol 2. Oxford University Press, New York, pp 410–411

Chamberlain CJ (1966) Gymnosperms: structure and evolution. Dover, New York

Chaney RW (1948) The bearing of the living *Metasequoia* on problems of Tertiary paleobotany. Proc Natl Acad Sci U S A 34(11):503–515

Chaney RW (1951) A revision of fossil *Sequoia* and *Taxodium* in western North America based on the recent discovery of *Metasequoia*. Trans Am Philos Soc II 40(3):171–263

Chen L-Z (ed) (1993) Biodiversity of China. Science Press, Beijing

Chen X-Y, Li Y-Y, Wu T-Y, Zhang X, Lu H-P (2003) Size-class differences in genetic structure of *Metasequoia glyptostroboides* Hu et Cheng (Taxodiaceae) plantations in Shanghai. Silvae Genetica 52:107–109

Chen L, Dong H, Peng H (2013) Diversity and distribution of higher plants in Yunnan, China. Biodivers Sci 21(3):359–363

Cheng WC, Chu ZX (1949) Forest status of *Metasequoia glyptostroboides* in Lichuan, Hubei. Science (China) 31(3):73–80

Choat B, Jansen S, Cochard H, Delzon S, Bhaskar R et al (2012) Global convergence in the vulnerability of forests to drought. Nature 491(7426):752–755

Christopher IE (2015) The gymnosperm database. *Thuja plicata* Donn ex D. Don. http://www.conifers.org/cu/Thuja.php. Retrieved 15 Jan 2015
Chu K-L, Cooper WS (1950) An ecological reconnaissance in the native home of *Metasequoia glyptostroboides*. Ecology 31(2):260–278
CPGC (Committee of Physical Geography of China) (1980) Physical geography of China. Science Press, Beijing
Crane P (2013) Ginkgo. Yale University Press, New Haven
Critchfield WB (1970) Shoot growth and heterophylly in *Ginkgo biloba*. Bot Gaz 131:150–162
Daniels LD, Marshall PL, Carter RE, Klinka K (1995) Age structure of *Thuja plicata* in the tree layer of old-growth stands near Vancouver, British Columbia. Northwest Sci 69(3):175–183
de Blois S, Bouchard A (1995) Dynamics of *Thuja occidentalis* in an agricultural landscape of southern Quebec. J Vegetation Sci 6(4):531–542
Del Tredici P, Ling H, Yang G (1992) The *Ginkgos* of Tian Mu Shan. Conserv Biol 6:202–209
Doweld AB (1998) Carpology, seed anatomy and taxonomic relationships of *Tetracentron* and *Trochodendron*. Ann Bot 82:413–443
EBAFC (Editorial Board of Annals of Fenggang County) (1994) Guizhou Province Annals of Fenggang County. Guizhou People's Press, Guiyang
EBANC (Editorial Board of Annals of Nanchuan County, Sichuan Province) (1991) Annals of Nanchuan County. Sichuan People's Press, Chengdu
EBAWGMAC (Editorial Board of Annals of Wuchuan Gelao and Miao Autonomous County) (2001) Guizhou Province. Annals of Wuchuan Gelao and Miao Autonomous County. Guizhou People's Press, Guiyang
Editorial Committee of Flora Reipublicae Popularis Sinicae (1959–2004) Flora Reipublicae Popularis Sinicae. Science Press, Beijing
eFloras (2014) Flora of China. Published on the Internet. Missouri Botanical Garden/Harvard University Herbaria, St. Louis/Cambridge. http://www.efloras.org. Retrieved 22 Aug 2014
Farjon A (2001) World checklist and bibliography of conifers. Royal Botanical Garden, Kew
Farjon A (2008) A natural history of conifers. Timber Press, Portland
Farjon A (2013) *Thuja occidentalis*. In: IUCN 2013. IUCN red list of threatened species. Version 2013.2. www.iucnredlist.org. Retrieved 25 Feb 2014
Farjon A, Page CN [compilers] (1999) Conifers. Status survey and conservation action plan. IUCN-SSC Conifer Specialist Group. IUCN, Gland
Florin R (1963) The distribution of conifer and taxad genera in time and space. Acta Horticulturae Bergiana 20(4):1–312
Forbe FB, Hemsley WB (1886) An enumeration of all the plants known from China Proper, Formosa, Hainan, Corea, the Luchu Archipelago, and the Island of Hongkong, together with their distribution and synonymy–part 1. J Linnean Soc Lond Bot 23:1–80
Franchet AR (1899) Plantarum sinensium ecloge tertia. J Bot (Morot) 13:253–266
Franklin JF, Debell DS (1988) Thirty-six years of tree population change in an old-growth *Pseudotsuga-Tsuga* forest. Can J Forest Res 8:633–639
Fu DZ, Bartholomew B (2001) *Tetracentraceae*. In: Wu CY, Raven PH (eds) Flora of China, vol 6. Science Press and Missouri Botanical Garden Press, Beijing/St. Louis, p 125
Fu L-K, Jin J-M (eds) (1992) China plant red data book–rare and endangered plants, vol 1. Science Press, Beijing
Futuyma DJ (2005) Evolution. Sinauer Associates, Sunderland
Gardener WH (1972) E. H. Wilson's first trip to China. Arnoldia 32(3):103–115; introduction by Gordon P. DeWolf, Jr
Givulescu R, Ghiurca V (1969) Flora Pliocene de la Chiuzbaia (Maramures) cu un studiu geologic introductiv. Memorii Institutul Geologic Comitetul de Stat al Geologiei 10:7–81
Gressitt JL (1953) The California Academy-Lingnan dawn redwood expedition. Proc Calif Acad Sci Fourth Series 28:25–58
Grime JP (1974) Vegetation classification by reference to strategies. Nature 250:26–31

Grímsson F, Denk T, Zetter R (2008) Pollen, fruits, and leaves of *Tetracentron* (Trochodendraceae) from the Cainozoic of Iceland and western North America and their palaeobiogeographic implications. Grana 47(1):1–14

Grubb PJ (1977) The maintenance of species-richness in plant communities: the importance of the regeneration niche. Biol Rev 52:107–145

Guan CY (2002) Investigation on geomantic forest in ancient China. Agric Archaeology 3:239–243

Hannah PR (2004) Stand structures and height growth patterns in northern white-cedar stands on wet sites in Vermont. Northern J Appl Forestry 21:173–179

Hao R-M, He S-A, Tang S-J, Wu S-P (1995) Geographical distribution of *Liriodendron chinense* in China and its significance. J Plant Res Environ 4:1–6

Harper RM (1914) Geography and vegetation of North Florida. In: Sellards EH (ed) Florida geological survey sixth annual report. E. O. Painter Printing Company, DeLand, pp 163–451

HDOSU (2014) Horticulture Department of Oregon State University Web site http://oregonstate.edu/dept/ldplants/megl.htm. Retrieved 16 Apr 2014

He SA, Hao RM (1999) Study on the natural population dynamics and the endangering habitat of *Liriodendron chinense* in China. Acta Phytoecologica Sin 23:87–95

Hobohm C (ed) (2014) Endemism in vascular plants, vol 9, Plants and vegetation. Springer, Dordrecht, p 348

Hofmeyer PV, Kenefic LS, Seymour RS (2009) Northern white-cedar ecology and silviculture in the Northeastern United States and Southeastern Canada: a synthesis of knowledge. Northern J Appl Forestry 26:21–27

Hori S, Hori T (1997) A culture history of *Ginkgo biloba* in Japan and the generic name Ginkgo. In: Hori T, Ridge RW, Tulecke W, Del Tredici P, Trémouillaux-Guille J, Tobe H (eds) *Ginkgo biloba*–a global treasure: from biology to medicine. Springer, Tokyo, pp 385–411

Hu HH (1946) Notes on a Palaeogene species of Metasequoia in China. Bull Geol Soc China 26:105–107

Hu HH (1948) How Metasequoia, the "living fossil" was discovered in China. J New York Bot Gard 49:201–202 (Excerpted in Arnoldia (1999) 59:4–7)

Hu HH, Cheng WC (1948) On the new family Metasequoiaceae and on *Metasequoia glyptostroboides*, a living species of the genus Metasequoia found in Szechuan and Hupeh. Bull Fan Mem Inst Biol New Ser 1(2):153–161

Huang W-L, Tu Y-L, Yang L (eds) (1988) The vegetation of Guizhou. Guizhong People's Press, Guiyang

Huzioka K, Uemura K (1973) The late Miocene Miyata flora of Akita Prefecture, northeast Honshu, Japan. Bull Natl Sci Museum Tokyo 18:661–738

Itihara M, Yoshikawa S, Inoue K, Hayashi T, Tateishi M, Nakajima K (1975) Stratigraphy of the Plio-Pleistocene Osaka Group in Sennan-Senpoku area, south of Osaka, Japan–a standard stratigraphy of the Osaka Group. J Geosci Osaka City University 19:1–29

IUCN (2006) Red list of threatened species. www.iucnredlist.org. Retrieved 25 Mar 2014

IUCN (2013) IUCN red list of threatened species, version 2013.2. www.iucnredlist.org. Retrieved 13 Feb 2014

Jacobson AL (2014) Sichuan arborvitae or *Thuja*, *Thuja sutchuenensis* Franch. Cupressaceae; Cypress Family. http://www.arthurleej.com. Retrieved 12 Oct 2014

Jetz W, Rahbek C, Colwell RK (2004) The coincidence of rarity and richness and the potential signature of history in centres of endemism. Ecol Lett 7:1180–1191

Jiang Y (2011) E. H. Wilson's Sichuan. Global Times Published in http://www.globaltimes.cn/content/657405.shtml. Retrieved 10/7/2014. Retrieved 20 May 2011

Jiang M, Jin Y, Zhang Q (1990) A preliminary study on *Ginkgo biloba* in the Dahongshan region, Hubei. J Wuhan Bot Res 8:191–193

Jiang M, Wu J, Ge J (2000) Studies on flora and ecological features of endangered plant communities in Songziyuan, the southern slope of Mt. Shengnongjia. J Wuhan Bot Res 18(5):368–374

Kelly PE, Cook ER, Larson DW (1992) Constrained growth, cambial mortality, and dendrochronology of ancient *Thuja occidentalis* on cliffs of the Niagara Escarpment: an eastern version of bristlecone pine? Int J Plant Sci 153:117–127

Kitamura A, Kimoto K (2004) Reconstruction of the southern channel of the Japan Sea at 3.9–1.0 Ma. Quaternary Res 43:417–434 (Tokyo)
Kokawa S (1965) Fossil endocarp of *Davidia* in Japan. J Biol Osaka City Univ 16:45–51, 4 pls
Kong WS, Watts D (1993) The plant geography of Korea with an emphasis on the alpine zones. Kluwer, Dordrecht
Kruckeberg AR (2002) Geology and plant life: the effects of landforms and rock type on plants. University of Washington Press, Seattle
Kuser JE, Sheely DL, Hendricks DR (1997) Genetic variation in two ex situ collections of the rare *Metasequoia glyptostroboides* (Cupressaceae). Silvae Genetica 46:258–264
Kwant C (2014) The *Ginkgo* Pages web site. http://kwanten.home.xs4all.nl/history.htm. Retrieved 24 Apr 2014
Lafon CW (2004) Stand dynamics of a yellow-poplar (*Liriodendron tulipifera* L.) forest in the Appalachian Mountains, Virginia, USA. Dendrochronologia 22:43–52
Larcher W (1976) Ecology of plants. Ulmer, Stuttgart
Larouche C, Kenefic LS, Ruel J-C (2010) Northern white-cedar regeneration dynamics on the Penobscot Experimental Forest in Maine: 40-year results. Northern J Appl Forestry 27:5–12
Larson DW, Kelly PE (1991) The extent of old-growth *Thuja occidentalis* on cliffs of the Niagara Escarpment. Can J Bot 69:1628–1636
Leigh A, Zwieniecki MA, Rockwell FE, Boyce CK, Nicotra AB, Holbrook NM (2011) Structural and hydraulic correlates of heterophylly in *Ginkgo biloba*. New Phytol 189:459–470
Leng Q, Fan SH, Li W, Yang H, Lai XL, Cheng DD, Ge JW, Shi GL, Jiang Q, Liu XQ (2007) Database of native *Metasequoia glyptostroboides* trees in China based on new Census surveys and expeditions. Bull Peabody Museum Nat Hist 48(2):185–233
Leonard N (2013) Refugia. http://www.eoearth.org/view/article/155685. Retrieved 25 Mar 2014
LePage BA (2003) A new species of *Thuja* (Cupressaceae) from the Late Cretaceous of Alaska: implications of being evergreen in a polar environment. Am J Bot 90:167–174
LePage BA, Yang H, Matsumoto M (2005) The evolution and biogeographic history of *Metasequoia*. In: LePage BA, Williams CJ, Yang H (eds) The geobiology and ecology of *Metasequoia*. Springer, Dordrecht, pp 3–114
Li YY (2006) *Phylogenetic position and recovery assessment of Metasequoia glyptostroboides*. Dissertation, East China Normal University, Shanghai
Li JH, Ban JD (1989) The water fir communities endemic to China. J Henan Norm Univ (Nat Sci Version) 4:49–55
Li J-H, Xiang Q-P (2005) Phylogeny and biogeography of *Thuja* L. (Cupressaceae), an eastern Asian and North American disjunct genus. J Integr Plant Biol 47(6):651–659
Li YY, Chen XY, Zhang X, Wu TY, Lu HP, Cai YW (2005) Genetic differences between wild and artificial populations of *Metasequoia glyptostroboides*: implications for species recovery. Conserv Biol 19:224–231
Li HF, Chaw SM, Du CM, Ren Y (2011) Vessel elements present in the secondary xylem of *Trochodendron* and *Tetracentron* (Trochodendraceae). Flora 206:95–600
Ling H, Zhang D (2004) Analysis for the origin of *Ginkgo* population in Tianmu Mountain. Sci Silvae Sinicae 40:28–31
Liu KB (1988) Quaternary history of the temperate forests of China. Quaternary Sci Rev 7:1–20
Liu H, Shen X, Ceng Y (1991) Genetic variations in morphological and growth characteristics of Chinese, American Tulip-trees and their hybrid. J Zhejinag Forestry Sci Technol 11(5):18–22
Liu ZY, Yang MH, Yi SR, Tang ZF, Li SL (2000) *Thuja sutchuenensis* is not extinct. Plant Mag 3:8–9
Liu J, Shi S, Chang E, Yang W, Jiang Z (2013) Genetic diversity of the critically endangered *Thuja sutchuenensis* revealed by ISSR markers and the implications for conservation. Int J Mol Sci 14:14860–14871
Lomolino MV, Riddle BR, Brown JH (2006) Biogeography. Sinauer, Sunderland
López-Pujol J, Zhang F-M, Sun H-Q, Ying T-S, Ge S (2011a) Centres of plant endemism in China: places for survival or for speciation? J Biogeogr 38:1267–1280

López-Pujol J, Zhang F-M, Sun H-Q, Ying T-S, Ge S (2011b) Mountains of Southern China as "plant museums" and "plant cradles": evolutionary and conservation insights. Mountain Res Dev 31(3):261–269

Luo J (1990) The rainstorm feature in the area of the Daba Mountains and its effect to the disastrous landforms of hillslopes. J Southwest Norm Univ 15(2):266–273

Mabberley DJ (1987) The plant-book. Cambridge University Press, Cambridge

Mädler K (1939) Die Pliozäne Flora von Frankfurt am Main. Abhandlungen der Senchenbergischen Naturforschenden Gesellschaft 446:1–202

Manchester SR (2002) Leaves and fruits of *Davidia* (Cornales) from the Paleocene of North America. Syst Bot 27(2):368–382, + cover photo

Manchester SR (2014a) *Davidia*-the dove tree and its fossil record. http://www.palaeobotany.org/page/living-fossils/davidia-involucrata. Retrieved 18 Apr 2014

Manchester SR (2014b) *Davidia involucrata*. http://www.palaeobotany.org/page/living-fossils/davidia-involucrata. Retrieved 18 Apr 2014

Manchester SR, Chen I (2006) *Tetracentron* fruits from the Miocene of Western North America. Int J Plant Sci 167:601–605

Manchester SR, McIntosh WC (2007) Late Eocene silicified fruits and seeds from the John Day Formation near Post, Oregon. PaleoBios 27(1):7–17

Manchester SR, Crane PR, Golovneva L (1999) An extinct genus with affinities to extant *Davidia and Camptotheca* (Cornales) from the Paleocene of North America and Eastern Asia. Int J Plant Sci 160:188–207

Matthes-Sears U, Larson DW (1999) Limitations to seedling growth and survival by the quantity and quality of rooting space: implications for the establishment of *Thuja occidentalis* on cliff faces. Int J Plant Sci 160:122–128

Matthes-Sears U, Nash CH, Larson DW (1995) Constrained growth of trees in a hostile environment: the role of water and nutrient availability for *Thuja occidentalis* on cliff faces. Int J Plant Sci 156:311–319

McIver EE, Basinger JF (1989) The morphology and relationships of *Thuja polaris* sp. *nov* (Cupressaceae) from the Early Tertiary, Ellesmere Island, Arctic Canada. Can J Bot 67:1903–1915

Médail F, Diadema K (2009) Glacial refugia influence plant diversity patterns in the Mediterranean Basin. J Biogeogr 36:1333–1345

Miki S (1938) On the change of flora of Japan since the upper Pliocene and the floral composition at present. Jpn J Bot 9:213–251, plates 3–4

Miki S (1941) On the change of flora in eastern Asia since Tertiary period (I). The clay or lignite beds flora in Japan with special reference to the Pinus trifolia beds in the central Hondo. Jpn J Bot 11:237–303

Miki S (1950) Taxodiaceae of Japan, with special reference to its remains. J Inst Polytechnics Osaka City Univ 1:63–77

Miki S (1956) Endocarp remains of Aiangiaceae, Cornaceae and Nyssaceae in Japan. J Inst Polytechnics Osaka City Univ 7:247

Miki S (1958) Gymnosperms in Japan, with special reference to the remains. J Ins Polytech Osaka City Univ D9:125–152

Minore D (1983) Western redcedar: a literature review. Pacific Northwest Forest and Range Experiment Station. USDA Forest Service General Technical Report PNW-150. Portland

Minore D (1990) Western redcedar (*Thuja plicata* Donn ex D. Don). In: Burns RM, Honkala BH (eds) Silvics of North America, vol 1, Conifers. USDA Forest Service Agriculture Handbook 654, Washington, DC, pp 590–600

Momohara A (1992) Late Pliocene plant biostratigraphy of the lowermost part of the Osaka Group, Southwest Japan, with reference to extinction of plants. Quaternary Res (Tokyo) 31:77–89

Momohara A (1994) Floral and paleoenvironmental history from the late Pliocene to middle Pleistocene in and around central Japan. Palaeogeogr Palaeoclimatol Palaeoecol 108:281–293

Momohara A (2005) Paleoecology and history of *Metasequoia* in Japan, with references to its extinction and survival in East Asia. In: LePage BA, Williams CJ, Yang H (eds) The geobiology and ecology of *Metasequoia*. Springer, Dordrecht, pp 115–136

Myers N, Mittermeier RA, Mittermeier CG, da Fonseca GAB, Kent J (2000) Biodiversity hotspots for conservation priorities. Nature 403:853–858

Nakamura Y, Krestov PV (2005) Coniferous forests of the temperate zone of Asia. In: Andersson F (ed) Coniferous forests. Elsevier, Amsterdam, pp 163–220

Nakane H, Takumi A, Tadaki Y (1988) Ecological succession around Mt. Yake, Kamikohchi. II Secondary forests and mature forests. J Faculty Sci Shinshu Univ 23:33–46

Nowacki GJ, Abrams MD (1994) Forest composition, structure, and disturbance history of the Alan Seeger Natural Area, Huntington County, Pennsylvania. Bull Torrey Bot Club 121:277–291

Ohsawa M (1987a) Vegetation zones in the Bhutan Himalaya. In: Ohsawa M (ed) Life zone ecology of the Bhutan Himalaya. Chiba University, Chiba, pp 1–71

Ohsawa M (1987b) Habitat differentiation and ecological niche in vegetation. Physiol Ecol Japan 24 (Special No.):15–27

Ohsawa M (1991) Montane evergreen broad-leaved forests of the Bhutan Himalaya. In: Ohsawa M (ed) Life zone ecology of the Bhutan Himalaya. Chiba University, Chiba

Olson DF (1969) Silvical characteristics of yellow-poplar (*Liriodendron tulipifera* L.). USDA Forest Service, Southeastern Forest Experiment Station. Research Paper SE-48. Asheville

Orme CDL, Davies RG, Burgess M, Eigenbrod F, Pickup N, Olson VA, Webster AJ, Ding T-S, Rasmussen PC, Ridgely RS, Stattersfield AJ, Bennett PM, Blackburn TM, Gaston KJ, Owens IPF (2005) Global hotspots of species richness are not congruent with endemism or threat. Nature 436:1016–1019

Ozaki K (1984) Two new fossil species of *Fortunearia and Davidia* from the upper Motojuku (Kabutoiwa) Formation in Central Japan. Memoirs of the Institute of Field Education, Yokohama National University 2:1–8

Parks CR, Wendel JF (1990) Molecular divergence between Asian and Northern American species of *Liriodendron* (Magnoliaceae) with implications for interpretation of fossil floras. Am J Bot 77:1243–1256

Peng D, Wang X-Q (2008) Reticulate evolution in *Thuja* inferred from multiple gene sequences: implications for the study of biogeographical disjunction between eastern Asia and North America. Mol Phylogenet Evol 47:1190–1202

Phillips R (1978) Trees of North America and Europe. Random House New York, New York

Pigg KB, Wehr WC, Ickert-Bond SM (2001) *Trochodendron* and *Nordenskioldia* (Trochodendraceae) from the Middle Eocene of Washington State, U.S.A. Int J Plant Sci 162:1187–1198

Pigg KB, Dillhoff RM, DeVore ML, Wehr WC (2007) New diversity among Trochodendraceae from the Early/Middle Eocene Okanogan Highlands of British Columbia, Canada, and Northeastern Washington State, United States. Int J Plant Sci 168:521–532

Pulle AA (1949) Over de *Ginkgo* alias Ginkyo. Jaarboek Nederlandsche Dendrologische Vereeniging 15:25–35

Qian H, Wang S, He J-S, Zhang J, Wang L, Wang X, Guo K (2006) Phytogeographical analysis of seed plant genera in China. Ann Bot 98:1073–1084

Qin HN (2004) Tetracentraceae. In: Wu CY (ed) Flora reipublicae popularis sinicae, vol 1. Science Press, Beijing, pp 767–769

Qin H, Phengklai C (2013) *Davidia involucrata*. Flora of China. Missouri Botanical Garden/Harvard University Herbaria, St. Louis/Cambridge. Retrieved 28 Apr 2013

Qiu Y-X, Fu C-X, Comes HP (2011) Plant molecular phylogeography in China and adjacent regions: tracing the genetic imprints of Quaternary climate and environmental change in the world's most diverse temperate flora. Mol Phylogenet Evol 59:225–244

RBGE (Royal Botanic Gardens Edinburgh) (2014) Himalayan geography. http://padme.rbge.org.uk/floraofnepal/?page=StaticRoot/Himalaya&edit=. Retrieved 5 Sept 2014

Ren Y, Chen L, Tian XH, Zhang XH, Lu AM (2007) Discovery of vessels in *Tetracentron* (Trochodendraceae) and its systematic significance. Plant Syst Evol 267:155–161

Royer DL, Hickey LJ, Wing SL (2003) Ecological conservatism in the "living fossil" *Ginkgo*. Paleobiology 29:84–104

Sargent CS (1903) *Liriodendron chinense*, Sarg. In: Sargent CS (ed) Trees and Shrubs: illustrations of new or little known ligneous plants/prepared chiefly from material at the Arnold Arboretum of Harvard University. Houghton Mifflin, Boston, pp 103–106

Sargent CS (ed) (1913) Plantae wilsonianae. An enumeration of the woody plants collected in western China for Arnold arboretum of Harvard University during the years 1907, 1908, and 1910/by E. H. Wilson. The University press, Cambridge

Sawada H, Kaji M, Igarashi Y, Omura K (1995) Study on the regeneration dynamics of natural forests in the Chichibu Mountains, Central Japan III. The Report of the University Forest of Faculty of Agriculture, University of Tokyo 93:21–40

Schlarbaum SE, Tsuchiyu T (1984) Cytotaxonomy and phylogeny in certain species of Taxodiaceae. Plant Syst Evol 147:29–54

Schweitzer HJ (1974) Die "Tertiaren" Koniferen Spitzbergens. Paleontogräphica 149B:1–89

Serbet R, Manchester SR, Aulenbach K, Braman D (2004) Nyssaceae among the dinosaurs: anatomically preserved fruits from the Upper Cretaceous Horseshoe Canyon Formation, Drumheller, Alberta, Canada. Botany 2004 Conference, Snowbird

Seward AC (1911) Links with the past in the plant world. Cambridge University Press, Cambridge

Shen L, Chen X-Y, Zhang X, Li Y-Y, Fu C-X, Qiu Y-X (2005) Genetic variation of *Ginkgo biloba* L. (Ginkgoaceae) based on cpDNA PCR-RF-LPs: inference of glacial refugia. Heredity 94:396–401

Shi YF, Ren BG, Wang JT, Derbyshire E (1986) Quaternary glaciation in China. Quaternary Sci Rev 5:503–507

Shi L, Zhang Y, Shi J (2009) Surface geohazards at the Qinling-Daba Mountains and impacts on Three Gorges Water Diversion Project. J Eng Geol 17:212–219

Smith AC (1945) A taxonomic review of *Trochodendron* and *Tetracentron*. J Arnold Arbor Harvard Univ 26:123–142

Stebbins GL, Major J (1965) Endemism and speciation in the California flora. Ecol Monogr 35:1–35

Suzuki M, Joshi L, Noshiro S (1991) *Tetracentron* wood from the Miocene of Noto Peninsula, central Japan, with a short revision of Homoxylic fossil woods. Bot Mag Tokyo 104:37–48

Taberlet P, Cheddadi R (2002) Quaternary refugia and persistence of biodiversity. Science 297:2009–2010

Tanai T (1967) Tertiary floral changes of Japan. Jubilee Publ. Commem., Professor Sasa, 60th birthday, Sapporo, pp 317–334

Tanai T, Huzioka K (1967) Climatic implications of Tertiary floras in Japan. In: Hatai (ed) Tertiary correlation and climatic changes in the Pacific. Sasaki Printing and Publishing Co., Sendai, pp 84–94

Tang CQ, Ohsawa M (1997) Zonal transition of evergreen, deciduous and coniferous forests along the altitudinal gradient on a humid subtropical mountain, Mt. Emei, Sichuan, China. Plant Ecol 133:63–78

Tang CQ, Ohsawa M (2002) Tertiary relict deciduous forests on a humid subtropical mountain, Mt. Emei, Sichuan, China. Folia Geobot 37:93–106

Tang CQ, Yang Y, Ohsawa M, Momohara A, Hara M, Cheng S, Fan S (2011) Population structure of relict *Metasequoia glyptostroboides* and its habitat fragmentation and degradation in south-central China. Biol Conserv 144:279–289

Tang CQ, Yang Y, Ohsawa M, Yi S-R, Momohara A, Su W-H, Wang H-C, Zhang Z-Y, Peng M-C, Wu Z-L (2012) Evidence for the persistence of wild *Ginkgo biloba* (Ginkgoaceae) populations in the Dalou Mountains, southwestern China. Am J Bot 99(8):1408–1414

Tang CQ, Peng M-C, He L-Y, Ohsawa M, Wang C-Y, Xie T-H, Li W-S, Li J-P, Zhang H-Y, Li Y, Yang X-M, Li G-S (2013a) Population persistence of a Tertiary relict tree *Tetracentron sinense* on the Ailao Mountains, Yunnan, China. J Plant Res 126(5):651–659

Tang CQ, Yang Y, Ohsawa M, Momohara A, Mu J, Robertson K (2013b) Survival of a Tertiary relict species: *Liriodendron chinense* (Magnoliaceae) in southern China, with special reference to village fengshui forests. Am J Bot 100(10):2112–2119

Tang CQ, Yang Y, Ohsawa M, Momohara A, Yi S-R, Robertson K, Song K, Zhang S, He L-Y (2015) Community structure and survival of Tertiary relict species *Thuja sutchuenensis* (Cupressaceae) in the subtropical Daba Mountains, southwestern China. PLoS One (accepted)

Tang CQ, Werger MJA, Ohsawa M, Yang Y (2014) Habitats of Tertiary relict trees in China. In: Hobohm C (ed) Endemism in vascular plants, vol 9, Plants and vegetation. Springer, Dordrecht, pp 289–308

Thompson JD (2005) Plant evolution in the Mediterranean. Oxford University Press, New York

Tralau H (1968) Evolutionary trends in the genus *Ginkgo*. Lethaia 1:63–101

Tsukagoshi M, Ono Y, Hashimoto T (1997) Fossil endocarp of *Davidia* from the early Pleistocene sediments of the Tokai Group in Gifu Prefecture, central Japan. Bull Osaka Museum Nat Hist 51:13–23

Tzedakis PC, Lawson IT, Frogley MR, Hewitt GM, Preece RC (2002) Buffered tree population changes in a Quaternary refugium: evolutionary implications. Science 297:2044–2047

Tzedakis PC, Hooghiemstra H, Paelike H (2006) The last 1.35 million years at Tenaghi Philippon: revised chronostratigraphy and long-term vegetation trends. Quaternary Sci Rev 25:3416–3430

Uemura K (1997) Cenozoic history of *Ginkgo* in East Asia. In: Hori T, Ridge RW, Tulecke W, Del Tredici P, Trémouillaux-Guiller J, Tobe H (eds) *Ginkgo biloba*–a global treasure from biology to medicine. Springer, Tokyo, pp 207–221

Wang CW (1961) The forests of China with a survey of grassland and desert vegetation, Maria Moors Cabot Foundation Publication. Ser. B. Harvard University, Cambridge

Wang H-S (1992) Phytogeography. Science Press, Beijing

Wang XQ, Ma LY, Guo BX, Fan SH, Tan JX (2005) Analysis on the changes of the original *Metasequoia glyptostroboides* population and its environment in Lichuan, Hubei from 1948 to 2003. Acta Ecologica Sin 25(5):972–977

Wang X-F, Guo QS, Hao JX, Ma C, Wan QX (2007) Study on the cone characteristics and seed output of a critically endangered plant, *Thuja sutchuenensis* in the world. Forest Res 20:673–677

Whetstone RD (2006) *Ginkgo biloba*. In: Flora of North America Editorial Committee (eds) Flora of North America, vol 2. Oxford University Press, New York/Oxford

Wikipedia (2014a) http://en.wikipedia.org/wiki/Davidiainvolucrata. Retrieved 18 Apr 2014

Wikipedia (2014b) http://en.wikipedia.org/wiki/ErnestHenry Wilson. Retrieved 4 Oct 2014

Wikipedia (2014c) Wikipedia hosted by Wikimedia Foundation. http://en.wikipedia.org/wiki/Liriodendron. Retrieved 23 Apr 2014

Williams CJ (2005) Ecological characteristics of *Metasequoia glyptostroboides*. In: LePage BA, Williams CJ, Yang H (eds) The geobiology and ecology of *Metasequoia*. Springer, Dordrecht, pp 285–304

Wilson EH (1905) The Chinese Tulip-Tree (*Liriodendron chinense*). Flora and Sylva 29:202–204

Wilson EH (1910) *Liriodendron chinense* China http://viacs.hul.harvard.edu/via/deliver/deepcontent?recordId=olvwork292744. Retrieved 24 Apr 2013

Wisheu IC, Keddy PA (1992) Competition and centrifugal organization of plant communities: theory and tests. J Veg Sci 3:147–156

Wolfe JA, Wehr WC (1987) Middle Eocene dicotyledonous plants from Republic, Northeastern Washington. U. S. Geol Surv Bull 1597:1–25

Wu ZY (ed) (1980) The vegetation of China. Science Press, Beijing

Xiang QP, Farjon A, Li ZY, Fu LK, Liu ZY (2002) *Thuja sutchuenensis*: a rediscovered species of the Cupressaceae. Bot J Linnean Soc 140:93

Yang H (1999) From fossil to molecules: the *Metasequoia* tale continues. Arnoldia 58(59):60–71

Yang H (2005) Biomolecules from living and fossil *Metasequoia*: biological and geological applications. In: LePage BA, Williams CJ, Yang H (eds) The geobiology and ecology of *Metasequoia*. Springer, Dordrecht, pp 253–279

Yang H, Jin JH (2000) Phytogeographic history and evolutionary stasis of *Metasequoia*: geological and genetic information contrasted. Acta Palaeontol Sin 39(Suppl):288–307

Yang Y-C, Li T-J (1989) Preliminary studies on the phytocommunity of *Davidia involucrata* in Emei mountain of Sichuan. Acta Phytotax Geobot Sin 13:270–276

Yang JM, Yang XY, Liang H (2004) The discovery of buried *Metasequoia* wood in Lichuan, Hubei, China, and its significance. Acta Palaeontol Sin 43(1):124–131

Yang W, Huang J, Ma K (2013) Taxonomic composition and spatial pattern of endemic seed plant species in Southwest China. In: Hobohm C (ed) Endemism in Vascular Plants. Plants and Vegetation, vol 9, Springer, Dordrecht pp 281–288

Ying T-S (2001) Species diversity and distribution pattern of seed plants in China. Biodivers Sci 9(4):393–398

Ying T-S, Zhang Z-S (1984) Endemism in the flora of China–studies on the endemic genera. Acta Phytotax Sin 22(4):259–268

Ying T-S, Zhang Y-L, Boufford DE (1993) The endemic genera of seed plants of China. Science Press, Beijing

You D, Wang Z, Lei Y, Ma G (2008) Community classification and dynamics succession of natural forest of *Metasequoia glyptostroboides* Hu et Cheng. Hubei Forestry Sci Technol 153:6–11

Yu YH (1995) Origin, evolution and distribution of the Taxodiaceae. Acta Phytotaxonomica Sin 33:362–389

Zhong ZQ, Boris DC (2007) Fengshui–a systematic research of vernacular sustainable development. Proceedings of the 7th annual general meeting on ancient China and its lessons for the future; 15 Sept 2007; UK-Chinese Association of Resources and Environment (UK CARE), Greenwich

Zhou Z, Zheng S (2003) Missing link in *Ginkgo* evolution. Nature 423:821–822

Chapter 8
Relations to Neighboring Regions

Abstract SW China represents a transition between the floras of western and eastern Asia, and forms a transitional zone between temperate eastern and tropical southeastern Asia. It is associated with the Sino-Himalayan, Sino-Japanese, African and Mediterranean, and Indo-Malesian floristic regions. The flora of SW China is influenced by its relation to diverse floristic regions and is especially reminiscent of the Sino-Himalayan (as on the Yunnan-Guizhou Plateaus and in western Sichuan), the Sino-Japanese (on the Yunnan-Guizhou Plateaus, in the Sichuan Basin and central and eastern Guizhou), and the Indo-Malesian floras (in southern, southeastern and southwestern Yunnan). Its flora has great value not only intrinsically, but also for the role it may play in the formulation of theories regarding the origin and evolution of the angiosperms and the patterns of plant migration and distribution. It is characterized by a high number of seed plant species, as well as endemic and Tertiary relict species. The floristic relationships of SW China to neighboring regions are discussed by comparing the geographic elements and are exemplified by the genera and species of the Fagaceae shared by these regions. The evergreen broad-leaved forests of SW China are compared with those of Taiwan, Japan, SE Asia and the eastern Himalayas. SW China's high floristic affinities with the neighboring regions are clearly expressed not only in floristic components, but also in the structure of the evergreen broad-leaved forest.

8.1 Phytogeographic Considerations

An understanding of the floristic relations of SW China to neighboring regions requires study of both the geological history (see Sect. 1.1.3) and the floristic regions. Good (1964) proposed a system of six floristic kingdoms (Australian, Antarctic, Boreal, Neotropical, Paleotropical, and South African) as the largest natural floristic units for the world's flowering plants. The kingdoms were subdivided into subkingdoms, provinces and regions. The Paleotropical kingdom was first divided into 3 subkingdoms, whereas the other five kingdoms were subdivided directly into a total of 37 provinces. Takhtajan (1986) extensively revised Good's system to produce a classification widely used today. Takhtajan recognized six kingdoms: Holarctic, Paleotropical, Neotropical, Cape, Australian and Antarctic (Fig. 8.1a). SW China

C.Q. Tang, *The Subtropical Vegetation of Southwestern China*,
Plant and Vegetation 11, DOI 10.1007/978-94-017-9741-2_8

lies at a transition zone between the two major phytogeographic realms (kingdoms): the Holarctic division and Paleotropical division. As described in Chap. 1, SW China includes the Sikang-Yunnan and central Chinese floras, and is closely related to the southeastern Chinese, Ryukyu, Japanese-Korean, the eastern Himalayan, northern Burmese and Khasi-Manipur floras (Fig. 8.1b) (Takhtajan 1986). In Asia, Wu and Wu (1998) expanded on Takhtajan's system by proposing refinements to the Eastern Asiatic Region. They promoted the Eastern Asiatic Region to Kingdom.

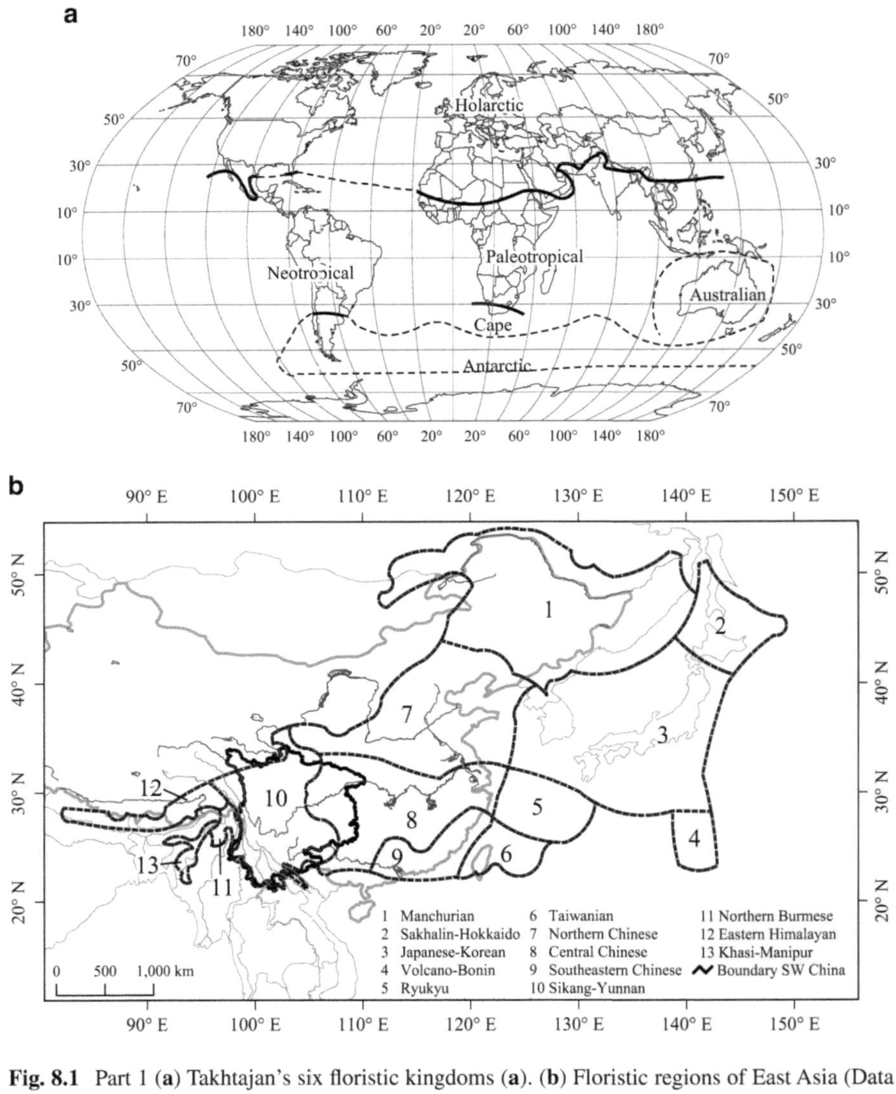

Fig. 8.1 Part 1 (**a**) Takhtajan's six floristic kingdoms (**a**). (**b**) Floristic regions of East Asia (Data source: Takhtajan (1986) for (a, b))

c

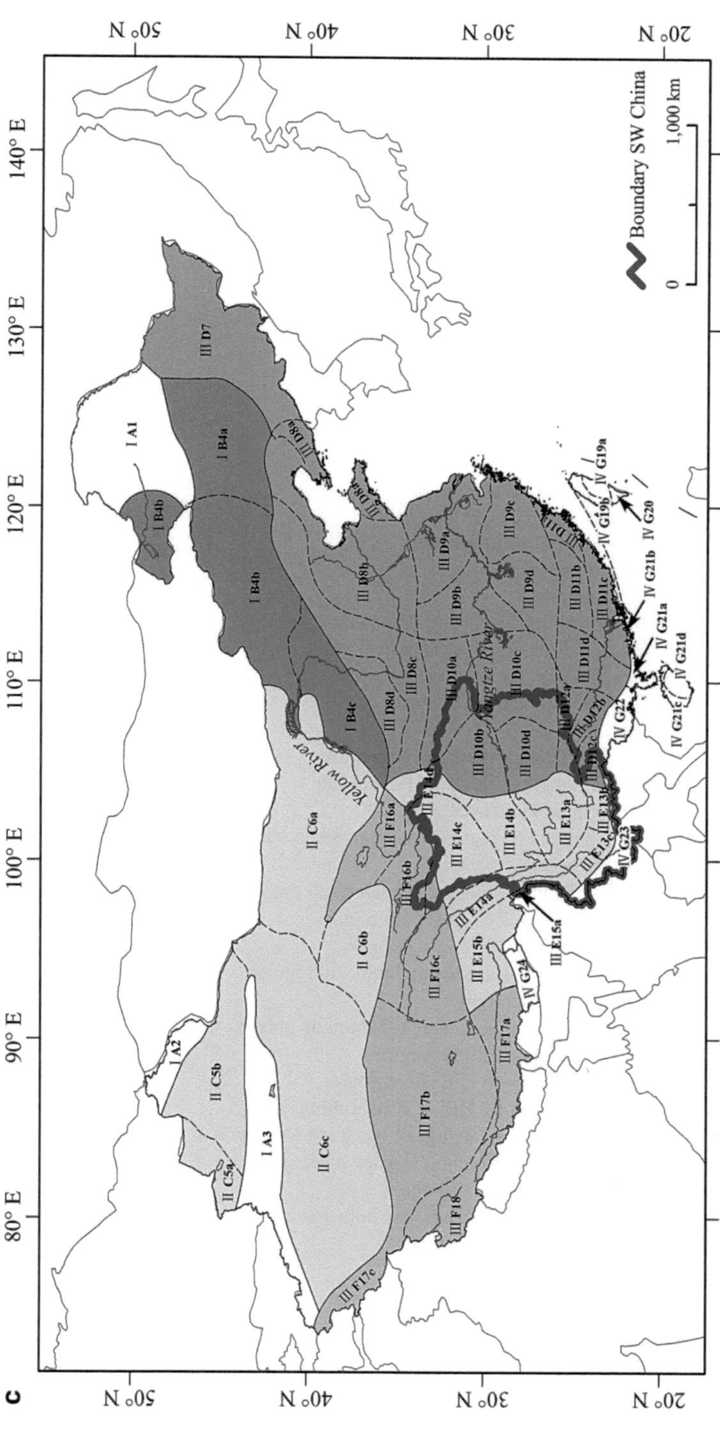

Fig. 8.1 (continued) Part 2 (**c**) The System of Chinese Floristic Regions (Data source: Wu and Wu (1998))

I. **Holarctic Kingdom**
  IA. **Eurasia Forest subkingdom**
    IA1. Great Hsingan region
    IA2. Altai region
    IA3. Tian Shan region
  IB. **Eurasia steppe subkingdom**
    IB4. Mongolian steppe region
      IB4a. NE China forest-steppe subregion
      IB4b. E Nei Mongol (Inner Mongolia) steppe subregion
      IB4c. Ordos, Shaanxi, Gansu and Ningxia desert steppe subregion
II. **Tethys kingdom**
  IIC. **Central Asiatic desert subkingdom**
    IIC5. Songaria region
      IIC5a. Tacheng-Ili (Tarbagatai-Ili) subregion
      IIC5b. Songaria subregion
    IIC6. Kashgar region
      IIC6a. Kashi region
      IIC6b. SW. Mongolia subregion
      IIC6c. Qaidam Basin subregion
III. **East Asiatic Kingdom**
  IIID. **Sino-Japanese Forest subkingdom**
    IIID7. NE. China region
    IIID8. N. China region
      IIID8a. Liaodong-Shandong Peninsula subregion
      IIID8b. N. China plain subregion
      IIID8c. N. China mountain subregion
      IIID8d. Loess plateau subregion
    IIID9. E. China region
      IIID9a. S. Zhejiang mountain subregion
      IIID9b. S. Jiangxi-E. Hunan Hill subregion
      IIID9c. Huaihe Plain subregion
      IIID9d. Jianghan (Yangtze-Hanjiang) Plain subregion
    IIID10. C. China region
      IIID10a. Sichuan, Hubei & Hunan border subregion
      IIID10b. Sichuan Red Basin subregion
      IIID10c. Guizhou Plateau subregion
      IIID10d. Qinling-Bashan Mountain subregion
    IIID11. S. China Mountain region
      IIID11a. Guangdong, Guangxi Mountain subregion
      IIID11b. S. Fujian Hill subregion
      IIID11c. N. Guangxi Hill subregion
      IIID11d. Guangdong Hill & Plain subregion
    IIID12. Yunnan, Guizhou & Guangxi Limestone Mountain & Hill region
      IIID12a. Guizhou-Guangxi Border subregion
      IIID12b. Hongshuihe subregion
      IIID12c. SE. Yunnan Limestone subregion
  IIIE. **Sino-Himalayan forest subkingdom**
    IIIE13. Yunnan Plateau region
      IIIE13a. C. Yunnan Plateau subregion
      IIIE13b. SE. Yunnan Mountain and Hill subregion
      IIIE13c. SW. Yunnan Mountain and Hill subregion
    IIIE14. Hengduan Mountain region

Fig. 8.1 (continued)

IIIE14a. Three River-Gorges subregion
IIIE14b. N. Hengduan Mountain subregion
IIIE14c. S. Hengduan Mountain subregion
IIIE14d. Taohe-Mingshan Mountain subregion
IIIE15. E. Himalayan region
IIIE15a. SE. Xizang subregion
IIIE15b. Taron-Taru (Irrawaddy) –N. Myanmar subregion
IIIF. **Qinghai-Xizang plateau subkingdom**
IIIF16. Tangut region
IIIF16a. Qilian Mountain subregion
IIIF16b. A'nyemaqen Mountain subregion
IIIF16c. Tanggula (Dangla) Mountain subregion
IIIF17. Xizang, Pamir & Kunlun region
IIIF17a. Upper and Middle Yarlun Zhanpo subregion
IIIF17b. Qiangtan (Changtang) subregion
IIIF17c. Pamir. Karakorum-Kunlun High Mountain & Plateau subregion
IIIF18. W. Himalaya region
IV. **Paleotropical Kingdom**
IVG. **Malesian subkingdom**
IVG19. N. Taiwan region
IVG19a. N. Taiwan High mountain subregion
IVG19b. N. Taiwan Hill subregion
IVG20. S. Taiwan Hill region
IVG21. South China Sea region
IVG21a. W. Guangdong-N. Hainan subregion
IVG21b. E. Guangdong Coastal Islands subregion
IVG21c. SW. Hainan subregion
IVG21d. C. Hainan subregion
IVG21e. South China Sea Coral-reef subregion
IVG22. Tonkin Bay region
IVG23 Yunnan, Myanmar & Thailand region
IVG24. S-Wing of E. Himalaya region

Fig. 8.1 (continued)

Figure 8.1c gives detailed information showing the floristic division of China and showing that China floristically comprises 4 kingdoms (Holarctic, Tethys, East Asiatic, Paleotropical) and 7 subkingdoms (Wu and Wu 1998). Tang (2000) reviewed the affinities between Chinese flora and those of other regions of the world as well as the role of Chinese flora in the origin of angiosperms. SW China represents the transition between the floras of western and eastern Asia, and is also a transitional zone between temperate eastern and tropical southeastern Asia (Takhtajan 1986; Wu and Wu 1998). The flora of SW China can be considered to be a mixture of Laurasian, Gondwanian and Tethyan elements (Wu 1988). It is closely related to the Sino-Japanese, Sino-Himalayan, African and Mediterranean, and Indo-Malesian floristic regions (Fig. 8.1a–c).

China is the only country in the world that supports vegetational continuity from tropical, subtropical, temperate to boreal forests (see Fig. 1.28). Most of SW China was never covered by ice sheets during the Last Glacial Maximum (LGM); it is now one of the world's most important Pleistocene refugia for lineages that evolved prior to the late Tertiary and Quaternary glaciations and became extinct elsewhere (Axelrod et al. 1998). The Hengduan Range is a complex series of high mountains and deep river valleys running north–south. Because of the varied and complex topography and climate, natural refugia have allowed the survival of many relict species in SW China (Wu 1998). In addition, the north–south oriented, highly rugged mountain ranges of SW China presumably reduced rates of extinction as scattered advances and retreats of Pleistocene glaciations forced plant species to migrate southward and northward. Furthermore, the Indian and Eurasian plates were driven together about 50 million years ago. The collision resulted in a near-doubling in thickness of the crust and the formation of the vast Himalayan range, with elevated and highly dissected landscapes in SW China, creating important refugia and centers of speciation and evolution for vascular plants (Axelrod et al. 1998; Qian 2002; López-Pujol et al. 2011).

Qian et al. (2006), following Wu (1991, 1993), assigned genera of seed plants in 28 provinces of China (see Fig. 1.27) to 14 floristic elements according to their worldwide geographic distributions (cosmopolitan, pantropical, tropical Asian–tropical American, paleotropical, tropical Asian–tropical Australian, tropical Asian–tropical African, tropical Asian, Holarctic, eastern Asian–North American, temperate Eurasian, temperate Asian, Mediterranean, western to central Asian, central Asian, eastern Asian) (Table 8.1). According to the data, China has 3,143 genera of native seed plants. Of these, 105 (3.3 %) are cosmopolitan, 1,512 (48.1 %) are tropical in nature (FE2–FE7 in Table 8.2) and 1,526 (48.6 %) are primarily temperate (FE8–FE14 in Table 8.2). In total, 590 (18.8 %) genera are restricted to East Asia, of which 270 are endemic to China. SW China, including Yunnan, Sichuan and Guizhou, has a high proportion of taxa belonging to the East Asia, pantropical, tropical Asia and Holarctic floristic elements (Table 8.2).

The subtropical vegetation of SW China is dominated by evergreen broad-leaved forests. But the region's topographic complexity implies that many different forest types occur at diverse elevations on slopes of varying aspect. SW China covers a wide altitudinal range, from 76.4 to 7,756 m. The climate ranges from tropical and subtropical in the lowlands to alpine and frigid at the highest elevations. In some gorge valleys along rivers such as Jinshajiang, Yuanjiang, Lancangjiang and Nujiang, the hot-dry or warm-dry or temperate-dry climate prevails. Despite this complexity, the overall forest zonation sequence follows a pattern that re-occurs in mountain ranges throughout subtropical East Asia. As elevation increases, evergreen broad-leaved forests are gradually replaced by deciduous hardwoods and conifers. In SW China, from low altitudes in the south to high altitudes in the north, the forests are dominated by various species. The tropical rain forests are dominated by Fabaceae, Meliaceae, Sapindaceae, Anacardiaceae, Burseraceae, Moraceae, or by Dipterocarpaceae, Tetramelaceae, Crypteroniaceae, etc. with a few gymnosperms such as pines, cycads and podocarps. The tropical montane evergreen broad-leaved forests and the subtropical evergreen broad-leaved forests are dominated by species of Fagaceae, Lauraceae, Hammaelidaceae, Theaceae and Magnoliaceae. The temperate coniferous and evergreen broad-leaved mixed forests are characterized by *Tsuga*, *Lithocarpus*, *Cyclobalanopsis* and *Castanopsis*. The temperate coniferous

Table 8.1 Definitions of floristic elements

Cosmopolitan (FE1)	Widely distributed across all or nearly all continents. In general, they do not have special distribution centers. Occasionally they have one or a few centers of high diversity but are distributed worldwide.
Pantropical (FE2)	Occurring in all three sectors of the tropical zone (i.e. America, Africa-Madagascar and Asia-Australia). Some of the genera in this category may have species extending into temperate regions.
Tropical Asian-tropical American (FE3)	Disjunct between tropical Asia and tropical America. Some of them may be sporadically distributed on the islands of the Pacific, or may extend into temperate regions.
Paleotropical (FE4)	Distributed in the tropics of the Old World: Asia, Australia and Africa (including Madagascar). This floristic element is also called the 'Old World Tropics' in the literature.
Tropical Asian-Tropical Australian (FE5)	Restricted to Asia and Australia and mainly distributed in the tropical regions of these two continents.
Tropical Asian-Tropical African (FE6)	Restricted to Asia and Africa and mainly distributed in the tropical regions of these two continents.
Tropical Asian (FE7)	Mainly restricted to tropical Southeast Asia, although some may extend northward into temperate regions.
Holarctic (FE8)	Primarily distributed in extratropical regions of all three continents in the Northern Hemisphere (Europe, Asia and North America), although a few extend their distributions to high elevations of tropical mountains or further south. This floristic element is referred to as 'North Temperate' in Wu (1980, 1991), and also called 'Eurasian-North American' and 'circumpolar' in the literature.
Eastern Asian-North American (FE9)	Having disjunct distributions between the temperate-subtropical regions of eastern Asia and North America, with a few extending into tropical regions of their respective continents.
Temperate Eurasian (FE10)	Widely distributed across the temperate zone of Europe and Asia. Some of them may extend into the northernmost part of Africa, but they do not occur in the New World. This floristic element is referred to as 'Old World Temperate' in Wu (1980, 1991).
Temperate Asian (FE11)	Restricted to temperate Asia. The distribution centers are mainly in temperate regions but they may occasionally occur in subtropical regions.
Mediterranean, western to central Asian (FE12)	Distributed mainly in the region across the Mediterranean, western Asia and central Asia. Some of the genera extend eastward into eastern Asia.
Central Asian (FE13)	Mainly distributed in dry areas of central Asia.
Eastern Asian (FE14)	Distributed mainly in warm temperate to subtropical regions of eastern Asia. This floristic element includes all genera in the categories 'eastern Asian' and 'Chinese endemic' of Wu (1991, 1993).

Data source: Qian et al. (2006)

and deciduous broad-leaved mixed forests are dominated by *Tsuga*, *Picea*, *Pinus*, *Acer* and *Betula*. These two types of mixed forests are ecotone between coniferous and broad-leaved forests. On some mountains *Taxus* can also be found in the ecotone including warm-temperate and cool-temperate zones. The cold-temperate

Table 8.2 Total number and proportion (%) of seed plant genera in each of the 14 floristic elements (FE1–FE14) in China as a whole and in 28 Chinese provinces

Flora	No. of genera	FE1	FE2	FE3	FE4	FE5	FE6	FE7	FE8	FE9	FE10	FE11	FE12	FE13	FE14
China	3,143	3.3	11.2	1.2	6.1	4.8	4.5	20.3	9.6	3.8	5.2	1.8	5.3	3.9	18.8 (8.6)
Yunnan	**2,069**	**4.5**	**14.2**	**1.3**	**7.0**	**5.2**	**5.7**	**24.1**	**11.2**	**4.0**	**4.3**	**0.9**	**0.6**	**0.6**	**16.4 (5.1)**
Sichuan	**1,384**	**5.6**	**13.9**	**1.3**	**5.5**	**3.0**	**3.9**	**10.2**	**17.2**	**5.7**	**6.7**	**1.7**	**1.2**	**1.6**	**22.6 (6.7)**
Guizhou	**1,249**	**5.9**	**16.7**	**1.5**	**7.5**	**4.6**	**4.6**	**18.8**	**12.4**	**5.1**	**3.8**	**1.0**	**0.2**	**0.1**	**17.7 (6.2)**
Anhui	886	8.5	15.8	1.2	5.0	3.4	2.5	6.0	18.1	8.8	7.8	1.6	0.6	0.2	20.7 (4.3)
Fujian	1,195	6.9	18.8	1.8	8.5	5.7	4.1	15.1	11.5	5.8	4.9	0.9	0.6	0.1	15.2 (3.4)
Gansu	753	9.7	8.0	0.8	2.1	1.1	1.5	1.9	28.4	5.8	10.6	3.3	6.8	3.9	16.2 (4.0)
Guangdong	1,460	5.8	19.0	1.8	9.8	7.2	5.5	19.0	9.0	4.3	3.6	0.8	0.5	0.1	13.6 (4.2)
Guangxi	1,693	5.0	16.2	1.7	8.3	6.6	5.7	24.0	9.1	4.1	3.8	0.5	0.3	0.1	14.6 (4.9)
Hainan	1,317	5.1	22.4	2.1	11.8	9.2	6.8	26.1	5.2	2.7	2.1	0.2	0.2	0.1	6.1 (1.7)
Hebei	719	11.8	11.1	0.3	2.5	1.9	1.3	1.3	30.9	6.8	11.7	4.3	4.0	1.4	10.7 (2.8)
Heilongjiang	559	14.7	7.7	0.0	1.8	0.5	0.7	0.5	39.0	7.7	12.5	5.2	2.0	0.7	7.0 (0.4)
Henan	505	11.1	10.1	1.4	3.4	2.6	1.2	3.4	27.5	7.9	8.9	3.0	1.6	0.6	17.4 (4.6)
Hubei	1,264	6.5	14.5	1.2	5.9	3.6	3.1	10.0	16.7	7.3	7.4	1.8	1.0	0.2	20.9 (6.3)
Hunan	1,156	6.5	16.2	1.3	6.7	4.2	3.8	13.1	14.4	7.0	5.8	1.1	0.3	0.1	19.6 (5.6)
Jiangsu	950	8.2	17.5	1.3	6.5	4.1	3.6	9.3	15.9	6.8	7.1	1.5	0.7	0.3	17.3 (3.2)

Jiangxi	1,086	7.4	16.7	1.3	6.7	4.1	4.1	12.3	14.3	7.2	6.0	1.0	0.4	0.3	18.2 (4.1)
Jilin	585	13.5	7.7	0.0	1.5	1.0	0.9	1.0	38.3	8.7	12.3	4.1	1.5	0.7	8.7 (0.7)
Liaoning	708	12.1	10.7	0.1	2.5	1.4	1.4	1.3	31.5	7.9	11.4	3.7	2.7	1.1	12.0 (2.5)
Neimenggu	647	12.2	6.6	0.0	1.5	0.8	0.9	0.5	34.9	5.1	13.0	5.7	7.4	4.3	7.0 (1.7)
Ningxia	413	14.0	6.1	0.2	1.2	0.2	0.5	0.5	35.6	5.1	13.3	3.6	7.0	4.1	8.5 (2.7)
Qinghai	581	11.9	5.7	0.2	1.0	0.0	0.5	0.7	33.2	3.4	13.4	4.1	6.4	5.9	13.6 (4.5)
Shaanxi	1,014	7.7	11.4	0.9	3.7	1.9	2.5	4.3	22.7	8.0	9.9	2.9	3.2	1.4	19.6 (5.1)
Shandong	586	13.3	14.0	0.5	4.1	2.4	1.7	2.4	27.0	7.3	10.6	2.7	1.9	0.7	11.4 (2.4)
Shanxi	678	10.6	9.6	0.6	1.8	1.9	1.2	1.6	32.0	6.0	11.8	4.1	3.4	1.2	14.2 (3.5)
Taiwan	1,206	7.0	21.8	2.0	9.9	7.3	4.2	15.1	12.4	5.1	4.2	0.6	0.2	0.1	10.1 (1.3)
Xinjiang	758	10.3	5.8	0.1	0.7	0.1	0.8	0.3	29.6	1.8	13.9	3.4	19.1	11.1	3.0 (0.9)
Xizang	1,395	5.8	14.0	1.4	5.9	3.8	4.3	14.7	16.1	4.5	5.9	1.6	3.1	2.3	16.6 (2.7)
Zhejiang	1,038	8.2	16.0	1.1	6.3	3.9	3.4	8.6	16.3	7.7	7.1	1.2	0.4	0.2	19.7 (4.3)

Data source: Qian et al. (2006)
Values in parentheses under FE14 are the proportions of the Chinese endemic genera. Codes for floristic elements are the same as in Table 8.1

coniferous forests (subalpine zone) are dominated by *Abies*, *Larix* and *Pinus*. And the sclerophyllous *Quercus* forests in the warm-, cool- and cold-temperate zones are dominated by evergreen *Quercus* tree species. In the temperate-dry and warm-dry valleys, small-leaved scrublands dominated by *Pinus*, *Quercus*, *Terminalia*, *Sophora* or *Desmodium*, and in the hot-dry valleys, *Woodfordia*, *Calotropis*, *Adinandra*, etc. are characteristics of the semi-savanna vegetation. Thus, there are different proportions of widely distributed East-Asian taxa, temperate zone genera, SE Asian, and tropical African elements. The floras of this vast region are extremely rich in taxa and diverse in origin and distribution, and include remarkable endemic families, genera and species. Yunnan and Sichuan are globally outstanding for their high number of seed plant species (Fig. 8.2a) and high degree of endemism (Fig. 8.2b).

Availability of refugia surely accounts for differences in species diversity. It has been reported that numerous taxa that occurred in all East Asia, Europe and North America in past geological times became extinct among the modern floras of Europe and North America due to Quaternary glacial events (Wolfe and Leopold 1967; Xiang et al. 1998). In contrast, during the Quaternary China was relatively warm and provided refugia for numerous temperate species, as evidenced by the many taxa once presumed extinct that are found there, notable examples being *Ginkgo biloba*, *Metasequoia glyptostroboides*, *Cathaya argyrophylla* and *Davidia involucrata* (in SW and central China), *Glyptostrobus pensilis* and *Pseudolarix amabilis* (in SE China), *Thuja sutchuenensis* (in the Sichuan Basin), *Taiwania cryptomerioides* (in SW and central China, and Taiwan), *Liriodendron chinense*, *Tetracentron sinense* and *Emmenopterys henryi* (in SW, central and eastern China, respectively), *Cunninghamia lanceolata* and *Nyssa sinensis* (in SW, southern, central and southeastern China), *Euptelea pleiosperma* (in SW, central, southeastern China, Hebei, Henan), *Eucommia ulmoides* (in SW, central and eastern China, Gansu, Henan), *Cercidiphyllum japonicum* (in SW and central China, Gansu). By contrast, some neighboring regions share some paleo-genera with China such as *Cercidiphyllum* (*C. japonicum*) and *Euptelea* (*E. polyandra*) in Japan, *Tetracentron* (*T. sinense*) and *Euptelea* (*E. pleiosperma*) in the eastern Himalayas, *Taiwania* (*T. cryptomerioides*) in Taiwan, Vietnam and Myanmar, and *Nyssa* (*N. sinensis*) in Vietnam.

Species of the Fagaceae occur as associates of different forest types in various parts of the Northern Hemisphere, as described in discussing the subtropical evergreen broad-leaved forests of Chap. 2 and the evergreen sclerophyllous *Quercus* forests of Chap. 3.

The Fagaceae family primarily includes trees, but counts certain shrubby plants among its members as well. They are not only distributed in China, but also in Japan, Korea, Mongolia, Russia (Siberia), Europe, North America, Indochina, and the Malaysian region including Malaysia, Indonesia and surrounding islands. Fagaceae species could be said to characterize the biodiversity and endemism of East and Southeast Asia. In the vast subtropical region of China, including the tropical mountainous regions, the Fagaceae are principal constituents of the forest vegetation. Many species of this family are dominants of the subtropical evergreen broad-leaved forests, of the mixed evergreen and deciduous broad-leaved forests in the transitional montane forest zone from subtropical EBLFs to temperate

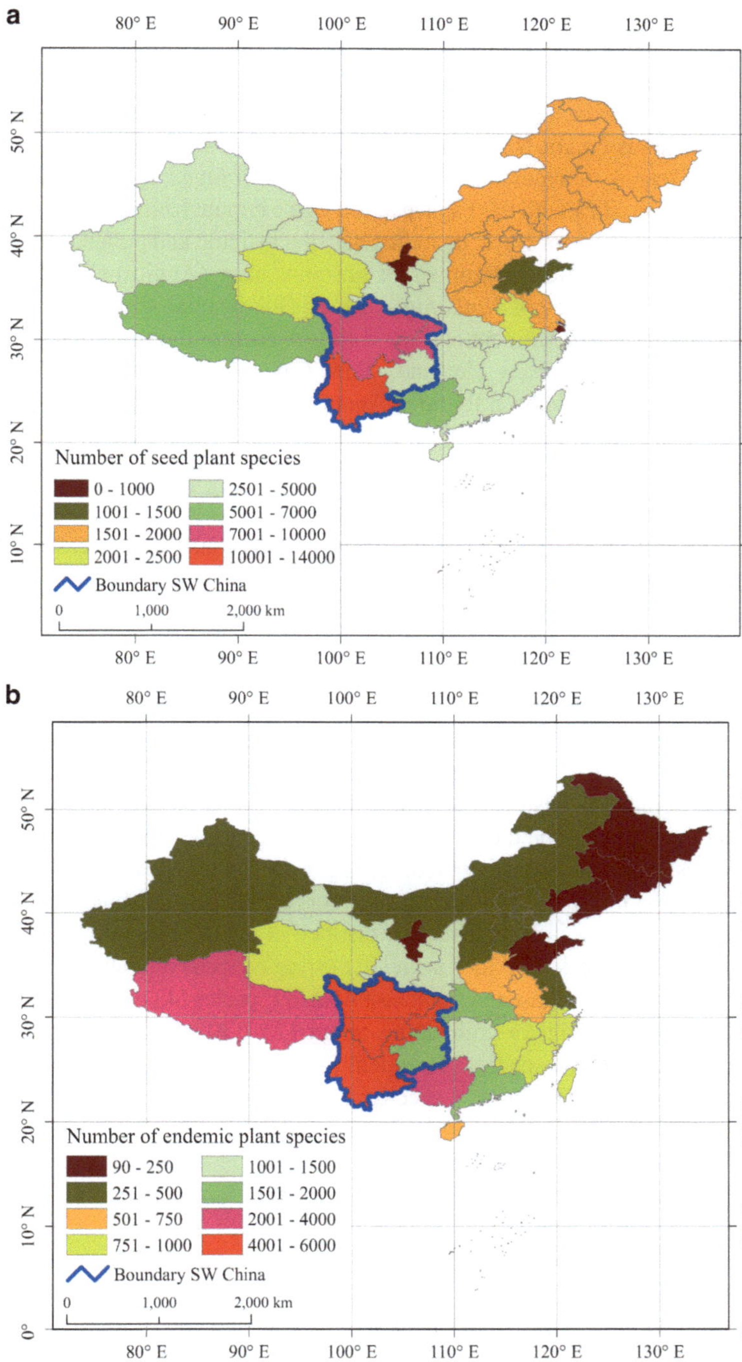

Fig. 8.2 Distribution pattern of the number of seed plant species (**a**), and distribution pattern of the number of endemic plant species in China (**b**) (Data source: Sang et al. (2011))

deciduous forests. The Fagaceae are generally regarded as including 9 genera (listed below). The southern Hemisphere genus *Nothofagus* (literally meaning "false beeches" or "bastard beeches") includes about 36 species of trees and shrubs in southern South America and Australasia, was formerly included in the Fagaceae as a sister to the genus *Fagus*, but is now treated as the separate family Nothofagaceae according to recent molecular evidence. While *Nothofagus* shares a number of common characteristics with the Fagaceae, such as a cupule fruit structure, it differs significantly in a number of ways, including a distinct stipule and pollen morphology, as well as having a different number of chromosomes (Angiosperm Phylogeny Group 2009). The species of *Nothofagus* are of Gondwanian distribution, having originated on that supercontinent and persisting in current-day East and Southeast Australia, New Guinea, New Zealand, New Caledonia, Argentina, and Chile. Recently, fossils also have been found in Antarctica (Li and Zhou 2007).

As regards the Fagaceae, the *Quercus* subgenus *Cyclobalanopsis* is treated as a distinct genus by the Flora of China, but as a subgenus by most taxonomists. The genera can be characterized as follows:

Fagus – beeches; about 10 species, deciduous, East Asia, Europe, eastern North America. Disjunct distribution between Europe and Asia, and Eurasia and North America. 4 species (3 endemic) in China.

Castanea – chestnuts; about 12 species, deciduous, Asia, Southeast Europe, eastern North America. Disjunct distribution between Europe and Asia, and Eurasia and North America. 3 (2 endemic) in China.

Quercus – (Latin "oak tree"); about 300 species deciduous and evergreen. The Northern Hemisphere: North Africa, Asia, Europe, North America, Central and South America (Colombia). Disjunct distribution between Eurasia and North Africa on the one hand and America on the other. North America contains the largest number of oak species, with approximately 90 occurring in the United States. Mexico has 160 species, of which 109 are endemic. 36 species (16 endemic) in China.

Cyclobalanopsis (here, it is treated as a distinctive genus following the Flora of China) – (Cyclo- Greek "Kyklo-", a circle); the ring-cupped oaks; about 150 species, evergreen, confined to East and Southeast Asia. 69 species (43 endemic) in China.

Castanopsis – chinquapins or chinkapins, about 125–130 species, evergreen, East and Southeast Asia. 60 species (31 endemic) in China.

Chrysolepis – golden chinkapins; 2 species, evergreen, western USA.

Lithocarpus – stone oaks; about 300 species, evergreen, East and Southeast Asia. 123 species (69 endemic) in China.

Notholithocarpus – Tanoaks; 1 species (formerly *Lithocarpus densiflorus*), evergreen, California and southwest Oregon, USA.

Trigonobalanus (*Formanodendron*) – 3 species, evergreen. *T. doichangensis*, south-western, southern Yunnan and northern Thailand., *T. verticillata*, tropical Indonesia and the Malay Peninsula as well as Hainan Island (South China), and *T. excelsa*, Colombia.

Of the above genera, deciduous species of *Fagus*, *Castanea* and *Quercus* are often ecologically dominant in northern temperate deciduous broad-leaved forests of the Northeastern United States (e.g. *Fagus grandifolia*, *Castanea dentata*, and

Quercus alba), of Europe (e.g. *Fagus sylvatica*, *Quercus robur* and *Quercus petraea*), and of East Asia (e.g. *Quercus acutissima*, *Quercus serrata*, *Quercus mongolica*, *Castanea crenata*, *Fagus crenata* and *Fagus japonica*). Some deciduous *Quercus* species (e.g. *Quercus variabilis* and *Quercus aliena*) are usually found in the secondary forests in the EBLF zone in subtropical China. *Fagus* species are found both in the transitional zone from subtropical evergreen to temperate deciduous broad-leaved forests and in the subtropical EBLF zone of SW China (e.g. *Fagus lucida*, *Fagus longipetiolata* on Fanjingshan, Guizhou, *Fagus longipetiolata* in Xichou, SE Yunnan, *Fagus hayatae* and *Fagus engleriana* on mountains including Dabashan, Sichuan), in central China (*Fagus engleriana* on Shennongjia, Hubei), and in northern Taiwan (*Fagus hayatae*). The altitudinal distribution of the *Fagus* species in China seems to be determined by the length of the dry period in spring before the arrival of monsoon rains (Guo and Werger 2010).

Various evergreen sclerophyllous *Quercus* species are widely distributed in the Mediterranean region, in North and Central America, in Colombia of South America, and in Asia. 16 evergreen sclerophyllous *Quercus* species are found in the Hengduan Mountains of China. Those species (such as *Q. aquifolioides*, *Q. pannosa* and *Q. longispica*) are morphologically and ecologically similar to Mediterranean *Q. suber* (cork oak) and *Q. ilex* (holm oak) (Li et al. 2000; Zhou et al. 2003). The sclerophyllous *Quercus* communities in NW Yunnan are ecologically similar to the *Q. ilex* community of NE Spain and the *Q. agrifolia* (coast live oak) community of central coastal California (Tang 2006).

Evergreen *Cyclobalanopsis*, *Castanopsis* and *Lithocarpus* are important dominants in the evergreen broad-leaved forests of tropical and subtropical Asia. SW China (especially Yunnan) and Indochina have the most diversity in these three genera. *Cyclobalanopsis* species, the ring-cupped oaks of Southeast and East Asia, are evergreen trees growing 10–40 m tall. They are distinct from the subgenus *Quercus* in having acorns with distinctive cups bearing concrescent rings of scales; they commonly also have densely clustered acorns, though this does not apply to all the roughly 150 species. Species of *Cyclobalanopsis* are common in the subtropical evergreen broad-leaved forests which extend from southern Japan, southern Korea (the narrow subtropical belt along the southern coast and Cheju Island) and Taiwan across southern China and northern Indochina to the eastern Himalayas, in association with trees of the genera *Castanopsis*, *Lithocarpus* and the family Lauraceae. Indochina, SW China and southern China have the highest species diversity (Fig. 8.3a, b; Table 8.3). A number of species in Malaysia are observed to have some plesiomorphic characteristics, and this region may be considered as a center of origin and a key area of endemism of this genus. On the island of Borneo, the area of Kalimantan is richest in endemic species, and fossil species records confirm this fact. In contrast, the endemic species from Hainan and Taiwan are more derivative. The endemic species of Yunnan represent a mixture of paleoendemics and neoendemics (Luo and Zhou 2001).

There are 125–130 species in the genus *Castanopsis* (chinquapins or chinkapins), and the species are mainly restricted to Southeast and East Asia. Yunnan (38 spp.) has the highest *Castanopsis* species diversity in China, with 11 species endemic to Yunnan. *Castanopsis* species occur in other areas of China, and further south, through Indochina to Indonesia, Malaysia, and also in Japan, southern Korea and the eastern Himalayas; with 82 species Indochina is especially rich (Fig. 8.3a, b; Table 8.4). In the

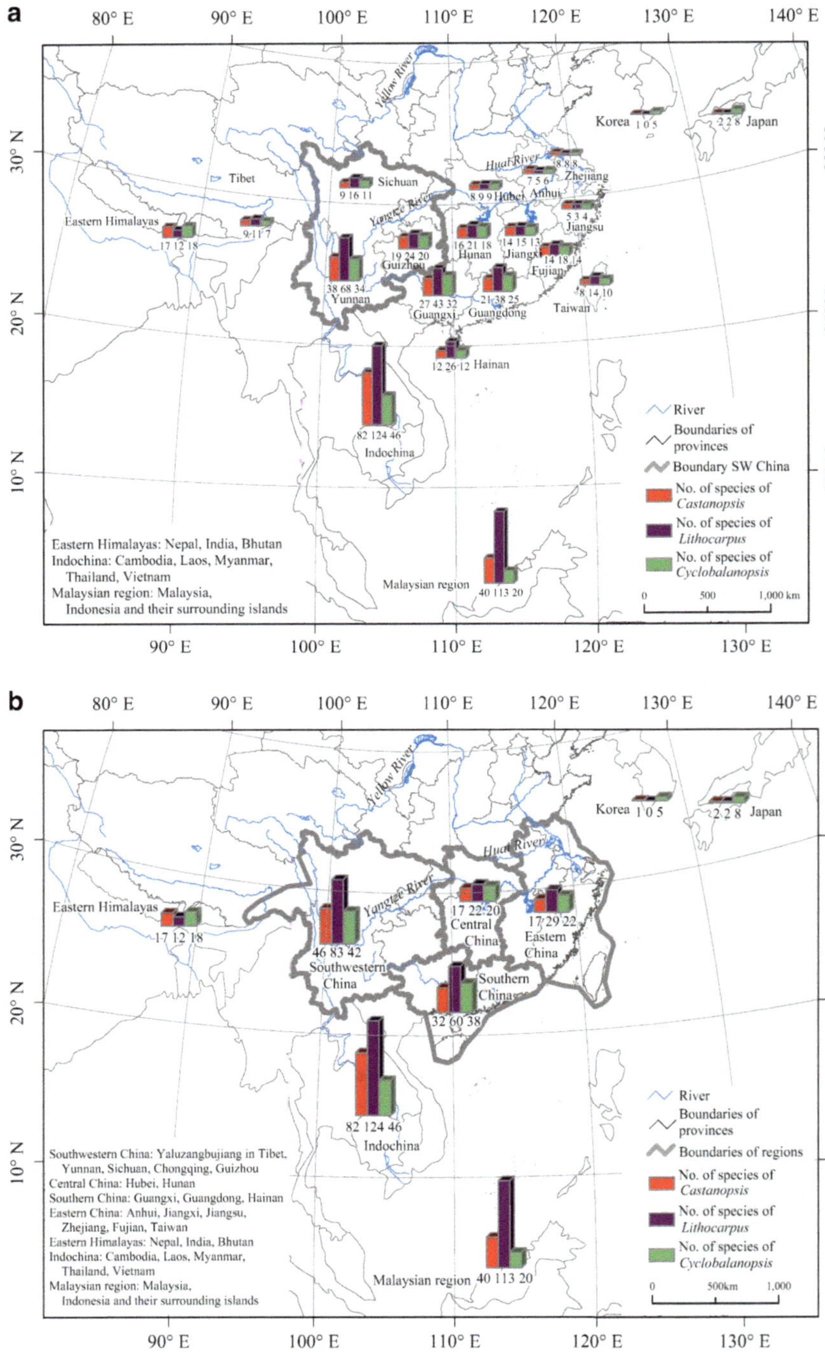

Fig. 8.3 Comparison of number of species of *Castanopsis*, *Lithocarpus* and *Cyclobalanopsis* among 15 provinces of China, the eastern Himalayas, Japan, Korea, Indochina and the Malaysian region (**a**), and between regions (**b**). The Malaysian region includes Malaysia, Indonesia and surrounding islands

Table 8.3 A comparison of *Cyclobalanopsis* species between SW China and other regions in East and Southeast Asia

Species	SW Ch.	S Ch.	C Ch.	E Ch.	Jap.	Kor.	E Him.	Indoch.	Mal. Reg.
Cyclobalanopsis annulata	*						*	*	
Cyclobalanopsis argyrotricha	*								
Cyclobalanopsis augustinii	*	*						*	
Cyclobalanopsis austrocochinchinensis	*							*	
Cyclobalanopsis austroglauca	*								
Cyclobalanopsis blakei	*	*						*	
Cyclobalanopsis camusiae	*							*	
Cyclobalanopsis championii	*	*		*					
Cyclobalanopsis chapensis	*							*	
Cyclobalanopsis chevalieri	*	*						*	
Cyclobalanopsis delavayi	*	*	*						
Cyclobalanopsis disciformis	*	*	*						
Cyclobalanopsis fleuryi	*	*	*	*				*	
Cyclobalanopsis gambleana	*		*				*		
Cyclobalanopsis gilva	*	*	*	*	*	*			
Cyclobalanopsis glauca	*	*	*	*	*	*	*	*	
Cyclobalanopsis glaucoides	*								
Cyclobalanopsis gracilis	*	*	*	*			*		
Cyclobalanopsis helferiana	*	*					*	*	
Cyclobalanopsis jenseniana	*	*	*	*					
Cyclobalanopsis jinpinensis	*								
Cyclobalanopsis kerrii	*	*						*	
Cyclobalanopsis kiukiangensis	*								
Cyclobalanopsis kouangsiensis	*	*	*						
Cyclobalanopsis lamellosa	*	*					*	*	

(continued)

Table 8.3 (continued)

Species	SW Ch.	S Ch.	C Ch.	E Ch.	Jap.	Kor.	E Him.	Indoch.	Mal. Reg.
Cyclobalanopsis lobbii	*						*	*	
Cyclobalanopsis lungmaiensis	*								
Cyclobalanopsis motuoensis	*								
Cyclobalanopsis multiervis	*	*	*	*					
Cyclobalanopsis myrsinifolia	*	*	*	*	*	*		*	
Cyclobalanopsis oxyodon	*	*	*	*			*	*	
Cyclobalanopsis pachyloma	*	*	*	*					
Cyclobalanopsis pentacycla	*								
Cyclobalanopsis rex	*						*	*	
Cyclobalanopsis sessilifolia	*	*	*	*	*			*	
Cyclobalanopsis sichourensis	*								
Cyclobalanopsis stewardiana	*	*	*	*					
Cyclobalanopsis tenuicupula	*								
Cyclobalanopsis thorelii	*	*						*	
Cyclobalanopsis tomentosinervis	*								
Cyclobalanopsis xanthotricha	*							*	
Cyclobalanopsis yingjiangensis	*							*	
Cyclobalanopsis albicaulis		*							
Cyclobalanopsis bella		*							
Cyclobalanopsis chungii		*	*	*					
Cyclobalanopsis daimingshanensis		*							
Cyclobalanopsis delicatula		*	*						
Cyclobalanopsis dinghuensis		*							
Cyclobalanopsis edithiae		*						*	
Cyclobalanopsis hui		*	*						
Cyclobalanopsis litoralis		*							

Cyclobalanopsis litseoides		*							
Cyclobalanopsis ningangensis		*	*	*					
Cyclobalanopsis obovatifolia		*	*	*					
Cyclobalanopsis patelliformis		*		*					
Cyclobalanopsis phanera		*							
Cyclobalanopsis poilanei		*							
Cyclobalanopsis tiaoloshanica		*							
Cyclobalanopsis elevaticostata				*					
Cyclobalanopsis hypophaea				*					
Cyclobalanopsis longinux				*					
Cyclobalanopsis morii				*					
Cyclobalanopsis stenophylloides				*					
Cyclobalanopsis yonganensis				*					
Cyclobalanopsis acuta					*	*			
Cyclobalanopsis hondae					*				
Cyclobalanopsis miyagii					*				
Cyclobalanopsis salicina					*				
Cyclobalanopsis stenophylla						*			
Cyclobalanopsis brandisiana							*	*	
Cyclobalanopsis eumorpha							*	*	
Cyclobalanopsis gomeziana							*	*	
Cyclobalanopsis hilldebrandii							*	*	
Cyclobalanopsis kamroopii							*		
Cyclobalanopsis mespilifolioides							*		
Cyclobalanopsis patkoiensis							*		
Cyclobalanopsis semiserrata							*	*	*
Cyclobalanopsis thomsoniana							*		

(continued)

Table 8.3 (continued)

Species	SW Ch.	S Ch.	C Ch.	E Ch.	Jap.	Kor.	E Him.	Indoch.	Mal. Reg.
Cyclobalanopsis arbutifolia								*	
Cyclobalanopsis asymmetrica								*	
Cyclobalanopsis bambusifolia								*	
Cyclobalanopsis braianensis								*	
Cyclobalanopsis brandisiana								*	
Cyclobalanopsis brevicalyx								*	
Cyclobalanopsis chrysocalyx								*	
Cyclobalanopsis dilacerata								*	
Cyclobalanopsis dussaudii								*	
Cyclobalanopsis fructiseptata								*	
Cyclobalanopsis langbianensis								*	
Cyclobalanopsis longistyla								*	
Cyclobalanopsis macrocalyx								*	
Cyclobalanopsis neglecta								*	
Cyclobalanopsis pilanei								*	
Cyclobalanopsis quangtriensis								*	
Cyclobalanopsis rupestris								*	
Cyclobalanopsis saravanensis								*	
Cyclobalanopsis xanthoclada								*	
Cyclobalanopsis argentata									*
Cyclobalanopsis chrysotricha									*
Cyclobalanopsis elmeri									*
Cyclobalanopsis gaharuensis									*

Cyclobalanopsis gamelliflora									*
Cyclobalanopsis gemelliflora									*
Cyclobalanopsis kerangasensis									*
Cyclobalanopsis lineata									*
Cyclobalanopsis lowii									*
Cyclobalanopsis merrillii									*
Cyclobalanopsis nivea									*
Cyclobalanopsis oidocarpa									*
Cyclobalanopsis percoriacea									*
Cyclobalanopsis pseudoverticillata									*
Cyclobalanopsis steenisii									*
Cyclobalanopsis subsericea									*
Cyclobalanopsis sumatrana									*
Cyclobalanopsis treubiana									*
Cyclobalanopsis valdinervosa									*

Data sources: Soepadmo (1972), eFloras (2014), Ohwi et al. (1965), Negi and Naithani (1995), Soepadmo et al. (2000), Luo and Zhou (2001), Wangda and Ohsawa (2006), Tang and Ohsawa (2009), Springer (2014), Wikipedia (2014b)

An asterisk indicates the presence. The list is not complete for Indochina and the Malaysian region

SW Ch. Southwestern China including the Yaluzangbujiang area and its surroundings in Tibet, Yunnan, Sichuan and Chongqing, and Guizhou, *S Ch*. Southern China including Guangxi, Guangdong, and Guizhou, *C Ch*. Central China including Hubei and Hunan, *E Ch*. Eastern China including Anhui, Jiangxi, Jiangsu, Zhejiang, Fujian, and Taiwan, *Jap*. Japan, *Kor*. Korea, *E Him*. Eastern Himalayas including Nepal, India, and Bhutan, *Indoch*. Indochina including Cambodia, Laos, Myanmar, Thailand, Vietnam, *Mal. Reg*. The Malaysian region including Malaysia; also Indonesia and surrounding islands

Table 8.4 A comparison of *Castanopsis* species between SW China and other regions in East and Southeast Asia

Species	SW Ch.	S Ch.	C Ch.	E Ch.	Jap.	Kor.	E Himl.	Indoch.	Mal. Reg.
Castanopsis argyrophylla	*						*	*	
Castanopsis boisii	*	*						*	
Castanopsis calathiformis	*								
Castanopsis carlesii	*	*	*	*				*	
Castanopsis ceratacantha	*		*					*	
Castanopsis cerebrina	*							*	
Castanopsis chinensis	*	*	*	*				*	
Castanopsis choboensis	*	*						*	
Castanopsis chunii	*	*	*	*					
Castanopsis clarkei	*						*	*	
Castanopsis delavayi	*	*							
Castanopsis densispinosa	*								
Castanopsis echinocarpa	*						*	*	
Castanopsis eyrei	*	*	*	*					
Castanopsis fabri	*	*	*	*				*	
Castanopsis fargesii	*	*	*	*					
Castanopsis ferox	*							*	
Castanopsis fissa	*	*	*	*				*	
Castanopsis fleuryi	*							*	
Castanopsis globigemmata	*								
Castanopsis hupehensis	*		*						
Castanopsis hystrix	*	*	*	*			*	*	
Castanopsis indica	*	*		*			*	*	
Castanopsis jinpingensis	*								
Castanopsis jucunda	*	*	*	*					

Castanopsis kweichowensis	*	*							
Castanopsis lamontii	*	*	*	*					
Castanopsis longispina	*							*	
Castanopsis malipoensis	*								
Castanopsis megaphylla	*								
Castanopsis mekongensis	*						*		
Castanopsis oblonga	*								
Castanopsis orthacantha	*								
Castanopsis ouonbiensis	*							*	
Castanopsis platyacantha	*								
Castanopsis remotidenticulata	*								
Castanopsis rockii	*							*	
Castanopsis rufotomentosa	*								
Castanopsis sclerophylla	*	*	*	*					
Castanopsis tcheponensis	*							*	
Castanopsis tessellata	*								
Castanopsis tibetana	*	*	*	*					
Castanopsis tonkinensis	*	*						*	
Castanopsis tribuloides	*						*		
Castanopsis wattii	*						*	*	
Castanopsis xichouensis	*								
Castanopsis amabilis		*							
Castanopsis concinna		*							
Castanopsis crassifolia		*						*	
Castanopsis damingshanensis		*							
Castanopsis fordii		*	*	*					

(continued)

Table 8.4 (continued)

Species	SW Ch.	S Ch.	C Ch.	E Ch.	Jap.	Kor.	E Himl.	Indoch.	Mal. Reg.
Castanopsis hainanensis		*							
Castanopsis jianfenglingensis		*							
Castanopsis kawakamii		*		*				*	
Castanopsis ledongensis		*							
Castanopsis longzhouica		*							
Castanopsis nigrescens		*	*	*					
Castanopsis subuliformis		*							
Castanopsis uraiana		*	*	*					
Castanopsis wenchangensis		*							
Castanopsis cuspidata var. *carlesii*				*					
Castanopsis cuspidata					*	*			
Castanopsis sieboldii					*				
Castanopsis argentea							*		*
Castanopsis armata							*	*	
Castanopsis castanicarpa							*		
Castanopsis clarkei							*		
Castanopsis kurzii							*		
Castanopsis lanceifolia							*	*	
Castanopsis purpurella							*	*	
Castanopsis wallichii							*	*	
Castanopsis calathiformis							*	*	
Castanopsis acuminatissima								*	*
Castanopsis annamensis								*	
Castanopsis arietina								*	
Castanopsis birmanica								*	

Castanopsis borneensi								*	
Castanopsis brevispinula								*	
Castanopsis cambodiana								*	
Castanopsis castanicarpa								*	
Castanopsis chapaensis								*	
Castanopsis chevalieri								*	
Castanopsis diversifolia								*	
Castanopsis dongchoensis								*	
Castanopsis gamblei								*	
Castanopsis griffithii								*	
Castanopsis guinieri								*	
Castanopsis harmandii								*	
Castanopsis jucunda								*	
Castanopsis lamantii								*	
Castanopsis lecomtei								*	
Castanopsis longipes								*	
Castanopsis longipetiolata								*	
Castanopsis malaccensis								*	*
Castanopsis megacarpa								*	*
Castanopsis microphylla								*	
Castanopsis namdinhensis								*	
Castanopsis nhatrangensis								*	
Castanopsis ninhhoensis								*	
Castanopsis phuthoensis								*	
Castanopsis pierrei								*	
Castanopsis piriformis								*	

(continued)

Table 8.4 (continued)

Species	SW Ch.	S Ch.	C Ch.	E Ch.	Jap.	Kor.	E Himl.	Indoch.	Mal. Reg.
Castanopsis poilanei								*	
Castanopsis purpurea								*	
Castanopsis siamensis								*	
Castanopsis symmetricupulata								*	
Castanopsis torulosa								*	
Castanopsis touranensis								*	
Castanopsis tranninhensis								*	
Castanopsis buruana									*
Castanopsis catappaefolia									*
Castanopsis clemensii									*
Castanopsis costata									*
Castanopsis curtisii									*
Castanopsis densinervia									*
Castanopsis endertii									*
Castanopsis evansii									*
Castanopsis foxworthyi									*
Castanopsis fulva									*
Castanopsis hypophoenicea									*
Castanopsis inermis									*
Castanopsis javanica									*
Castanopsis johorensis									*
Castanopsis lucida									*
Castanopsis motleyana									*
Castanopsis nephelioides									*
Castanopsis oligoneura									*

Castanopsis oviformis									*
Castanopsis paucispina									*
Castanopsis pedunculata									*
Castanopsis philipensis									*
Castanopsis psilophylla									*
Castanopsis rhamnifolia									*
Castanopsis schefferiana									*
Castanopsis scortechinii									*
Castanopsis tungurrut									*

Data sources: Soepadmo (1972), eFloras (2014), Ohwi et al. (1965), Ohsawa (1991), Negi and Naithani (1995), Soepadmo et al. (2000), Liu and Zhou (2006), Tang and Ohsawa (2009), Springer (2014)

An asterisk indicates the presence. The list is not complete for Indochina and the Malaysian region

SW Ch. Southwestern China including the Yaluzangbujiang area and its surroundings in Tibet, Yunnan, Sichuan and Chongqing, and Guizhou, *S Ch*. Southern China including Guangxi, Guangdong, and Guizhou, *C Ch*. Central China including Hubei and Hunan, *E Ch*. Eastern China including Anhui, Jiangxi, Jiangsu, Zhejiang, Fujian, and Taiwan, *Jap*. Japan, *Kor*. Korea, *E Him*. Eastern Himalayas including Nepal, India, and Bhutan, *Indoch*. Indochina including Cambodia, Laos, Myanmar, Thailand, Vietnam, *Mal. Reg*. The Malaysian region including Malaysia; also Indonesia and surrounding islands

Malaysian region, among 40 species there are 29 species endemic to Malaysia, the richest endemism region of this genus. This fact supports the results from molecular research suggesting that floristic migration in this region is limited. Indochina and Malaysia may be considered as two modern centers of diversity of this genus (Liu and Zhou 2006). The English name chinkapins is shared with other related plants, including the golden chinkapins of the Pacific United States, which are sometimes included within *Castanopsis* but are more often considered a separate but very closely related genus, *Chrysolepis*. If the genus *Chrysolepis* is accepted, the distribution type of *Castanopsis* is only of tropical Asia. The earliest reliable records of *Castanopsis* go back to the Eocene, and subfamily fossil records found from the Paleocene indicate that the appearance of *Castanopsis* is not later than the Paleocene. All fossil records of Fagaceae including *Castanopsis* are reported from the Northern Hemisphere, as are their main modern distribution regions (Liu and Zhou 2006). *Castanopsis* species are able to inhabit a wide range of tropical to subtropical locations, extending to temperate habitats, and are often keystone species in their ecosystems. They are plentiful in ecotones as diverse as Borneo montane rain forests and Chinese subtropical evergreen broad-leaved forests. Generally they are common in lauro-fagaceous-dominated subtropical/warm-temperate evergreen broad-leaved forests, and in upper montane forests. In the former, they are characteristic elements of the climax vegetation in their entire continental Asian range, as well as in Japan.

The tropical and subtropical genus *Lithocarpus* (stone oak) represents a major diversification within the globally dominant Fagaceae. *Lithocarpus* is very closely related to *Quercus*. The key difference is that *Lithocarpus* species have erect flowering spikes and *Quercus* species have trailing ones. *Lithocarpus* species are confined to Southeast and East Asia, where two centers of biodiversity exist: one in Indochina and one in the Malaysian region (Fig. 8.3a, b; Table 8.5). The highest levels of genetic endemism were observed in western Borneo (Cannon and Manos 2003). These trees dominate the tropical forests of Indochina and subtropical China (especially Yunnan), the eastern Himalayas and Japan. *Lithocarpus* has as many as 124 species in Indochina, 113 in the Malaysian region, 83 in SW China, 60 in southern China, 12 in the eastern Himalayas, and 2 in Japan.

Morphological characteristics of some species of Fagaceae are shown in Fig. 8.4a–n.

Evidently, the Fagaceae are widely distributed across the Northern Hemisphere. Genus-level diversity is concentrated in Southeast and East Asia (*Cyclobalanopsis*, *Lithocarpus* and *Castanopsis* are limited to those regions.). Species-level diversity of the Fagaceae is greatest in Mexico, Central America and Southeast Asia. Fossil remains of *Quercus* are common and have been discovered in the three continents. The earliest fossils of *Quercus* in North America (Bones 1979; Manchester 1994), China (Jiang 1993), and Europe (Kvaček and Walther 1989) are from the Eocene. The almost simultaneous appearance of the genus on the three continents, as witnessed by fossil remains, has raised the question of the geographic origin and radiation of the genus (Quercus Portal 2014).

Zhou (1992) suggests that the genus *Quercus* appeared in Southeast Asia deriving from a sister genus *Trigonobalanus* during the Paleocene, and migrated to Europe

Table 8.5 A comparison of *Lithocarpus* species between SW China and other regions in East and Southeast Asia

Species	SW Ch.	S Ch.	C Ch.	E Ch.	Jap.	Kor.	E Himl.	Indoch.	Mal. Reg.
Lithocarpus amoenus	*	*	*	*					
Lithocarpus apricus	*								
Lithocarpus areca	*	*							
Lithocarpus bacgiangensis	*	*							
Lithocarpus balansae	*								
Lithocarpus bonnetii	*	*							
Lithocarpus brevicaudatus	*	*	*	*					
Lithocarpus calolepis	*								
Lithocarpus calophyllus	*	*	*	*					
Lithocarpus chifui	*	*							
Lithocarpus cinereus	*	*							
Lithocarpus cleistocarpus	*		*	*					
Lithocarpus collettii	*								
Lithocarpus confinis	*								
Lithocarpus corneus	*	*	*	*					
Lithocarpus craibianus	*								
Lithocarpus crassifolius	*							*	
Lithocarpus cryptocarpus	*								
Lithocarpus dealbatus	*						*		
Lithocarpus echinophorus	*							*	
Lithocarpus echinotholus	*							*	
Lithocarpus elizabethae	*	*		*					
Lithocarpus eriobotryoides	*		*						
Lithocarpus fangii	*								
Lithocarpus farinulentus	*								

(continued)

Table 8.5 (continued)

Species	SW Ch.	S Ch.	C Ch.	E Ch.	Jap.	Kor.	E Himl.	Indoch.	Mal. Reg.
Lithocarpus fenestratus	*	*					*		
Lithocarpus fohaiensis	*								
Lithocarpus fordianus	*								
Lithocarpus gaoligongensis	*								
Lithocarpus garrettianus	*								
Lithocarpus glaber	*	*	*	*	*				
Lithocarpus grandifolius	*							*	
Lithocarpus gymnocarpus	*	*							
Lithocarpus haipinii	*	*	*						
Lithocarpus hancei	*	*	*	*					
Lithocarpus henryi	*		*	*				*	
Lithocarpus hypoglaucus	*								
Lithocarpus jenkinsii	*								
Lithocarpus laetus	*								
Lithocarpus laoticus	*								
Lithocarpus leucodermis	*								
Lithocarpus levis	*								
Lithocarpus listeri	*						*		
Lithocarpus litseifolius	*	*	*	*				*	
Lithocarpus longanoides	*	*							
Lithocarpus longipedicellatus	*	*							
Lithocarpus lycoperdon	*	*							
Lithocarpus magneinii	*								
Lithocarpus mairei	*								
Lithocarpus megalophyllus	*	*	*						
Lithocarpus mekongensis	*								

Lithocarpus mianningensis	*								
Lithocarpus microspermus	*								
Lithocarpus nitidinux	*								
Lithocarpus oblanceolatus	*								
Lithocarpus obscurus	*								
Lithocarpus pachylepis	*	*							
Lithocarpus pachyphyllus	*						*		
Lithocarpus pakhaensis	*								
Lithocarpus pasania	*								
Lithocarpus petelotii	*	*	*						
Lithocarpus propinquus	*								
Lithocarpus pseudoreinwardtii	*								
subsp. *dakhaensis*									
Lithocarpus pseudovestitus	*	*							
Lithocarpus qinzhouicus	*	*							
Lithocarpus rhabdostachyus	*	*							
Lithocarpus rosthornii	*	*	*						
Lithocarpus silvicolarum	*	*							
Lithocarpus skanianus	*	*	*	*					
Lithocarpus sphaerocarpus	*	*							
Lithocarpus tabularis	*								
Lithocarpus taitoensis	*	*	*	*					
Lithocarpus talangensis	*								
Lithocarpus tenuilimbus	*	*							
Lithocarpus tephrocarpus	*								
Lithocarpus thomsonii	*						*		
Lithocarpus trachycarpus	*								

(continued)

Table 8.5 (continued)

Species	SW Ch.	S Ch.	C Ch.	E Ch.	Jap.	Kor.	E Himl.	Indoch.	Mal. Reg.
Lithocarpus triqueter	*								
Lithocarpus truncatus	*						*		
Lithocarpus tubulosus	*								
Lithocarpus variolosus	*							*	
Lithocarpus xizangensis	*								
Lithocarpus xylocarpus	*						*	*	
Lithocarpus amygdalifolius		*		*				*	
Lithocarpus attenuatus		*							
Lithocarpus brachystachyus		*							
Lithocarpus caudatilimbus		*							
Lithocarpus chiungchungensis		*							
Lithocarpus chrysocomus		*	*						
Lithocarpus cucullatus		*	*						
Lithocarpus cyrtocarpus		*						*	
Lithocarpus damiaoshanicus		*							
Lithocarpus elaeagnifolius		*							
Lithocarpus elmerrillii		*							
Lithocarpus fenzelianus		*							
Lithocarpus floccosus		*		*					
Lithocarpus handelianus		*							
Lithocarpus harlandii		*	*	*					
Lithocarpus howii		*							
Lithocarpus irwinii		*		*					
Lithocarpus iteaphyllus		*	*	*					
Lithocarpus ithyphyllus		*							
Lithocarpus konishii		*		*					

Lithocarpus macilentus		*							
Lithocarpus melanochromus		*							
Lithocarpus naiadarum		*							
Lithocarpus obovatilimbus		*							
Lithocarpus paihengii		*	*	*					
Lithocarpus paniculatus		*	*	*					
Lithocarpus phansipanensis		*							
Lithocarpus quercifolius		*		*					
Lithocarpus uvariifolius		*		*					
Lithocarpus dodonaeifolius				*					
Lithocarpus formosanus				*					
Lithocarpus kawakamii				*					
Lithocarpus lepidocarpus				*					
Lithocarpus nantoensis				*					
Lithocarpus shinsuiensis				*					
Lithocarpus yongfuensis				*					
Lithocarpus edulis					*				
Lithocarpus elegans							*		*
Lithocarpus kamengensis							*		
Lithocarpus milroyi							*		
Lithocarpus polystachya							*		
Lithocarpus wrayi							*		*
Lithocarpus annamensis								*	
Lithocarpus bacgiangensis								*	
Lithocarpus ceriferus								*	
Lithocarpus curtisii								*	*
Lithocarpus ducampii								*	

(continued)

Table 8.5 (continued)

Species	SW Ch.	S Ch.	C Ch.	E Ch.	Jap.	Kor.	E Himl.	Indoch.	Mal. Reg.
Lithocarpus echinops								*	
Lithocarpus encleisacarpus								*	*
Lithocarpus guinieri								*	
Lithocarpus harmandii								*	
Lithocarpus leptogyne								*	
Lithocarpus licentii								*	
Lithocarpus lindleyanus								*	
Lithocarpus longipes								*	
Lithocarpus orbicarpus								*	
Lithocarpus orneus								*	
Lithocarpus polistachyus								*	
Lithocarpus pseudomoluccus								*	
Lithocarpus recurvatus								*	
Lithocarpus sundaicus								*	
Lithocarpus tijsmannii								*	
Lithocarpus touranensis								*	
Lithocarpus urceolaris								*	
Lithocarpus vestitus								*	
Lithocarpus wallichianus								*	*
Lithocarpus bancanus									*
Lithocarpus beccarianus									*
Lithocarpus bennettii									*
Lithocarpus blumeanus									*
Lithocarpus bullatus									*
Lithocarpus burkillii									*
Lithocarpus cantleyanus									*

Lithocarpus clementianus									*
Lithocarpus conocarpus									*
Lithocarpus coopertus									*
Lithocarpus crassinervius									*
Lithocarpus dasystachyus									*
Lithocarpus echinifer									*
Lithocarpus erythrocarpus									*
Lithocarpus ewyckii									*
Lithocarpus ferrugineus									*
Lithocarpus gracilis									*
Lithocarpus hallieri									*
Lithocarpus hatusimae									*
Lithocarpus havilandii									*
Lithocarpus hendersonianus									*
Lithocarpus indutus									*
Lithocarpus jacobsii									*
Lithocarpus kalkmanii									*
Lithocarpus keningauensis									*
Lithocarpus kingianus									*
Lithocarpus kingii									*
Lithocarpus kostermansii									*
Lithocarpus kunstleri									*
Lithocarpus lampadarius									*
Lithocarpus lucidus									*
Lithocarpus luteus									*
Lithocarpus maingayi									*

(continued)

Table 8.5 (continued)

Species	SW Ch.	S Ch.	C Ch.	E Ch.	Jap.	Kor.	E Himl.	Indoch.	Mal. Reg.
Lithocarpus meijeri									*
Lithocarpus neorobinsonii									*
Lithocarpus nieuwenhuisii									*
Lithocarpus ovalis									*
Lithocarpus palungensis									*
Lithocarpus papillifer									*
Lithocarpus platycarpus									*
Lithocarpus porcatus									*
Lithocarpus pseudokunstleri									*
Lithocarpus pulcher									*
Lithocarpus pulongtauensis									*
Lithocarpus rassa									*
Lithocarpus revolutus									*
Lithocarpus rotundatus									*
Lithocarpus ruminatus									*
Lithocarpus sericobalanus									*
Lithocarpus turbinatus									*

Data sources: Soepadmo (1972), eFloras (2014), Ohwi et al. (1965), Negi and Naithani (1995), Soepadmo et al. (2000), Tang and Ohsawa (2009), Springer (2014), Wikipedia (2014a), Terrestrial Biozones (2014)

An asterisk indicates the presence

SW Ch. Southwestern China including the Yaluzangbujiang area and its surroundings in Tibet, Yunnan, Sichuan and Chongqing, and Guizhou, *S Ch.* Southern China including Guangxi, Guangdong, and Guizhou, *C Ch.* Central China including Hubei and Hunan, *E Ch.* Eastern China including Anhui, Jiangxi, Jiangsu, Zhejiang, Fujian, and Taiwan, *Jap.* Japan, *Kor.* Korea, *E Him.* Eastern Himalayas including Nepal, India, and Bhutan, *Indoch.* Indochina including Cambodia, Laos, Myanmar, Thailand, Vietnam, *Mal. Reg.* The Malaysian region including Malaysia; also Indonesia and surrounding islands

Fig. 8.4 Part 1 (**a**) Fruits with short peduncles (0.5–1.5 cm) of *Fagus lucida* (deciduous). (**b**) Fruits with long peduncles (1–10 cm) of *Fagus longipetiolata* (deciduous)

Fig. 8.4 (continued) Part 2 (**c**) Fruits of *Castanea mollissima* (deciduous). Cupule densely covered with pubescent spinelike bracts (Photograph courtesy of Huan-Chong Wang). (**d**) A fruit of *Quercus variabilis* (deciduous). Cupule cupular, enclosing 2/3 of nut; bracts long and subulate, inflexed, pilose (Photograph courtesy of Cheng-Tao Lin)

Fig. 8.4 (continued) Part 3 (**e**) Foliage of *Quercus pannosa* (evergreen sclerophyllous). Leaf blade oblong, elliptic, ovate, or obovate, abaxially with yellowish-brown stellate hairs, hairy along midvein, base rounded, margin with spiniform teeth, apex obtuse to mucronate; secondary veins 5–12 on each side of midvein; tertiary veins often hidden by indumentum (Photograph © Jan De Langhe, Ghent University Botanical Garden). (**f**) Fruits of *Quercus tatakaensis* (evergreen sclerophyllous). Cupule cupular, enclosing 1/4–1/2 of nut; bracts triangular to triangular-ovate, crowded, gray pubescent except for apex. Nut ovoid to ellipsoid (Photograph courtesy of Huan-Ching Lin)

Fig. 8.4 (continued) Part 4 (**g**) Fruits of *Cyclobalanopsis glauca*. Cupule bowl-shaped, enclosing 1/3–1/2 of nut, outside white puberulent or glabrous, inside white sericeous; bracts in 5 or 6 rings, crowded, margin entire or denticulate. Nut ovoid, oblong-ovoid, or ellipsoid, glabrous or rarely hairy; flat or slightly convex (Photograph courtesy of Cheng-Tao Lin). (**h**) Fruits of *Cyclobalanopsis stenophylloides*. Cupule cupular, enclosing about 1/2 or less of nut, outside and inside grayish velutinous; bracts in 6 or 8 rings, margin dentate. Nut ellipsoid, glabrous; stylopodium persistent, conspicuous, 3-ringed (Photograph courtesy of Cheng-Tao Lin)

Fig. 8.4 (continued) Part 5 (**i**) Fruits of *Castanopsis orthacantha*. Cupule subglobose, bracts pubescent, spinelike, basally conate and transversely united to 4–6-cristate rings (Photograph courtesy of Long-Yuan He). (**j**) Foliage of *Castanopsis orthacantha*. Leaf blade ovate, base often oblique. (**k**) Foliage and fruits of *Castanopsis fargesii*. Leaf blades covered with rust-colored trichomes. Cupule globose to broadly ovoid; bracts spinelike, connate and in bundles basally (Photograph courtesy of Cheng-Tao Lin)

Fig. 8.4 (continued) Part 6 (**l**) Fruits of *Lithocarpus mairei*. Cupule bowl-shaped, enclosing about 1/3 of nut; Nut broadly conical to depressed globose (Photograph courtesy of Long-Yuan He). (**m**) Fruits of *Lithocarpus lepidocarpus*. Cupules in clusters, rarely solitary. Cupule subglobose, completely enclosing nut, apically flat; bracts imbricate to spirally arranged apically on cupule, rhomboid, squamose, with grayish waxy scalelike trichomes (Photograph courtesy of Cheng-Tao Lin). (**n**) Fruits of *Lithocarpus dealbatus*. Cupule cupular, enclosing 1/2 to most of nut; bracts imbricate, triangular, appressed or a few spreading. Nut depressed globose to subglobose, apex rounded or flat

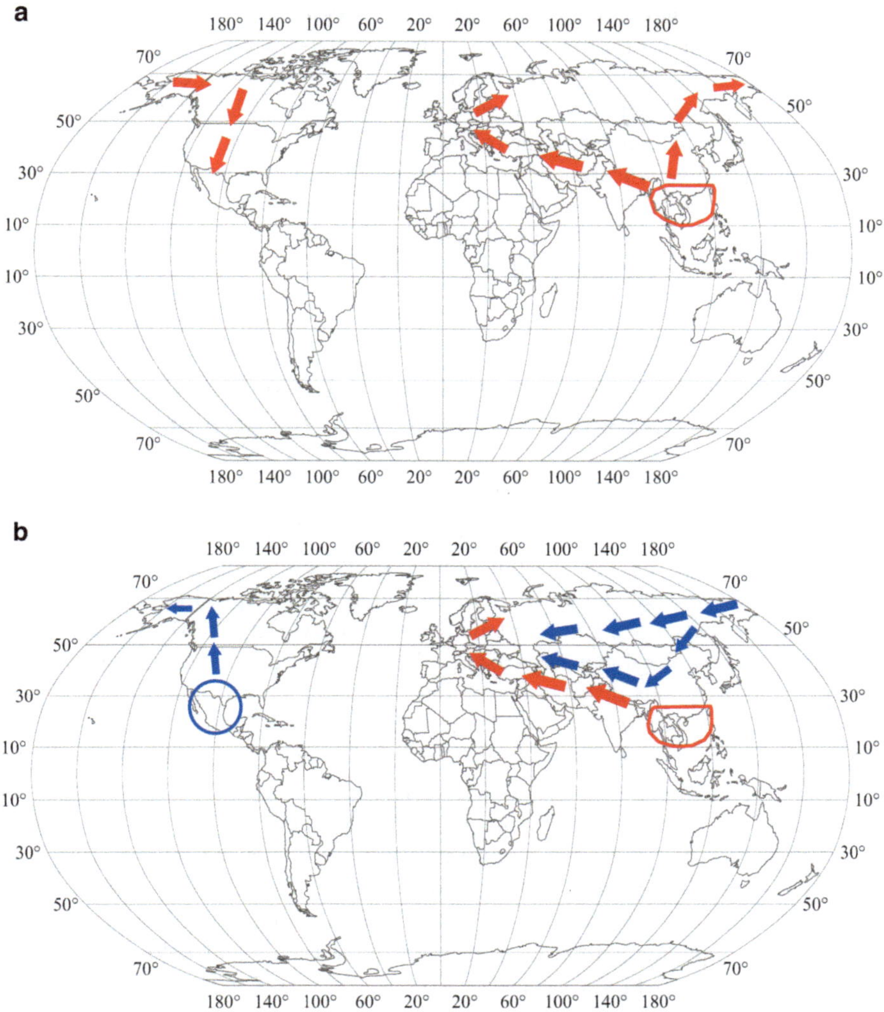

Fig. 8.5 (**a**) Origin and migration of *Quercus* by Zhou (1992). (**b**) Origin and migration of *Quercus* by Trelease (1924) and Manos and Stanford (2001)

and North America via the North Atlantic Land Bridge before the Eocene, and via the Asia-Bering land bridge after the Miocene (Fig. 8.5a). However, Trelease (1924) and Manos and Stanford (2001) suppose that the genus *Quercus* derived from the widely distributed boreal-tropical deciduous forest that extended throughout the Northern Hemisphere at the beginning of the Tertiary, and differentiated further as the separation between the continents became more pronounced. Accordingly, *Quercus* species developed on the different continents through vicariance between Asia and North America. In Asia the "ancestral group" differentiated into the subgenus *Cyclobalanopsis* and the section Cerris of the genus *Quercus*. And species of the Cerris group later migrated westward to Europe. In America, the "ancestral group" differentiated into the sections Lepidobalanus (white oaks), Protobalanus and Erythrobalanus (red oaks) with subsequent migration of white oaks from North America to Asia and further to Europe (Fig. 8.5b) (Quercus Portal 2014).

Despite the disagreement on the origin of the genus, paleobotanists agree on the extremely rapid diversification of the genus during the Oligocene and Miocene as a response to marked climatic changes. Most fossil remains of that period are similar to extant samples. Hence it is believed that most of the extant species existed already in the mid Miocene (Trelease 1924; Axelrod 1983).

8.2 The Subtropical Evergreen Broad-Leaved Forests of SW China as compared with Those in Other Regions

The subtropical/warm-temperate evergreen broad-leaved forest (EBLF) is the most diverse in East Asia. It is widely distributed in southwestern, southern, central and southeastern mainland China, and in Taiwan, southwestern Japan, southern Korea, and northern Myanmar, India, Bhutan, Nepal and some tropical mountain regions such as in Vietnam, Laos and Thailand. Li (1997) divided the EBLF of East Asia into four regions: the Japanese, eastern Chinese, western Chinese, and western Chinese sclerophyllous distribution regions. In China it corresponds to the region from 23° to 33° N and from 95° to 122° E (Fig. 8.6). The subtropics from 33° to 35° N constitute a transitional zone from EBLFs to deciduous broad-leaved forests. The differences between the western and eastern EBLFs within China result mainly from the East Asian monsoon climate. In general, Chinese regions east of c. 103° E (referred to here as the eastern region) are strongly influenced by the East Asian monsoon, while the regions to the west are affected by the Indian monsoon (see Fig. 1.15). Figure 8.7a–f offers the general information on the climate of the EBLFs. The annual temperature difference between the warmest and coldest months is smaller in the west (10–14 °C) than in the east (14–27 °C) (see Fig. 1.17d). In Holdridge's life zone system (1947, 1967), bio-temperature (the sum of temperatures >0 °C, if mean monthly temperature >30 °C is calculated as 30 °C) ranges from 12 to 24 °C for the EBLF. In the subtropical EBLF region of China, the bio-temperature is 15–22 °C (Fig. 8.7a). Kira (1945) has proposed a warmth index (WI) and a coldness index (CI) for vegetation zones in East Asia:

$$\mathrm{WI}\ (^\circ\mathrm{C}) = \sum (t-5) \cdot \mathrm{months};\ \mathrm{CI}\ (^\circ\mathrm{C}) = \sum (5-t') \cdot \mathrm{months}$$

Where t is the mean monthly temperature ≥5 °C and t′ is the mean monthly temperature ≤5 °C.

In Japan, a WI of 85–180 °C·months and a CI of −10 to 0 °C·months correspond to the EBLF distribution, and a CI of −10 °C is a decisive northern limit for the EBLF (Kira 1945; Hattori and Nakanishi 1985). In China, a WI of 125 °C and a CI of −5 °C determine the northern limit of the EBLF distribution; a WI of 135 °C,

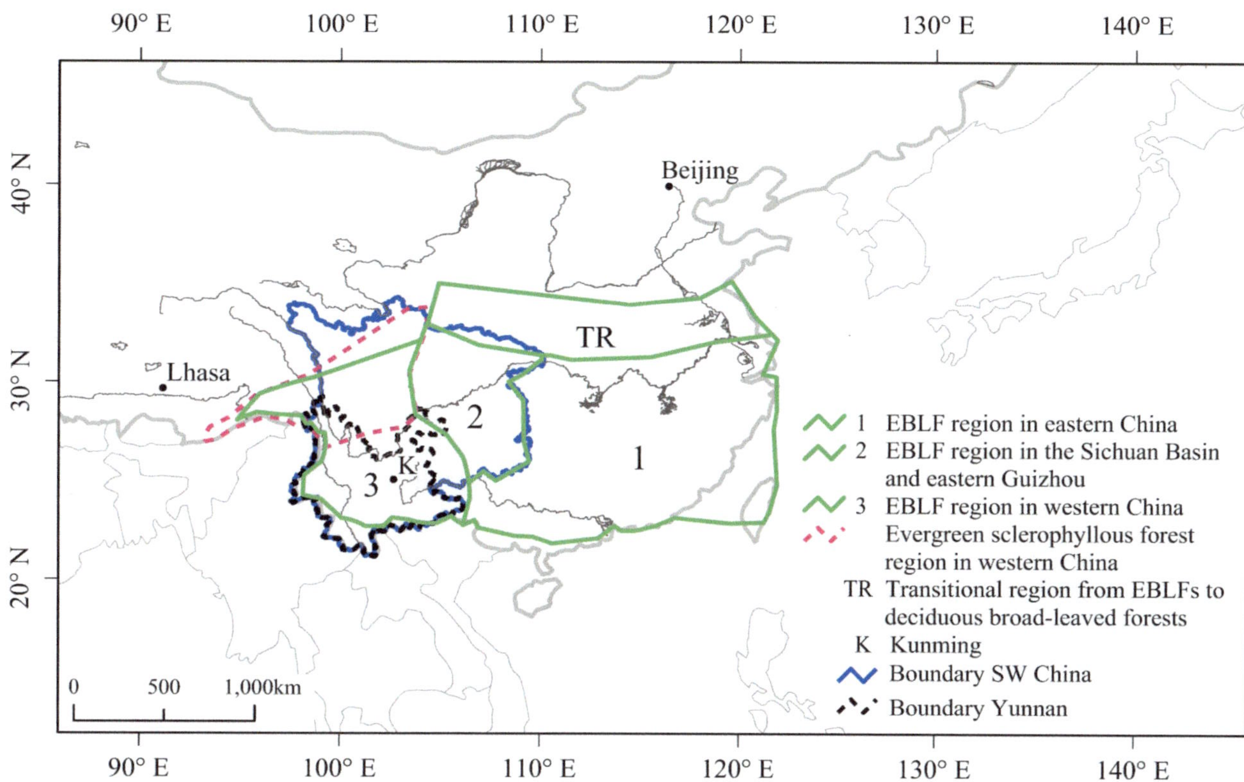

Fig. 8.6 Distribution of subtropical evergreen broad-leaved forests in China. Note: Chayu and Motou, in Yaluzangbujiang, southern Tibet, are included in the subtropical evergreen broad-leaved forest, although in deep gorges at 100–600 m asl, very limited patches of the tropical rain forest are found, dominated by *Dipterocarpus* (Data source: Tang (2015))

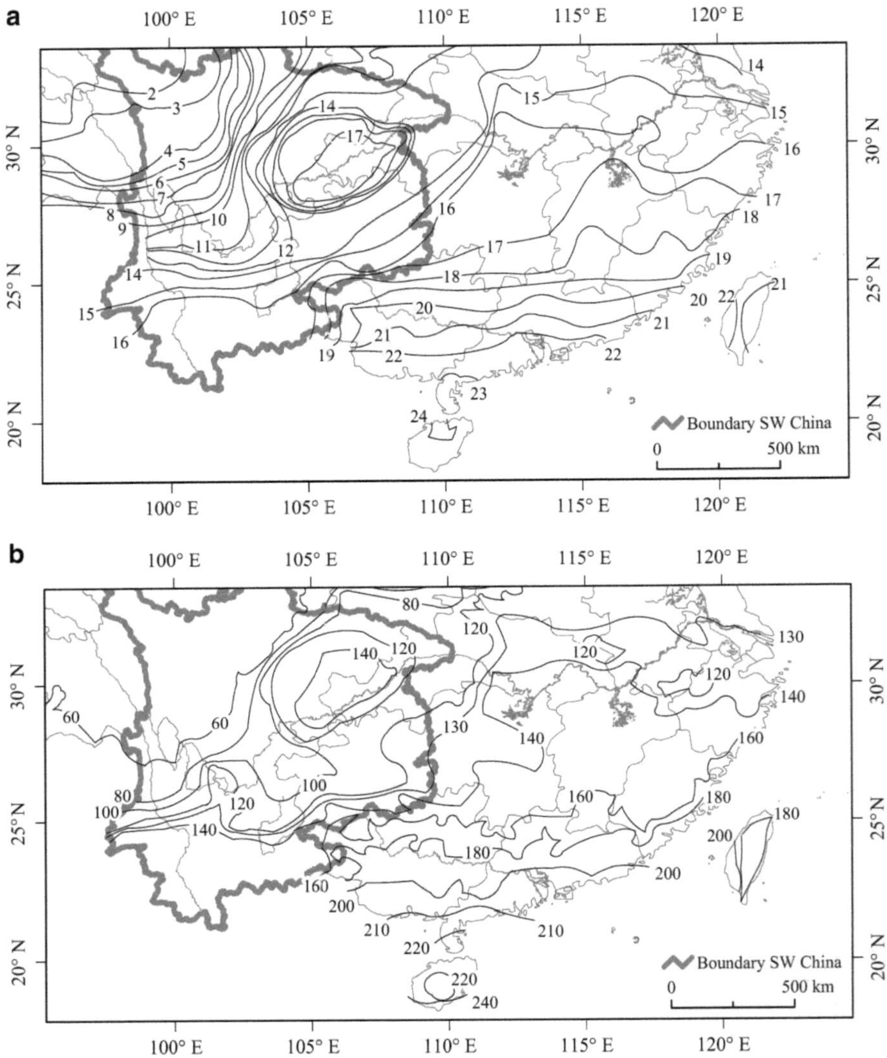

Fig. 8.7 Part 1 Climate in the EBLF region of China. (**a**) Annual mean bio-temperature (°C), (**b**) warmth index (WI)

and a CI of −2 °C correspond to the northern limit of the EBLF zone (Song 2013) (Fig. 8.7b, c).

In China, the annual precipitation for the EBLF in the eastern region ranges from 1,100 to 1,700 mm, and is between 800 and 1,000 mm in most areas of the western region (such as central and northern Yunnan and northwestern Guizhou) (Fig. 8.7d). However, due to the complexity of the topography and varying elevations, the annual rainfall is c. 1,500–1,800 mm in southern Yunnan, though it reaches 1,500–2,800 mm

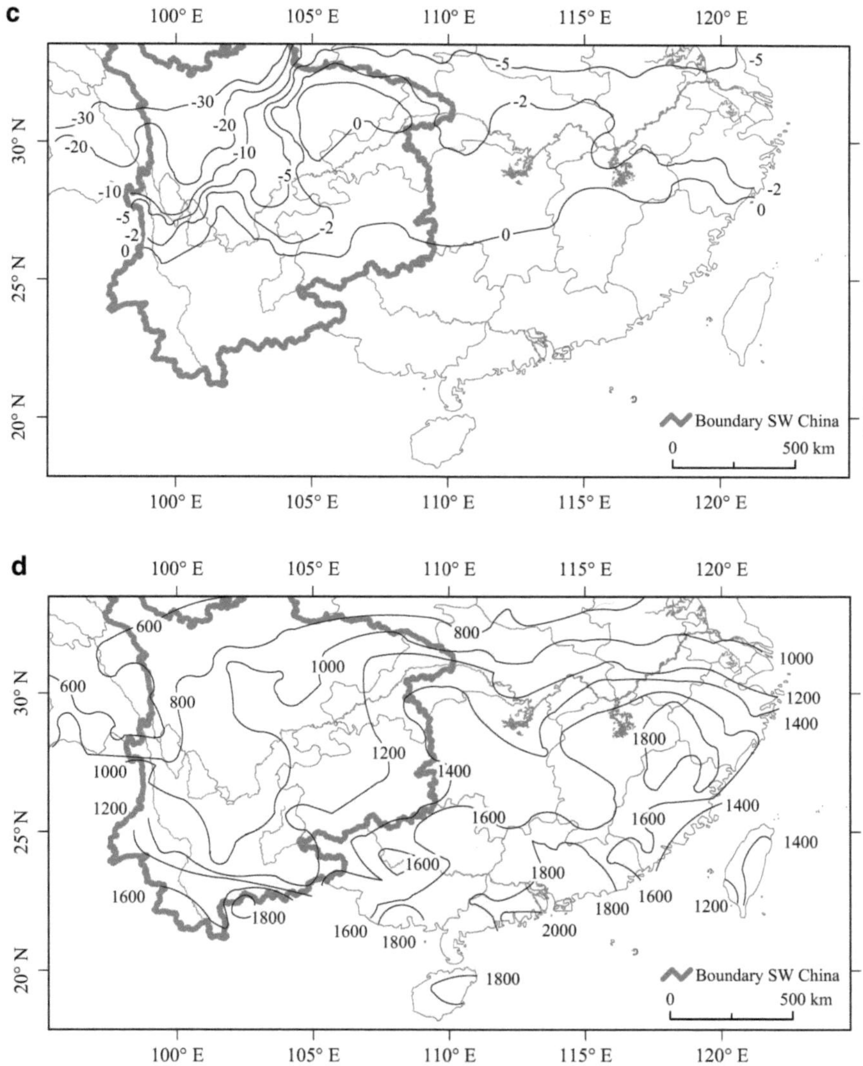

Fig. 8.7 (continued) Part 2 (**c**) Cold index (CI). (**d**) Annual precipitation (mm)

(but 3673 mm in Bapo, Dulongjiang) in some localities of western Yunnan, but in deep river valleys and on lee slopes it is just c. 600–800 mm. In the western region, 80–90 % of the annual precipitation falls from May to October, and the dry season is much more pronounced from November to April as compared with the east (Fig. 8.7e). The mean annual potential evapotranspiration for the subtropical EBLF region is c. 800–1,200 mm, and it is greater in the west than in the east. The Sichuan Basin area has the lowest evapotranspiration (Fig. 8.7f).

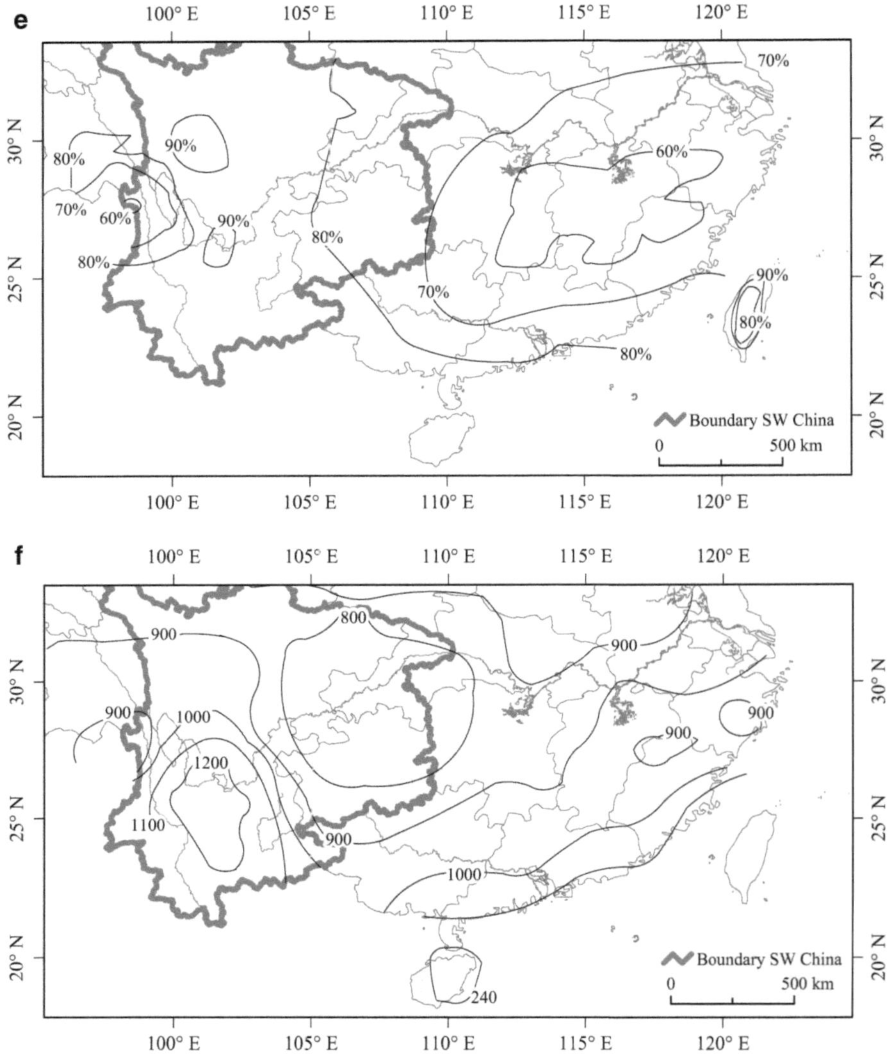

Fig. 8.7 (continued) Part 3 (**e**) the seasonal share (May–October) of precipitation (mm). (**f**) Annual potential evapotraspiration (Data sources: Song (2013) for **a–e**, Gao et al. (2006) for **f**)

For the subtropical EBLFs of SW China, four representative types are summarized as follows:

1. the semi-humid EBLF in Yunnan, western Sichuan and western Guizhou.
2. the monsoon EBLF in Yunnan.
3. the mid-montane moist EBLF in Yunnan, which I combine with the montane mossy EBLF as classified by former researchers (e.g. Wu 1980; Wu et al. 1987) in view of their physiognomy and ecological phenomena, including various epiphytes and lianas on tree trunks and branches.

4. the humid EBLF in the Sichuan Basin (also including Chongqing) and eastern Guizhou.

The montane mossy dwarf forest, dominated by species of *Rhododendron* along with some dwarf trees of the Fagaceae, grows in special topo-habitats such as near mountaintops. This is not considered a typical type of the subtropical EBLF. The evergreen broad-leaved sclerophyllous forest is also considered apart from the subtropical EBLF.

8.2.1 As compared with Other Chinese Regions

8.2.1.1 Forest Communities, Species Distribution and Composition

In this section, for convenience I define eastern China as all the Chinese regions to the east of SW China. Both climate and biogeography produce similarities and differences in floristic composition between southwestern and eastern China. In the southwest the climate depends more on altitude than on latitude, as do the species. The humid EBLF in the Sichuan Basin and eastern Guizhou is considered as the same type as the typical humid EBLF of eastern China. If we compare the EBLFs in Fig. 2.6 (in Chap. 2) with those in Fig. 8.8, representing 23 mountains, we find that the Sichuan Basin, eastern Guizhou and eastern China share some dominant canopy tree species of Fagaceae (e.g. *Castanopsis carlesii*, *Castanopsis fargesii*, *Castanopsis eyrei*, *Castanopsis tibetana*, *Lithocarpus glaber*, *Cyclobalanopsis glauca*, *Cyclobalanopsis myrsinifolia*), Lauraceae (e.g. *Phoebe neurantha*), Theaceae (e.g. *Schima superba*) and Elaeocarpaceae (e.g. *Elaeocarpus sylvestris*). In the understory, they often share some genera including *Ilex* (Aquifoliaceae), *Lindera*, *Litsea* (Lauraceae), *Camellia*, *Eurya* (Theaceae), *Symplocos* (Symplocaceae), and Vaccinium (Ericaceae). However, these forests lack flourishing epiphytes and lianas on tree trunks and branches, as compared to the mid-montane moist EBLF of Yunnan. Additionally, in the Yunnan area there are more types of plant communities dominated by *Lithocarpus* or *Cyclobalanopsis* than in eastern China, where the chief dominant genus is *Castanopsis*. The upper limits of the EBLFs are mainly 600–1,500 m in eastern China (but 2,500 m in Taiwan), much lower than those in western Yunnan and western Sichuan (2,400–2,800 m) and in the southern and eastern fringes of SW China (1,000–1,500 m) (Fig. 8.9).

Among the evergreen genera of the Fagaceae, evergreen *Quercus* has a wide altitudinal distribution ranging from low to high altitudes, such as 500–4,800 m in Yunnan and Sichuan, and 1,300–4,300 m in southeastern Tibet. In general, *Castanopsis* is distributed at lower altitudes, *Lithocarpus* and *Cyclobalanopsis* at higher altitudes; sometimes they exist in the same altitudinal range and co-dominate the EBLF. Figure 8.10 shows an example from SE Tibet.

Now I would like to give an example to support drawing a comparison between Yunnan and Taiwan. Taiwan is situated between the southernmost part (Ryukyu Islands) of the Japanese Archipelago and the Asian continent and runs in a north–south direction east of mainland China, north of the Philippines, and is traversed

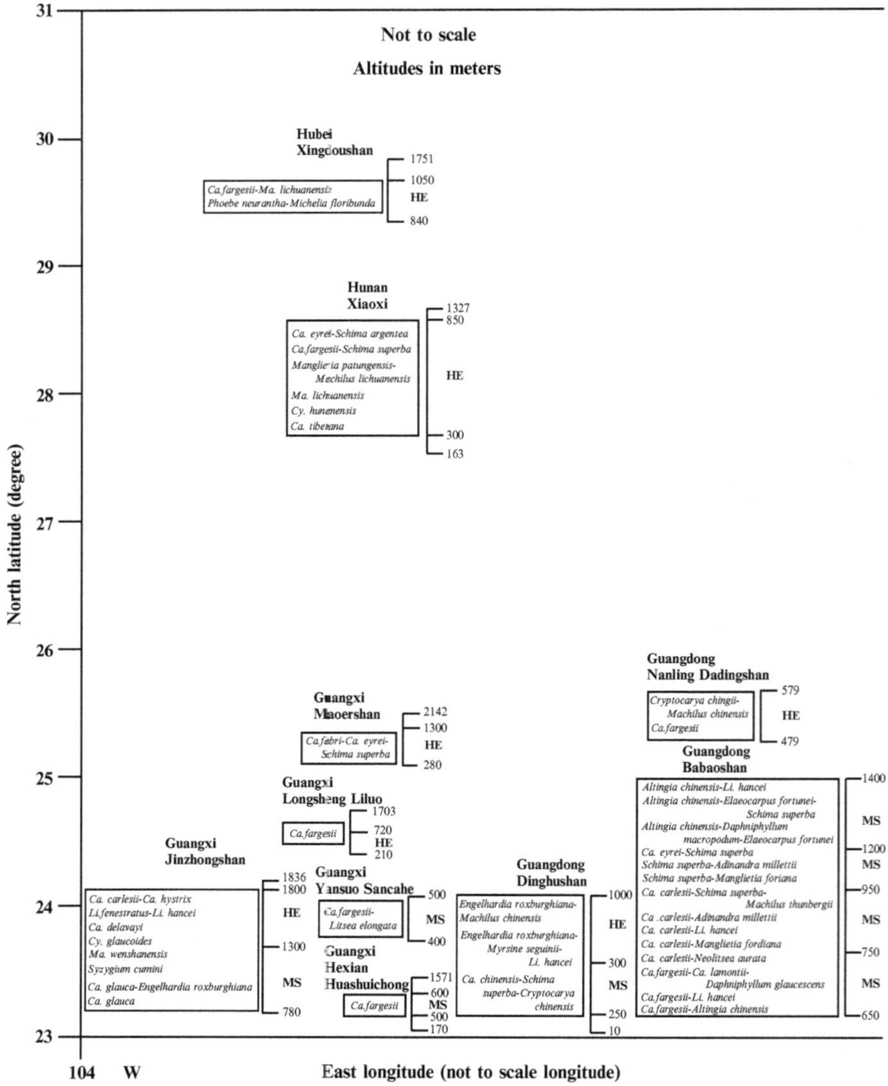

Fig. 8.8 The spatial distribution patterns of dominant evergreen broad-leaved communities along latitudes, longitudes and altitudes in eastern China. *MS* Monsoon EBLF, *HE* humid EBLF, *Cy Cyclobalanopsis*, *Ca Castanopsis*, *Li Lithocarpus, Ma Machilus* (After Tang (2015). Data sources: Chen and Su (1995), Huang et al. (1998), Hu et al. (2003), Da et al. (2005), Xia and Nie (2004), Zheng et al. (2004), Zhu et al. (2004), Peng et al. (2006)), Huang et al. (2006), Wan (2007), Deng et al. (2008), Wang et al. (2009), Qiu et al. (2010), Chang et al. (2013), Tang et al. (2013), Yang et al. (2013))

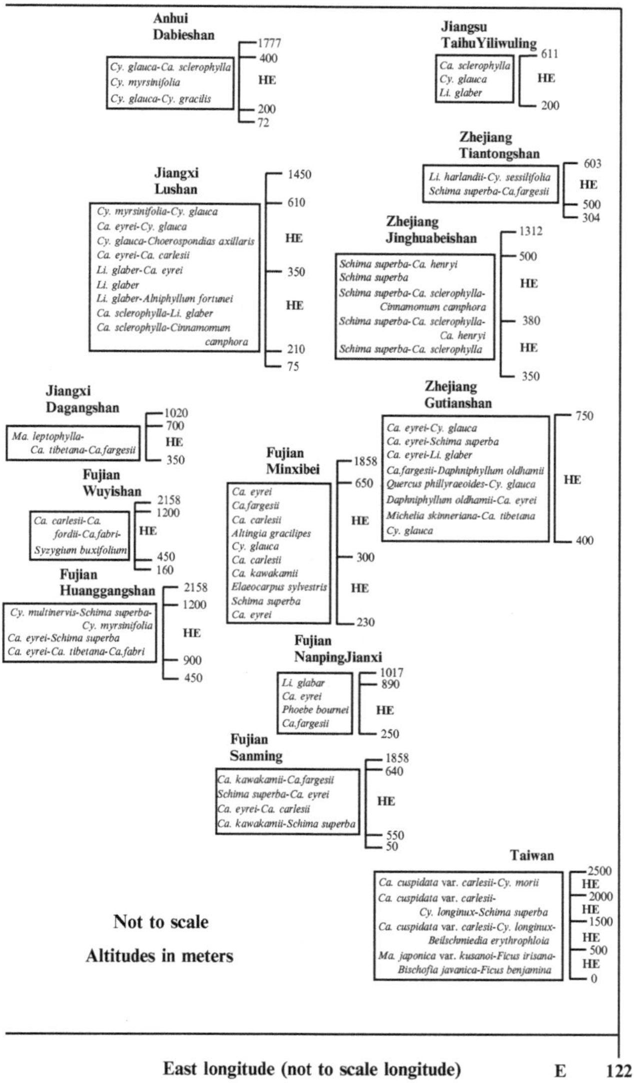

Fig. 8.8 (continued)

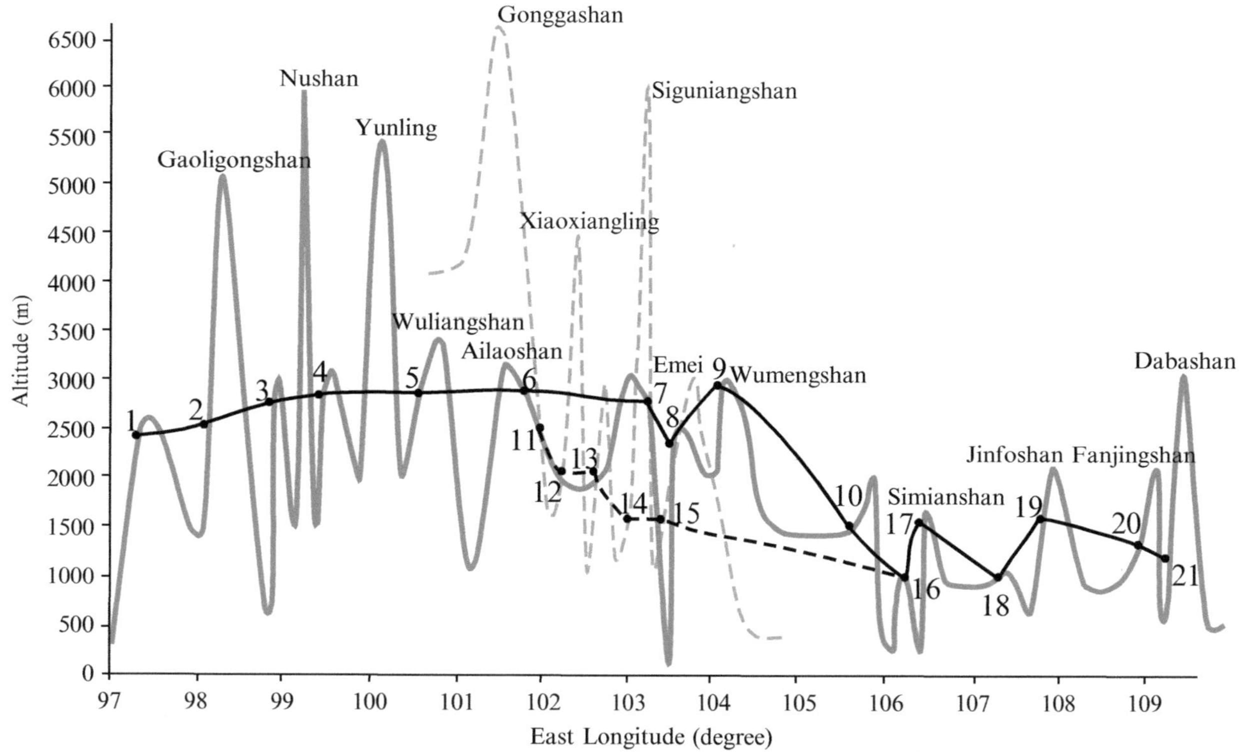

Fig. 8.9 The upper limits of EBLFs in SW China. A *dot* indicates the upper limit of the EBLF on its respective mountain. *1* Tongbiguan, *2* Dulongjiang, *3* Xiaoheshan in Longling, *4* Yongdedaxueshan, *5* Wuliangshan, *6* Ailaoshan, *7* Xilongshan, *8* Daweishan, *9* Wumengshan, *10* Tianlongshan in Guizhou, *11* Gonggashan, *12* Xiaoxiangling in Sichuan, *14* Siguliangshan, *15* Emeishan, *16* Jingyunshan, *17* Simianshan, *18* Maolan in Guizhou, *19* Jinfoshan, *20* Fanjingshan, *21* Kanghe on Dabashan (Data source: Tang (2015))

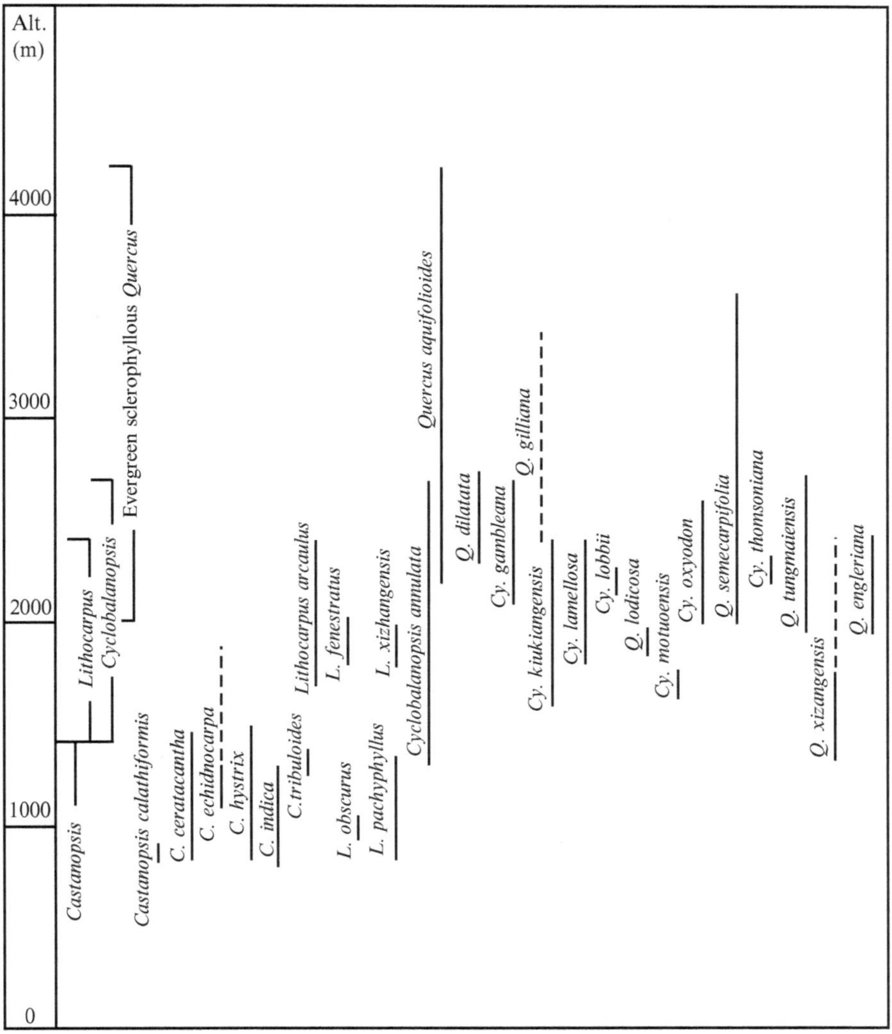

Fig. 8.10 Altitudinal distribution for the evergreen genera of the Fagaceae in Tibet (Data source: Chen (1988))

by the Tropic of Cancer. A land bridge connected southeastern China, Taiwan, and the Ryukyu Islands during the early Pleistocene about 1.5 million years ago, and it may have lasted into the middle Pleistocene (about one million years: Kizaki and Oshiro 1977). Species from southwestern China may have migrated from the Yunnan-Guizhou Plateaus to regions further east, eventually reaching Taiwan by the land bridge (Huang 2011), as evidenced by the island flora, which includes elements from the Sino-Himalaya, Sino-Japanese, Japanese, and Philippine regions.

Taiwan is dominated by forested mountains, with 30 % of its area strongly undulating between 1,000 and 3,000 m asl. More than 200 peaks are higher than 3,000 m; all of these are located within the Central Ridge. This lofty range runs basically along the north–south axis of the island (Fig. 8.11). Taiwan has a distinctive oceanic subtropical climate (Hsieh et al. 1997). The annual rainfall ranges from 1,000 to >4,500 mm. With respect to the EBLF zone, annual precipitation is approximately 1,800–2,500 mm at altitudes of <500 m, 2,350–3,000 mm at altitudes of 500–1,500 m, and approximately 2,400–2,900 mm at altitudes of 1,500–2,500 m. Mean monthly temperatures in the lowlands range from 15 to 28 °C. Typhoons are not infrequent during the summer and early autumn, but have less impact in western Taiwan, which is sheltered by the Central Ridge. The climate ranges from (tropical) subtropical in the lowlands to alpine at the highest altitudes.

The EBLFs of Taiwan cover a wide altitudinal range from sea level to 2,500 m asl. They can be classified into four altitudinal vegetation zones: *Ficus–Machilus* zone (<500 m), *Machilus–Castanopsis* zone (500–1,500 m), lower *Cyclobalanopsis* zone (1,500–2,000 m), and upper *Cyclobalanopsis* zone (2,000–2,500 m) (Su 1984). Most of the lowland forests have been cleared for agriculture and industry, with only a few modified remnants remaining. Major tree species (RBA $\geq$ 1.5 %) in each altitudinal forest zone are shown in Table 8.6. The dominant tree species in the *Ficus–Machilus* zone at low altitudes are *Machilus*, *Ficus*, and *Bischofia* (Euphorbiaceae). In the *Machilus–Castanopsis* zone at middle altitudes, the dominant tree species are those of *Castanopsis* and *Cyclobalanopsis* and of *Machilus* and *Beilschmiedia*. In the lower *Cyclobalanopsis* zone at middle-high altitudes, trees of *Castanopsis* and *Cyclobalanopsis*, *Schima* and *Machilus* are dominant, and in the upper *Cyclobalanopsis* zone at high altitudes, species of *Castanopsis*, *Cyclobalanopsis*, *Trochodendron* (Trochodendraceae), and *Schima* are dominant. These data clearly reveal that in Taiwan *Castanopsis cuspidata* var. *carlesii* is distributed widely along the altitude zones, while the low altitude zone (below 500 m) is distinguished by several species of *Ficus* (e.g. *Ficus irisana*, *Ficus benjamina*, *Ficus microcarpa*, *Ficus nervosa*, and *Ficus virgata*), the middle zone (500–1,500 m) by several species of *Machilus* (e.g. *Machilus japonica*, *Machilus thunbergii*, *Machilus konishii* and *Machilus zuihoensis* var. *mushaensis*), the middle-high zone (1,500–2,000 m) by *Cyclobalanopsis longinux*, and the high altitude zone (2,000–2,500 m) by *Cyclobalanopsis morii*. The typical physiognomy of EBLFs in Taiwan is shown in Fig. 8.12a, b. For Yunnan, it can be seen in Fig. 2.3a–f.

8.2.1.2 Effects of Environmental Factors on Species Diversity

On the basis of data collected from 105 to 223 plots (20×20 m each) for Yunnan and Taiwan, respectively, Tang et al. (2013) report that the richness of taxa of the EBLFs as a whole in Yunnan at the family (89) and genus (242) levels have values similar to those in Taiwan (82 families, 221 genera), while at species level the Yunnan EBLFs have a higher value (610) of richness of taxa than those of Taiwan (533). The EBLFs

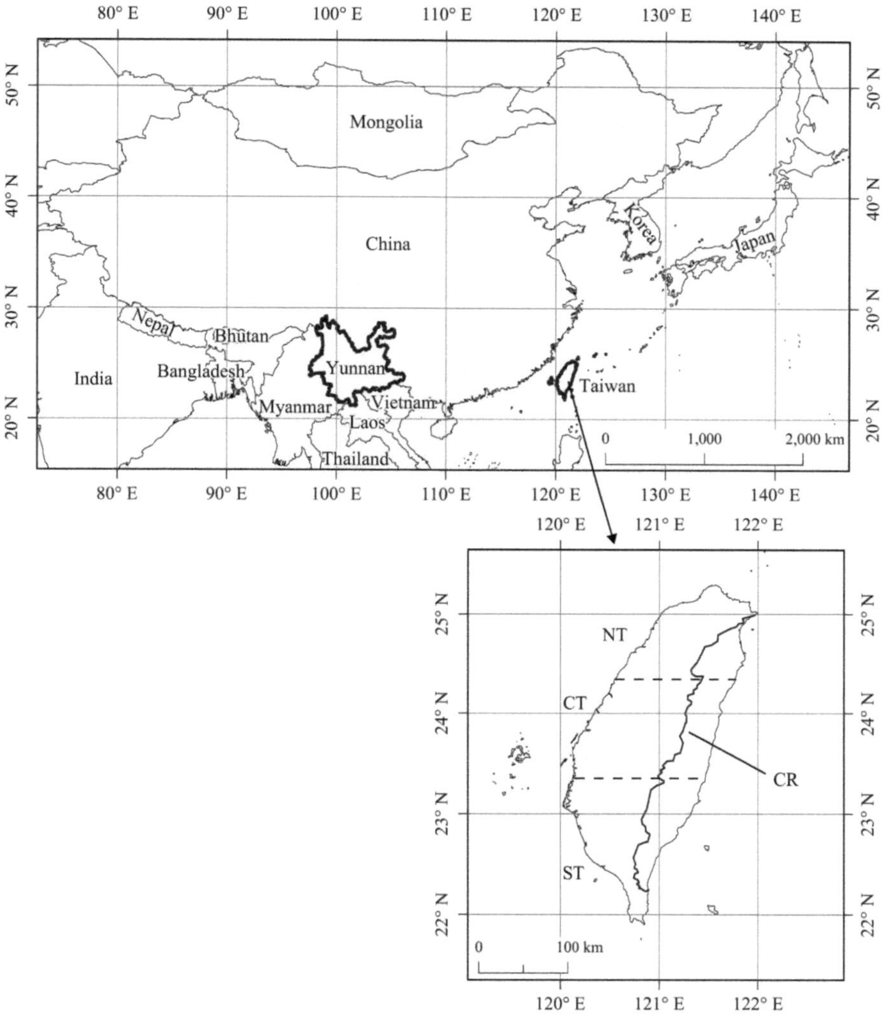

Fig. 8.11 Map showing the location of Taiwan. *NT* northern Taiwan, *CT* central Taiwan, *ST* southern Taiwan, *CR* Central Ridge

in Yunnan and Taiwan had 30 species, 96 genera, and 59 families in common. The Sørensen's similarity index (beta diversity) at the family level between the EBLFs of Yunnan and Taiwan is 0.69, at the genus level 0.41, and at the species level 0.05.

Species diversity models comparing the two EBLF ecosystems of Yunnan and Taiwan are proposed in Fig. 8.13a, b. In Yunnan, the monsoon EBLF in the mesic-humid sites tends to have higher species richness and diversity (Fig. 8.13a). Along the altitudinal gradients in Taiwan, higher diversity is found in the *Machilus–Castanopsis* zone at the middle altitudes (500–1,500) and in the lower *Cyclobalanopsis* zone at the mid-high altitude (1,500–2,000 m) (Fig. 8.13b). Species diversity, as influenced by temperature and precipitation, varies with altitude. Thus, altitude in the EBLFs of Taiwan may be a proxy for temperature,

Table 8.6 Major tree species (RBA ≥ 1.5 %) in each altitudinal forest zone of Taiwan

Altitudinal forest zone	FM	MCa	LCy	UCy
Species	RBA (%)	RBA (%)	RBA (%)	RBA (%)
Machilus japonica var. *kusanoi*	21.13	1.72		
Ficus irisana	15.62			
Bischofia javanica	14.74			
Ficus benjamina	13.08			
Ficus microcarpa	7.91			
Neolitsea konishii	4.37			
Schefflera octophylla	3.10	2.13		
Ficus nervosa	1.88			
Ficus virgata	1.51			
Castanopsis cuspidata var. *carlesii*		8.14	12.04	26.13
Cyclobalanopsis longinux		7.42	10.61	2.90
Beilschmiedia erythrophloia		6.87	3.49	
Quercus variabilis		5.99		
Schima superba		4.94	7.92	5.52
Machilus japonica		3.53	6.18	
Machilus thunbergii		3.38	6.78	2.91
Machilus konishii		3.00		
Castanopsis fabri		2.71		
Pasania shinsuiensis		2.36		
Litsea acuminata		2.30	4.52	
Machilus zuihoensis var. *mushaensis*		2.18		
Sloanea formosana		2.15		
Carpinus kawakamii		2.10		
Pasania kawakamii		1.62		
Acer albopurpurascens		1.51		
Machilus zuihoensis			5.43	
Elaeocarpus japonicus			3.51	
Michelia compressa			3.20	
Cyclobalanopsis stenophylloides			2.03	4.71
Elaeocarpus sylvestris			1.81	
Cyclobalanopsis morii			1.53	19.84
Lithocarpus amygdalifolius			1.47	2.33
Trochodendron aralioides				6.90
Cyclobalanopsis sessilifolia				2.98
Neolitsea acuminatissima				1.62

Data source: Tang et al. (2013)
FM Ficus-Machilus zone, *MCa Machilus-Castanopsis* zone, *LCa* Lower *Cyclobalanopsis* zone, *UCa* Upper *Cyclobalanopsis* zone, *RBA* Relative basal area

because the annual rainfall is abundant (1,800–3,000 mm) throughout the EBLF zone. In short, site moisture is the best predictor of species composition and diversity in the EBLFs of Yunnan, while temperature along altitudes plays a decisive role in Taiwan. For forest ecosystems, various temporal trends in diversity have been observed during succession, including increases, decreases, and one or more peaks

Fig. 8.12 (**a**) An evergreen broad-leaved forest dominated by *Castanopsis cuspidata* var. *carlesii*, *Castanopsis kawakamii*, *Pasania* (*Lithocarpus*) *nantoensis* and *Cyclobalanopsis pachyloma* at 550–850 m asl in Lianhuachi, central Taiwan. (**b**) An evergreen broad-leaved forest dominated by *Castanopsis cuspidata* var. *carlesii*, *Pasania* (*Lithocarpus*) *kawakamii* and *Pasania* (*Lithocarpus*) *harlandii* around 1,500 m asl in Guanwu, mid-northern Taiwan (Photogrphs courtesy of Cheng-Tao Lin)

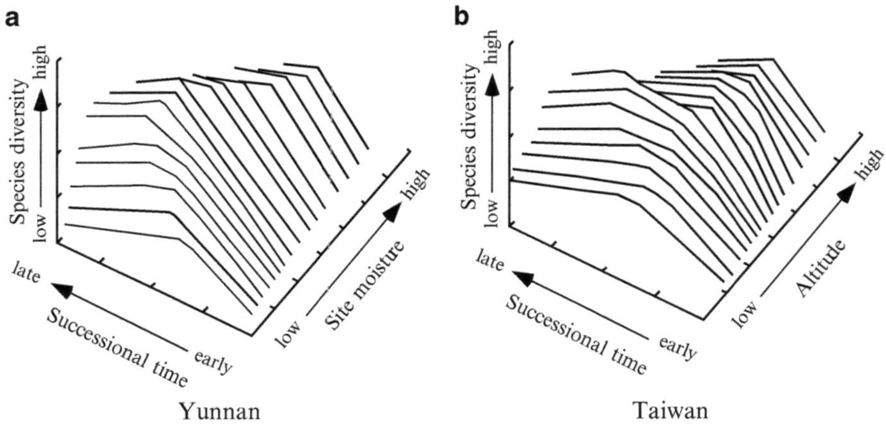

Fig. 8.13 (**a**) Changes in species diversity with site moisture and successional time in the EBLFs of Yunnan. (**b**) Changes in species diversity with altitude and successional time in the EBLFs of Taiwan (Data source: Tang et al. (2013))

over time (e.g., Monk 1967; Habeck 1968; Auclair and Goff 1971; Peet 1978; Halpern and Spies 1995; Turner et al. 1997). Relationships between diversity and succession are complex and should be seen as system-specific. Secondary forests, under varying degrees of human disturbance, such as coppice woods or those growing on abandoned farmland, exhibit a range of stages of plant succession in both the Yunnan and Taiwan sites. The peak in diversity in the stands at the middle successional stage of each study site in Yunnan and Taiwan (Fig. 8.13a, b) reflects the persistence of species with varying attributes, including fast growth, slow growth, light demands, limited shade-tolerance, and greater shade-tolerance, contributing to significantly higher species diversities as compared with the early successional stage. In general, as a forest develops, species diversity in the late stage tends to decrease with the disappearance of short-lived light-demanding pioneer species (e.g., *Pinus*, *Alnus*, *Populus*, *Carpinus*, *Acacia*, *Mallotus*, and *Diospyros*); however, in both Yunnan and Taiwan the differences in species diversity between the middle and late successional stages are not significant, which reflects the recruitment of shade-tolerant evergreen trees (e.g., *Cyclobalanopsis*, *Castanopsis*, *Machilus*, *Ficus*, *Manglietia*, and *Michelia*) into the overstory.

The effects of moisture, altitude, and disturbance/succession on the plant species diversity patterns can be seen in both Yunnan and Taiwan. The characteristics seen in the two forest ecosystems are summarized in Table 8.7.

8.2.2 As compared with Japan

8.2.2.1 Forest Communities and Species Diversity

Parts of Japan are situated along the northern border of the EBLF zone in East Asia. The Japanese Archipelago was connected to the Asian continent in the last glacial period, some 20,000–15,000 years ago. The northern temperate region of the

Table 8.7 Characteristics of the subtropical EBLFs of Yunnan and Taiwan

Region	Forest types/zones	Altitude (m)	AMT (°C)	AMP (mm)	Dominants	Species diversity	Drivers of the species diversity patterns
Yunnan	Semi-humid EBLF	1,500–2,500	8.1–13.5	900–1,200	*Cyclobalanopsis*, *Castanopsis*	Medium	Moisture, disturbance/succession
	Monsoon EBLF	800–2,000	13.3–20	1,100–2,500	*Castanopsis*, *Lithocarpus*, *Schima*, *Machilus*	Very high	Moisture, disturbance/succession
	Mid-montane EBLF	2,200–2,900	7.8–14.6	1,700–3,700	*Lithocarpus*, *Castanopsis*, *Cyclobalanopsis* *Manglietia*, *Alcimandra*, *Michelia*	High	Moisture, disturbance/succession
Taiwan	*Ficus-Machilus* zone	<500	15–28	1,800–2,500	*Ficus*, *Machilus*, *Bischofia*	High	Altitude, disturbance/succession
	Machilus-Castanopsis zone	500–1,500	17–23	2,350–3,000	*Machilus*, *Castanopsis*, *Beilschmiedia*, *Cyclobalanopsis*	Very high	Altitude, disturbance/succession
	Lower *Cyclobalanopsis* zone	1,500–2,000	14–17	2,400–2,800	*Cyclobalanopsis*, *Castanopsis*, *Schima*, *Machilus*	High	Altitude, disturbance/succession
	Upper *Cyclobalanopsis* zone	2,000–2,500	11–14	2,400–2,900	*Cyclobalanopsis*, *Castanopsis*, *Trochodendron*, *Schima*	High	Altitude, disturbance/succession

Revised from Tang et al. (2013)
AMT annual mean temperature, *AMP* annual mean precipitation

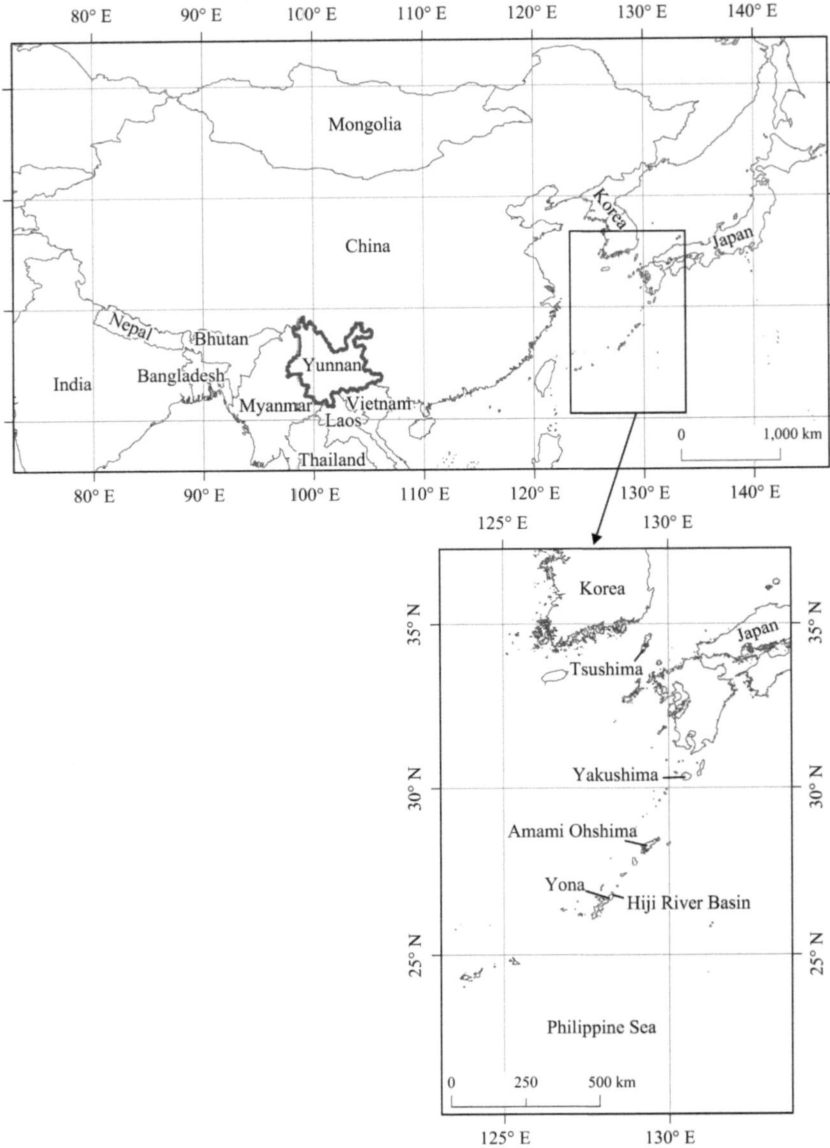

Fig. 8.14 Map showing the location of SW Japan

archipelago supports a summer-green forest, while an EBLF is found in the warm southern part (Numata 1974; Itow 1988). In SW Japan including Okinawa Island, Amami Ohshima and Yakushima (Fig. 8.14), the forests are dominated by *Castanopsis* or *Distylium* along with *Cyclobalanopsis* species (Suzuki 1979; Miyawaki 1980, 1981; Miyawaki and Sasaki 1985). The annual rainfall is generally 2,100–3,000 mm and annual mean temperatures vary from 14 to 22 °C in the localities of the forests.

Kira's warmth index (1945) ranges from 79 to 202 °C·months, Holdridge's annual biotemperature (1947, 1967) ranges from 10.8 to 21.8 °C for the forests of SW Japan. The temperature limits of the EBLF in SW Japan correspond roughly with both the Kira (85–180 °C·months) and the Holdridge (12–24 °C) values.

Tang and Ohsawa (2009) used the top dominant species as dominance types to obtain groups among the plots of Yunnan (47 plots) and Japan (39 plots), though most of the studied forests are co-dominated by several species. At 9 % floristic similarity, the forests fell into two groups: the Yunnan forests and the SW Japan forests (Fig. 8.15). The similarity dendrogram shows three forest types at 14 % of similarity for the Yunnan area, namely the subtropical mid-montane moist EBLF, subtropical semi-humid EBLF, and subtropical monsoon EBLF. At 63 % similarity, the forests fell into 19 dominant types. In the mid-montane moist EBLF, 10 dominance types are identified. Their altitudinal distribution is 2,200–2,800 m on the Ailao Mountains, and 1,880–2,400 m on the Gaoligong Mountains. Nine types are dominated by Fagaceae species (*Castanopsis wattii*, *Lithocarpus xylocarpus*, *Castanopsis remotidenticulata*, *Lithocarpus pachyphylloides* (*Lithocarpus crassifolius*), *Lithocarpus hancei*, *Castanopsis fleuryi*, *Cyclobalanopsis myrsinifolia*, *Cyclobalanopsis lamellosa*, *Castanopsis echinocarpa*), while one by a Magnoliaceae species (*Alcimandra cathcartii*). Four dominance types including *Castanopsis delavayi*, *Castanopsis orthacantha*, *Cyclobalanopsis glaucoides* and *Cyclobalanopsis delavayi* are found in the semi-humid EBLF in Xishan, Qiongzhusi, and Luquan of the Kunming area and Mt. Huafu of Chuxiong, central Yunnan. Five dominance types including *Castanopsis fabri*, *Castanopsis calathiformis*, *Machilus longipedicellata*, *Machilus gamblei* and *Castanopsis rufotomentosa* are in the monsoon EBLF of Fadou and Xiaoqiaogou, Xichou, and Wenshan, SE Yunnan. The dendrogram for SW Japan divided the subtropical/warm temperate evergreen broad-leaved rain forests (hereafter subtropical/warm temperate rain EBLF) into two groups at 40 % floristic similarity: the forests in Okinawa and Amami Ohshima are separated from those in Aya, Yakushima and Tsushima. At 63 % similarity, the dendrogram shows six dominance types: the *Distylium racemosum* type on Mt. Omori, Aya, Miyazaki and in Yakushima. The *Castanopsis sieboldii* subsp. *lutchuensis* type in Yona and the Hiji River basin, Okinawa and Amami Ohshima. The *Castanopsis cuspidata* type in Yakushima. The *Cyclobalanopsis acuta* and *Cyclobalanopsis gilva* types on Mt. Omori, Aya. And the *Castanopsis sieboldii* subsp. *sieboldii* type in Tsushima. The typical physiognomy of the EBLFs in Japan is shown in Fig. 8.16a, b. For Yunnan, it can be seen in Fig. 2.3a–f.

Table 8.8 shows the tree species with RBA $\geq$ 10 % in each plot of Yunnan and SW Japan. These genera and species are East Asian elements and indicate a very close relationship between the flora of Yunnan and SW Japan.

Tang and Ohsawa (2009) applied diversity indices to woody species (DBH $\geq$ 5 cm) to compare those forests in Yunnan and SW Japan (Fig. 8.17a–e). The monsoon EBLF forest of Yunnan had significantly higher values on the Shannon–Wiener H' (3.0) and Fisher's alpha (17.0) indices than the other forests. The mid-montane moist EBLF in Yunnan showed similar values on the Shannon–Wiener H' (2.43), Equitability J' (0.82), Simpson D (0.87) and Fisher's alpha (7.44) indices to the

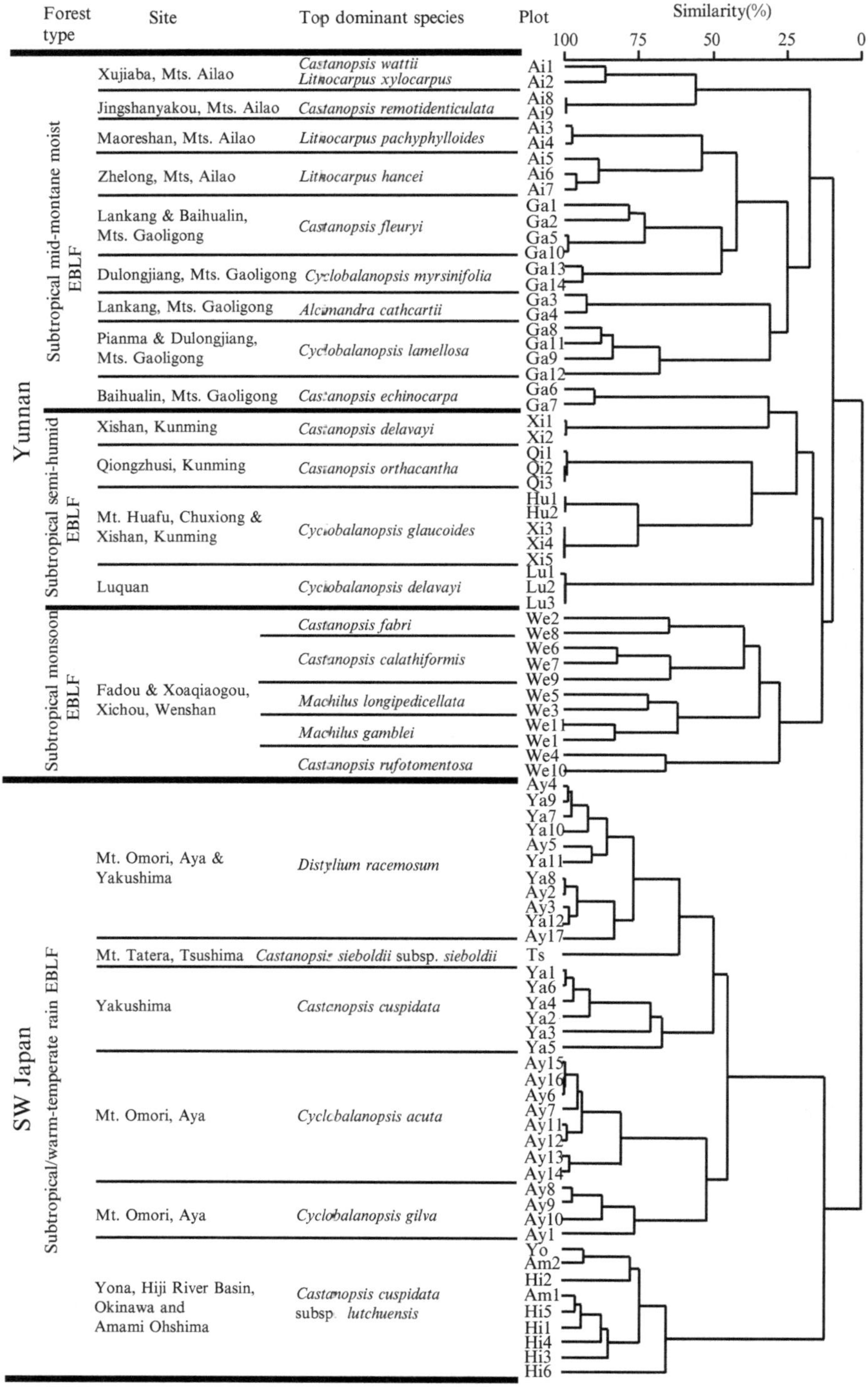

Fig. 8.15 Dendrogram showing floristic similarity between the EBLFs of Yunnan and SW Japan. Abbreviations are for plots (Data source: Tang and Ohsawa (2009))

Fig. 8.16 (**a**) An evergreen broad-leaved forest dominated by *Cyclobalanopsis acuta*, *Castanopsis cuspidata*, *Cyclobalanopsis gilva* and *Distylium racemosum* at 400–700 m asl in Aya, southwestern Japan. Photograph courtesy of Masahiko Ohsawa. (**b**) An evergreen broad-leaved forest dominated by *Castanopsis cuspidata*, *Cyclobalanopsis acuta*, *Cyclobalanopsis salicina* and *Distylium racemosum* at 500–850 m asl in Yakushima, southwestern Japan

Table 8.8 Tree species with relative basal area ≥10 % occurring in each plot of the EBLFs in Yunnan and SW Japan

Yunnan			SW Japan
MM	MS	SH	SWR
Lithocarpus xylocarpus	*Castanopsis fabri*	*Castanopsis delavayi*	*Distylium racemosum*
Castanopsis wattii	*Castanopsis calathiformis*	*Castanopsis orthacantha*	*Castanopsis cuspidata*
Alcimandra cathcartii	*Lithocarpus fenestratus*	*Cyclobalanopsis delavayi*	*Castanopsis siebodii* subsp. *sieboldii*
Cyclobalanopsis myrsinifolia	*Cyclobalanopsis glaucoides*	*Cyclobalanopsis glaucoides*	*Castanopsis sieboldii* subsp. *lutchuensis*
Cinnamomum iners	*Castanopsis rufotomentosa*	*Lithocarpus dealbatus*	*Cyclobalanopsis acuta*
Castanopsis orthacantha	*Bretschneidera sinensis*	*Lithocarpus confinis*	*Cyclobalanopsis gilva*
Castanopsis delavayi	*Cyclobalanopsis chevalieri*		*Cyclobalanopsis glauca*
Castanopsis fabri	*Exbucklandia populnea*		*Cyclobalanopsis miyagii*
Lithocarpus hancei	*Lithocarpus dealbatus*		*Cyclobalanopsis salicina*
Castanopsis fleuryi	*Machilus gamblei*		*Machilus japonica*
Cyclobalanopsis lamellosa	*Machilus longipedicellata*		*Machilus thunbergii*
Cyclobalanopsis oxyodon	*Machilus velutina*		*Ardisia sieboldii*
Lithocarpus craibianus	*Manglietiastrum sinicum*		*Camellia japonica*
Elaeocarpus lanceaefolius	*Neocinnamomum delavayi*		*Myrsine seguinii*
Lithocarpus confinis	*Ormosia henryi*		*Neolitsea aciculata*
Lithocarpus pachyphyllus	*Parakmeria yunnanensis*		*Diospyros morrisiana*
Hartia sinensis	*Phoebe angustifolia*		*Litsea acuminata*
Ilex forrestii	*Rhoiptelea chiliantha*		*Rhododendron tashiroi*
Castanopsis remotidenticulata	*Sloanea sinensis*		*Schima wallichii*
Lithocarpus pachyphylloides			*Symplocos prunifolia*
Castanopsis echinocarpa			*Ternstroemia gymnanthera*
Machilus longipedicellata			
Machilus viridis			
Manglietia insignis			
Michelia floribunda			

(continued)

Table 8.8 (continued)

Yunnan			SW Japan
MM	MS	SH	SWR
Phoebe legendrei			
Rhododendron irroratum			
Schima argentea			
Schima khasiana			
Schima noronhae			
Skimmia arborescens			
Symplocos dryophila			
Symplocos racemosa			

Revised from Tang and Ohsawa (2009)
MM subtropical mid-montane moist EBLF, *MS* subtropical monsoon EBLF, *SH* subtropical semi-humid EBLF, *SWR* subtropical/warm-temperate rain EBLF

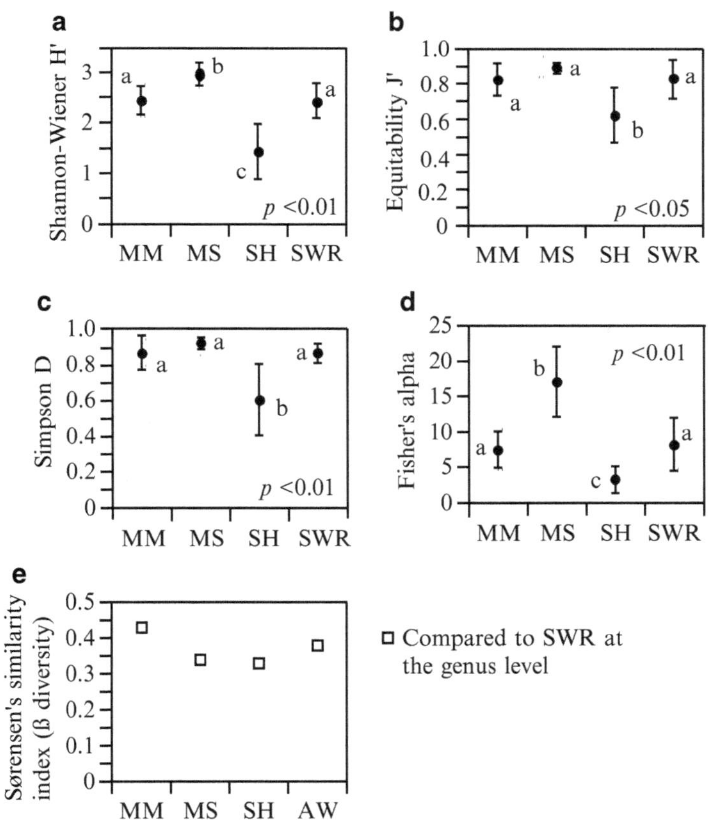

Fig. 8.17 Diversity indices for woody species [diameter at breast height (DBH) ≥ 5 cm] (**a**–**e**). *MM* Subtropical mid-montane moist evergreen broad-leaved forest (EBLF) in Yunnan, *MS* subtropical monsoon EBLF in Yunnan, *SH* subtropical semi-humid EBLF in Yunnan, *SWR* subtropical/warm-temperate rain EBLF in SW Japan, *AW* subtropical EBLF as a whole in Yunnan. Significantly different values are indicated by different letters. Forests sharing the same letters do not differ significantly by ANOVA, $p<0.01$ for a, c and d, $p<0.05$ for b. Error bars Standard deviation (Data source: Tang and Ohsawa (2009))

subtropical/warm-temperate rain EBLF of SW Japan (2.4 for H′, 0.83 for J′, 0.87 for D and 8.15 for Fisher's alpha). The semi-humid EBLF in Yunnan had significantly lower values on the Shannon–Wiener H′ (1.43), Equitability J′ (0.62), Simpson D (0.61) and Fisher's alpha (3.29) indices as compared to the other forests. The Sørensen's similarity indices (beta diversity) at the genus level between the Yunnan EBLF and the rain EBLF of SW Japan ranged from 0.33 to 0.43.

The Yunnan subtropical lucidophyllous forests contain more diverse types, including the mid-montane moist EBLF, monsoon EBLF and semi-humid EBLF. This is due primarily to the diversity in climates and landscapes in Yunnan. The higher species diversity of the monsoon EBLF forest of Yunnan may be attributable to the lower altitudes and southern geographical location, which lead to intermingling with the flora of tropical SE Asia. The rain EBLF in SW Japan displays a richness of taxa similar to that of the monsoon EBLF of Yunnan and diversity indices similar to those of the mid-montane moist EBLF of Yunnan. The beta diversity indices (Fig. 8.17e) confirm that the generic disjunctive distribution phenomenon (Good 1964; Wu 1983) between Yunnan and Japan is a most prominent feature that can indicate the degree of similarity of the two East Asian areas. Among the 151 genera in Yunnan and 79 genera in SW Japan, 44 shared genera are found within the studied forests. The genera and species of the forests of Yunnan and SW Japan are East Asian elements and show a very close relationship, though different assemblages of species occur in the two regions, as Fig. 8.15 shows, the Yunnan area and SW Japan being characterized by only 9 % similarity in species composition.

8.2.2.2 Forest Structure and Dynamics

The vertical structure of primary subtropical EBLF in Yunnan is multilayered, including canopy, subcanopy, shrub layer and understory, and the canopy height of the forest ranged from 23 to 34 m, much like the old-growth subtropical/warm-temperate rain EBLF (20–30 m) on Mt. Tatera, Tsushima (Manabe et al. 2000) and Mt. Omori, Aya, Miyazaki (Tanouchi and Yamamoto 1995), SW Japan. Most of the dominant canopy tree species of the forests had the sporadic type of size distribution frequency (Fig. 2.12 and Fig. 8.18). The sporadic type, with several peaks for the intermediate size classes, indicates gap regeneration with a poor seedling/sapling bank, as in the dominant species in the mid-montane moist EBLF and the monsoon EBLF as well as the *Castanopsis cuspidata*-, *Cyclobalanopsis acuta*- and *Cyclobalanopsis gilva*-dominated types in the rain forest of SW Japan. A few species were of the inverse-J type, such as *Distylium racemosum* in SW Japan, indicating strong regeneration. In the semi-humid EBLF on Mt. Huafu, *Cyclobalanopsis glaucoides*, *Castanopsis orthacantha* and *Castanopsis delavayi* have a great number of stems below 20 cm DBH, most of them being sprouts from the base of trunks. In old-growth stands of the rain forest in SW Japan, the canopy tree species *Castanopsis cuspidata*, *Castanopsis sieboldii* subsp. *sieboldii* and *Machilus thunbergii* can regenerate by base sprouting (Yamamoto 1997; Miura and Yamamoto 2003).

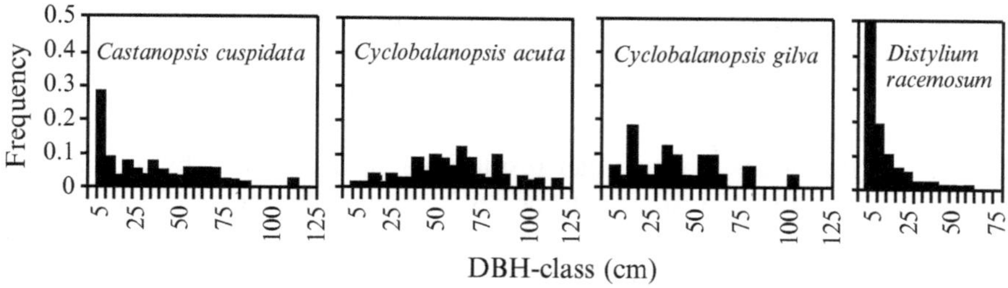

Fig. 8.18 DBH-class frequency distribution of some dominant canopy tree species of representative old-growth forests on Mt. Omori, Aya, Miyazaki, SW Japan (Data source: Tanouchi and Yamamoto (1995))

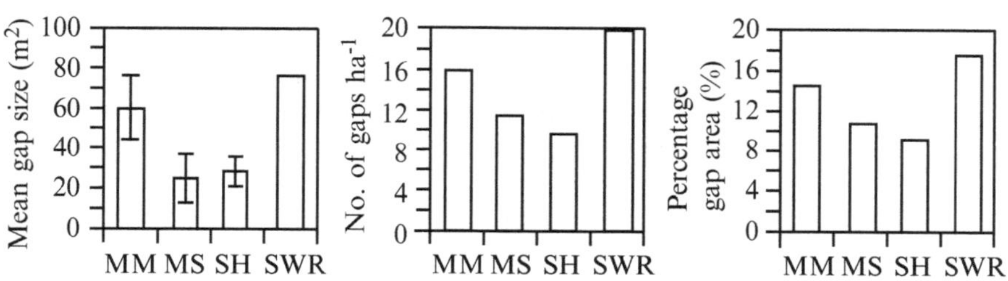

Fig. 8.19 Mean gap size, gap density and percentage of total gap area to total forested area surveyed in Yunnan and SW Japan. *MM* Subtropical mid-montane moist EBLF in Yunnan, *MS* subtropical monsoon EBLF in Yunnan, *SH* subtropical semi-humid EBLF in Yunnan, *SWR* subtropical/warm-temperate rain EBLF in Japan. *Error bars* are standard deviations (Data sources: Tang and Ohsawa (2009) for Yunnan, Yamamoto (1997) for SWR)

Regeneration in canopy gaps is an important mechanism maintaining the composition and structure of mature stands. Figure 8.19 indicates gap-distribution regimes by parameters including percentage of gap size, gap density and gap area. In SW Japan, where typhoons are frequent, gaps are more active as agents of natural disturbance affecting the forest dynamics (Yamamoto 1992, 1997; Miura et al. 2001; Saito 2002) than in the Yunnan forests. In Yunnan, the dense bamboo coverage in the understory of the mid-montane moist EBLF affects tree regeneration, which may be synchronously related to the bamboo flowering event occurring at roughly 55–60-year intervals (see Sec. 2.4 in Chap. 2). Dwarf bamboo was rare in the monsoon EBLF and the semi-humid EBLF of Yunnan, and is not reported in the rain forest of SW Japan.

Some species of *Castanopsis* and *Cyclobalanopsis* are found to regenerate from saplings already existing at the time of gap formation. But the regeneration mechanisms of some species, such as *Lithocarpus xylocarpus*, *Castanopsis remotidenticulata*, *Cyclobalanopsis myrsinifolia*, *Castanopsis fabri*, *Castanopsis rufotomentosa*, *Castanopsis delavayi* and *Cyclobalanopsis delavayi* in Yunnan, and *Cyclobalanopsis acuta* and *Cyclobalanopsis salicina* in Japan, remain unknown. Some studies have suggested that large disturbances may be required for successful recruitment of *Cyclobalanopsis* spp. (e.g., Yamamoto 1992; Tanouchi et al. 1994). The poor natural regeneration found in species of *Castanopsis* and *Cyclobalanopsis* may be due to damage to seed by insects and vertebrates, or grazing of seedlings by deer, decreasing the probabilities of seedling success (Fukumoto and Kajimura 2001). But the opposite effect can result from seed dispersal by animals, on which the regeneration of some oak species is strongly dependent (Chambers and Macmahon 1994). The dynamics of the early stages, from acorn fall to establishment of the seedling and sapling bank, need further study.

8.2.2.3 Forest Succession

Changes in land use are among the most important environmental events in both China and Japan. Secondary forests, such as coppice woods or those growing on abandoned farmland or plantations, are common in the subtropical/warm-temperate zones of both areas. Forest succession as exemplified by central Yunnan and the Hiji River basin of SW Japan is shown in Fig. 8.20a–c. The forest succession

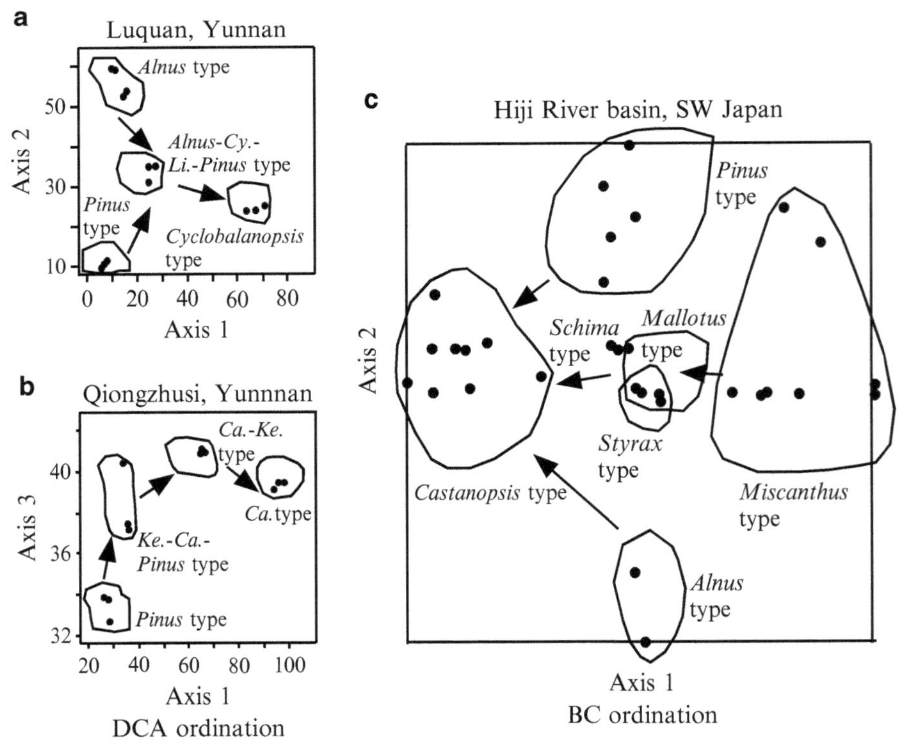

Fig. 8.20 Succession patterns in Luquan (**a**) and Qiongzhusi (**b**) in Yunnan, and in the Hiji River. basin (**c**), Okinawa, SW Japan. *Dots* represent plots. *Cy Cyclobalanopsis*, *Li Lithocarpus*, *Ca. Castanopsis*, *Ke Keteleeria* (**a–c** are redrawn from Tang and Ohsawa (2009))

sequence and the distribution of species are similar in Luquan, Yunnan and the Hiji River basin, SW Japan: from the pioneer *Pinus* or *Alnus* communities to the seral mixed community of various tree species having different traits, such as fast or slow growth, shade-tolerance or intolerance, and finally to late-successional *Castanopsis* and *Cyclobalanopsis* communities (Fig. 8.20a, c). *Keteleeria evelyniana* in Qiongzhusi, Yunnan (Fig. 8.20b), has unique traits, being somewhat shade-intolerant, requiring relatively moist soil, and having a longer life span and slower growth rate than *Pinus*. Those traits are similar to those of the Japanese *Cryptomeria japonica*, which produces no seedlings under a closed canopy, requiring semi-open habitats such as scree slopes, gaps and roadsides. *Keteleeria davidiana* var. *formosana* in Taiwan is a shade-intolerant species which is usually found in association with some evergreen broad-leaved trees such as species of *Machilus* and *Castanopsis* in open places (Lin et al. 2003). Late successional tree species of *Cyclobalanopsis* and *Castanopsis* have slow growth rates and are able to germinate and establish in some degree of shade. *Pinus* invades dry and poor soil habitats while *Alnus* is found in moist and rich soil habitats. Certain fungi living on the roots of *Alnus nepalensis* have the function of nitrification, which can enrich soils. Pioneer *Alnus* thus probably improves soil conditions, supporting the regeneration of late-successional species. In the subtropical area of Yunnan, most disturbances take the form of forest clearance through cutting for firewood. The intensity of disturbance has often been severe, with mature forest being cleared, but with many stumps remaining and then being allowed to grow back. Resprouting is a key mechanism in the secondary succession following a subtropical EBLF forest clearance. The significance of resprouting has been recognized for *Castanopsis cuspidata* in the secondary succession in Yakushima, southwestern Japan (Aiba et al. 2001). Additionally, bamboo stands are a major secondary vegetation type in southern Japan. Bamboo stands have not replaced primary EBLF in the subtropical zone of Yunnan, although bamboo stands do commonly replace the subtropical EBLFs in the Sichuan Basin and eastern Guizhou, and the tropical rain EBLF in the tropical zone (e.g. Xishuangbanna) of southern Yunnan.

Thus, similarities between the EBLFs in Yunnan and SW Japan are evident in the dominance of particular genera (rather than species), ecological attributes of pertinent species, structural diversity, and succession.

8.2.3 As compared with Southeast Asia

8.2.3.1 Characteristics of Flora

The South China, Indochina and East Malaya geoblocks are interpreted to have formed part of the India–Australian margin of Gondwana in the Early Palaeozoic and to have rifted and separated from Gondwana by the opening of the Paleo-Tethys ocean in the Early Devonian (Metcalfe 2005, 2006). About 50 million years ago, the Indian plate moved rapidly northward, colliding with and crowding into the

Laurasian plate. The collision with Eurasia had significant impact on the shaping, climate, and biogeography of southern China and Southeast Asia (including Myanmar, Thailand, Laos, Vietnam, Cambodia, Malaysia, Singapore, Brunei, the Philippines, and Indonesia). There was no geographical barrier between Yunnan, mainland SE Asia, and west Malaysia during most of the Tertiary (Morley 1998), Malesia being a floristic region that includes the Peninsula Malaysia, Sumatra, Java, Bali, and Borneo, Philippines, Wallacea, New Guinea and the Bismark Archipelago. Tropical-subtropical Yunnan of SW China is where the Holarctic comes into contact with tropical flora. The topography and climate of the region have been strongly affected by the uplift of the Himalayas and the formation of the eastern monsoon climate (Shi et al. 1998). Biogeographically, the region is located in a transitional zone with tropical SE Asia to the south, subtropical East Asia to the north, the Sino-Japanese floristic region to the east, and the Sino-Himalayan floristic region to the west (Zhu and Yan 2002). Zhu (2013a) considered a line at c. 22°30′ N to be the northern biogeographic boundary of the tropical zone in southern China. Southern, southeastern and southwestern Yunnan are areas on the transition from the tropical to the subtropical zone (Fig. 8.21). Southern Yunnan (Xishuangbanna including the three counties of Jinghong, Menghai and Mengla, with an area of 19,690 km^2), lying between 21°09′–22°32′ N and 99°58′–101° 50′ E, borders on Myanmar and Laos, and has a mountainous topography with the mountain ridges running north–south with elevation diminishing toward the south. Its altitude varies from 480 m at the bottom of the lowest valley in the south (Lancangjiang: the upper reach of the Mekong River) to 2,430 m at the highest mountain top in the north. The area in southeastern Yunnan described in this section is located between the Tropic of Cancer and the Yunnan-Vietnam border, 22°26′–23°26′ N and 104°27′–108°48′ E, known as tropical southeastern Yunnan. It borders Vietnam to the south, and also has a mountainous topography with the mountain ridges running in a north–south direction, decreasing in elevation southeastward. The region, including the six counties of Pingbian, Hekou, Jingping, Lüchun, Yuanyang and Honghe, is c. 14,389 km^2 in area (Southwestern Yunnan will be described later in this section.). Here the soil is mostly limestone-based. The karst landscape in this region makes the flora more peculiar. Its altitude ranges from 74.6 m at the bottom of the lowest valley in the south (Hekou) to 3,047.3 m at the top of the Xilong Mountains in Jingping County (Zhu and Yan 2009). The two regions have similar monsoonal climates and lowland tropical rain forest vegetation. Zhu and his colleagues (e.g. Zhu and Roos 2001; Zhu and Yan 2002, 2009; Zhu et al. 2006; Zhu 2013a, b) have studied the floristic elements of southern and southeastern Yunnan in great detail. The floras of both regions are dominated by tropical floristic elements, and both belong to the Indo-Malesian flora at the northern margin of tropical Asia. However, some temperate East Asian characteristic genera are well represented in southeast Yunnan, while genera characteristic of tropical SE Asia are well represented in southern Yunnan (Table 8.9).

Tables 8.10 and 8.11 show the floristic similarities between southern Yunnan, Hainan, Vietnam, the Malay Peninsula, and Brunei Darussalam. The floristic similarity between the flora of southern Yunnan and those of the Malay Peninsula

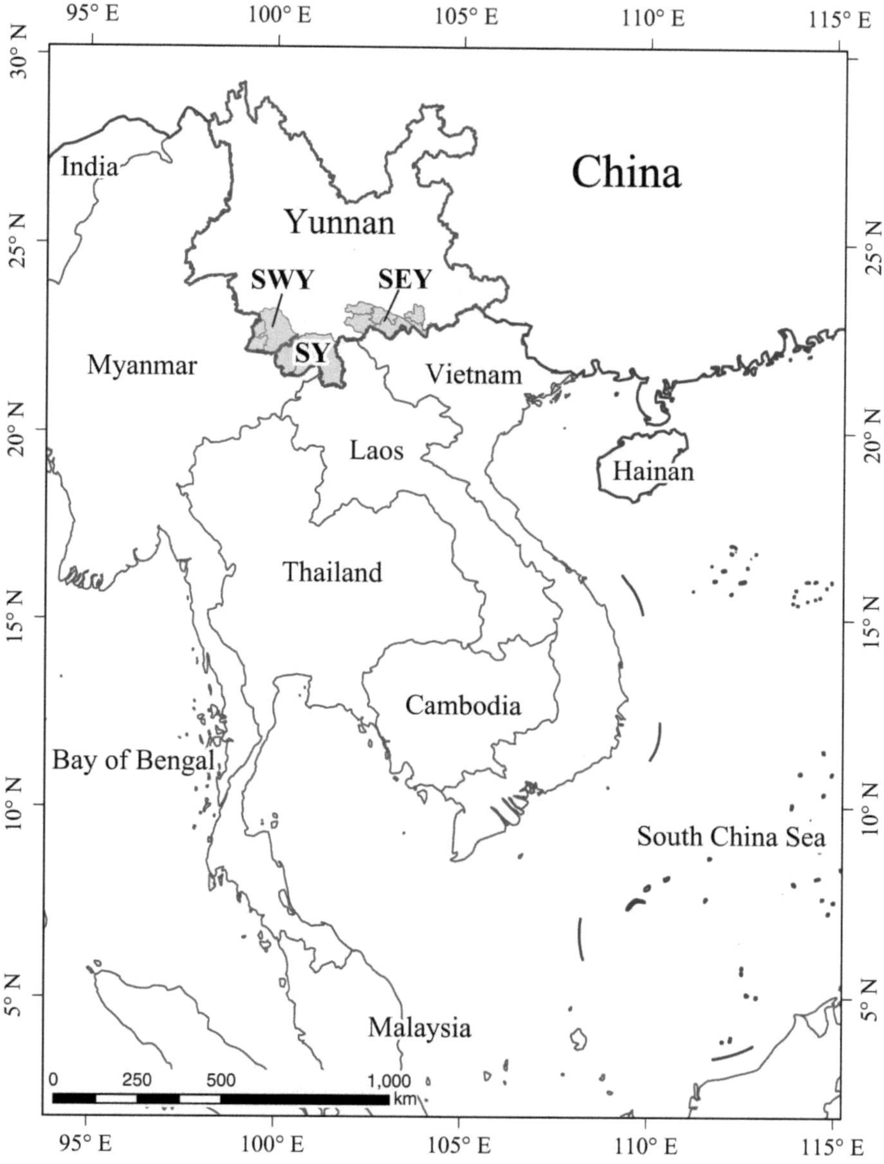

Fig. 8.21 The location of southern (*SY*), southeastern (*SEY*) and southwestern (*SWY*) Yunnan near 22°30′ Northern Latitude

and Brunei Darussalam is more than 72 % at the family level and more than 47 % at the genus level, while between the flora of southern Yunnan and those of Hainan and Vietnam it is over 84 % at the family level and over 65 % at the genus level. This suggests a high floristic affinity between southern Yunnan and Hainan and Indo-Malesia (Indochinese and Malesian floristic regions).

Table 8.9 Generic-level diversity (in total numbers and percentages) in southern and southeastern Yunnan, respectively, across geographic elements

Geographical element at generic level	Genera only in southern Yunnan	%	Genera only in southeastern Yunnan	%
1. Cosmopolitan	8	3.42	6	1.72
2. Pantropical	47	20.09	27	7.74
3. Tropical Asia and Tropical America disjunct	5	2.14	6	1.72
4. Old World Tropical	16	6.84	9	2.58
5. Tropical Asia to Tropical Australia	28	11.97	14	4.01
6. Tropical Asia to Tropical Africa	19	8.12	23	6.59
7. Tropical Asia	82	35.04	98	28.08
8. North Temperate	5	2.14	53	15.19
9. East Asia and North America disjunct	2	0.85	17	4.87
10. Old World Temperate	7	2.99	14	4.01
11. Temperate Asia	1	0.43	2	0.57
12. Mediterranean, Western Asia to Central Asia	1	0.43	1	0.29
13. Central Asia	2	0.85	0	0.00
14. East Asia	7	2.99	57	16.33
15. Endemic to China	4	1.71	22	6.30
Total	234	100.00	349	100.00

Data source: Zhu (2013b)

Table 8.10 Comparison of floristic similarities at the family level between southern Yunnan, Hainan, Vietnam, Malay Peninsula and Brunei Darussalam

	No. of families	Southern Yunnan Shared/S.C.	Hainan	Vietnam Shared/S.C.	Malay Peninsula Shared/S.C.	Brunei Darussalam Shared/S.C.
Southern Yunnan	207	100/100				
Hainan	206	181/87.86	100/100			
Vietnam	249	179/84.43	187/90.78	100/100		
Malay Peninsula	215	173/83.98	179/86.89	202/93.95	100/100	
Brunei Darussalam	164	119/72.56	115/70.17	147/89.63	161/98.17	100/100

Data source: Zhu and Roos (2001)
Similarity coefficient between A and B = the number of taxa shared by both A and B divided by the lower number of taxa of A or B, multiplied by 100
Note: *S.C.* Similarity coefficient (%)

Floras of southern and southeastern Yunnan are comprised of almost the same primary species-rich families, such as Orchidaceae, Fabaceae, Rubiaceae, Poaceae, Asteraceae, Euphorbiaceae, Lauraceae and Urticaceae. However, among the secondary species-rich families, tropical families, such as Zingiberaceae,

Table 8.11 Comparison of floristic similarities at the genus level between southern Yunnan, Hainan, Vietnam, Malay Peninsula and Brunei Darussalam

	No. of genera	Southern Yunnan Shared/S.C.	Hainan	Vietnam Shared/S.C.	Malay Peninsula Shared/S.C.	Brunei Darussalam Shared/S.C.
Southern Yunnan	1,218	100/100				
Hainan	1,237	803/65.93	100/100			
Vietnam	1,885	125/69.06	1,237/85.69	100/100		
Malay Peninsula	1,547	786/64.53	844/68.23	1,547/69.36	100/100	
Brunei Darussalam	917	446/47.80	480/52.34	917/65.98	789/86.04	100/100

Data source: Zhu and Roos (2001)
Similarity coefficient between A and B = the number of taxa shared by both A and B divided by the lower number of taxa of A or B, multiplied by 100
Note: *S.C.* Similarity coefficient (%)

Asclepiadaceae, Apocynaceae and Annonaceae are more dominant in the flora of southern Yunnan, while those with a mainly temperate distribution, such as Ericaceae, Theaceae, Liliaceae and Araliaceae, dominated more in the flora of southeastern Yunnan. Most dominant families from the two regions of Yunnan are also dominant in Indo-Malaysia (Table 8.12). For example, Orchidaceae, Rubiaceae, Euphorbiaceae, Lauraceae, Moraceae, Asclepiadaceae, Apocynaceae, Annonaceae, and Zingiberaceae are the large tropical families shared by southern and southeastern Yunnan and Indo-Malaysia. The family Dipterocarpaceae is the most characteristic and a dominant family in the Malesian flora (here: the floras of the Malay Peninsula and Brunei Darussalam). It has only a small number of species in Yunnan. Other dominant families in the Malesian flora, such as the Clusiaceae, Meliaceae, Melastomataceae and Myristicaceae, show a similar pattern to the Dipterocarpaceae. Although the dominant families Urticaceae, Fagaceae, Theaceae and Rosaceae in the tropical flora of southern and southeastern Yunnan are not among those dominant in the Malesian flora, the tropical floras of southern and southeastern Yunnan are basically of Indo-Malesian nature. While the floras of southern and southeastern Yunnan, as mentioned already, occur on the margin of tropical Asia, the families and genera of strictly tropical distribution are underrepresented as compared to the Malesian flora. Many Malesian elements reach their northern limits in southern and southeastern Yunnan (Zhu et al. 2006; Zhu and Yan 2009). The seasonal rain forest in the limestone mountains around Monglun has some unique species, such as *Tetrameles nudiflora*. Below 800 m many rubber (*Hevea brasiliensis*) plantations are found in both south and southeastern Yunnan.

The rivers and mountains of southwestern Yunnan (see Fig. 8.21), i.e., the southwestern part of the Linchang and Dehong autonomous prefectures located in southern Nushan and Gaoligongshan, with an area of 13,720 km^2 bordering Myanmar, support tropical-subtropical flora. The tropical element *Dipterocarpus*,

Table 8.12 The 20 families with highest species richness among the floras of southern, southeastern Yunnan and several parts of Southeast Asia

Flora of southern Yunnan		Flora of tropical SE Yunnan		Flora of Vietnam		Flora of Malay Peninsula		Flora of Brunei Darussalam	
Family	No. sp.	Family	No. sp.	Family	No. sp.	Name of family	No. sp.	Name of family	No. sp.
Orchidaceae	377	Fabaceae	285	Fabaceae	628	Orchidaceae	853	Rubiaceae	269
Fabaceae	261	Orchidaceae	276	Orchidaceae	621	Rubiaceae	562	Euphorbiaceae	199
Rubiaceae	201	Rubiaceae	235	Poaceae	516	Euphorbiaceae	368	Dipterocarpaceae	177
Poaceae	189	Poaceae	219	Rubiaceae	425	Fabaceae	298	Orchidaceae	145
Euphorbiaceae	148	Asteraceae	180	Euphorbiaceae	405	Poaceae	238	Arecaceae	132
Lamiaceae	139	Lamiaceae	151	Cyperaceae	325	Myrtaceae	215	Araceae	120
Asteraceae	137	Lauraceae	141	Asteraceae	293	Lauraceae	214	Melastomataceae	115
Lauraceae	105	Urticaceae	134	Lauraceae	244	Annonaceae	202	Fabaceae	110
Urticaceae	84	Euphorbiaceae	126	Fagaceae	213	Arecaceae	198	Annonaceae	96
Zingiberaceae	84	Rosaceae	124	Acanthaceae	195	Gesneriaceae	189	Zingiberaceae	88
Moraceae	83	Fagaceae	109	Annonaceae	168	Melastomataceae	172	Moraceae	84
Acanthaceae	77	Moraceae	104	Apocynaceae	155	Cyperaceae	162	Lauraceae	82
Asclepiadaceae	66	Ericaceae	96	Myrsinaceae	140	Acanthaceae	158	Myrtaceae	78
Cyperaceae	63	Cyperaceae	87	Moraceae	138	Dipterocarpaceae	156	Cyperaceae	73
Cucurbitaceae	60	Acanthaceae	85	Verbenaceae	131	Zingiberaceae	150	Clusiaceae	70
Fagaceae	60	Theaceae	81	Lamiaceae	128	Araceae	141	Poaceae	68
Rosaceae	59	Gesneriaceae	79	Araceae	128	Moraceae	138	Meliaceae	58
Annonaceae	57	Celastraceae	65	Rosaceae	124	Clusiaceae	120	Myristicaceae	56
Apocynaceae	56	Liliaceae	64	Zingiberaceae	124	Apocynaceae	119	Anacardiaceae	53
Vitaceae	56	Araliaceae	64	Scrophulariaceae	117	Asclepiadaceae	116	Gesneriaceae	53

Data sources: Zhu and Roos (2001), Zhu (2013b)

typical of the SE Asian rain forest, is found in the tropical monsoon rain forest below 1,000 m in southwestern Yunnan, for instance *Dipterocarpus retusus* in the Tonbiguan Nature Reserve (97°31′–97°46′ N, 23°54′–24°51′ E). Tropical/subtropical *Dimocarpus yunnanensis*, *Duabanga grandiflora*, *Garuga forrestii*, *Eriolaena spectabilis*, *Syzygium tetragonum*, *Pometia pinnata* and *Terminalia myriocarpa*, *Markhamia stipulata*, *Sterculia villosa* and *Hydnocarpus anthelminthicus* are found in the scattered seasonal rain forests of this area. On the limestone hills there are patchy seasonal rain forests including tropical *Tetrameles nudiflora*. Above 1,000 m, tropical/subtropical *Lithocarpus grandifolius*, *Castanopsis calathiformis*, *Castanopsis hystrix*, *Castanopsis indica*, *Lindera tonkinensis* and *Schima wallichii* predominate in the monsoon EBLF. Furthermore, *Trigonobalanus doichangensis* of tropical SE Asian origin is found in the monsoon EBLF in the Nangunhe Nature Reserve (98°54′–99°05′ N, 23°12′–23°19′ E). In sum, the flora in southwestern Yunnan clearly shows affinities with tropical SE Asia.

8.2.3.2 Altitudinal Vegetation Zones

The basic pattern of altitudinal vegetation zonation on tropical mountains is different from that on temperate mountains. The most obvious feature of tropical as compared with temperate zonation is the lack of both the boreal coniferous forests and the temperate deciduous broad-leaved forests (Troll 1948). These two forest types reach their southern limit in East Asia on subtropical high mountains in the Himalayas and Taiwan. But EBLFs or tropical coniferous forests develop at propitious altitudes on the tropical mountains of East Asia (Richards 1952; Whitmore 1975; Ohsawa 1985).

Figure 8.22 depicts the zonation on the Xilong Mountains (22°26′–22°46′ N, 102°32′–103°04′ N; 3,074 m) in southeastern Yunnan and Mt. Kerinci (1°43′ S, 101°15′ E; 3,800 m) in Sumatra. Since increasing elevation causes a distribution of habitat types similar to that of increasing latitude, the flora and the dominance of the vegetation can be categorized along altitudinal gradients. The vegetation on the two mountains can be divided into three main zones (tropical, subtropical and warm-temperate) and five types, as follows.

On Mts. Xilong, southeastern Yunnan, the tropical scrubby community (<8 m in height) is dominated by *Schima khasiana* and *Cratoxylum formosum* ssp. *pruniflorum* at 500–700 (800) m (Fig. 8.22). The montane rain forest, which can only be found in patch-mosaic forest at 800–1,300 m (the tropical zone) due to human activities at low altitudes, is a type of tropical Asian rain forest. It is characterized by the tree species *Dipterocarpus retusus*, *Altingia excelsa*, *Cyclobalanopsis austrocochinchinensis*, *Castanopsis indica*, *Terminalia myricarpa*, *Duabanga grandiflora* and *Schefflera octophylla*. Sometimes, deciduous pioneer species such as *Sapium discolor* and *Rhus sinensis* invade in canopy gaps and at forest edges. At the lower part (1,300–1,800 m) of the subtropical zone disturbed secondary forests resulting from human activity are common. They are codominated by evergreen and deciduous broad-leaved trees, such as *Castanopsis calathiformis*,

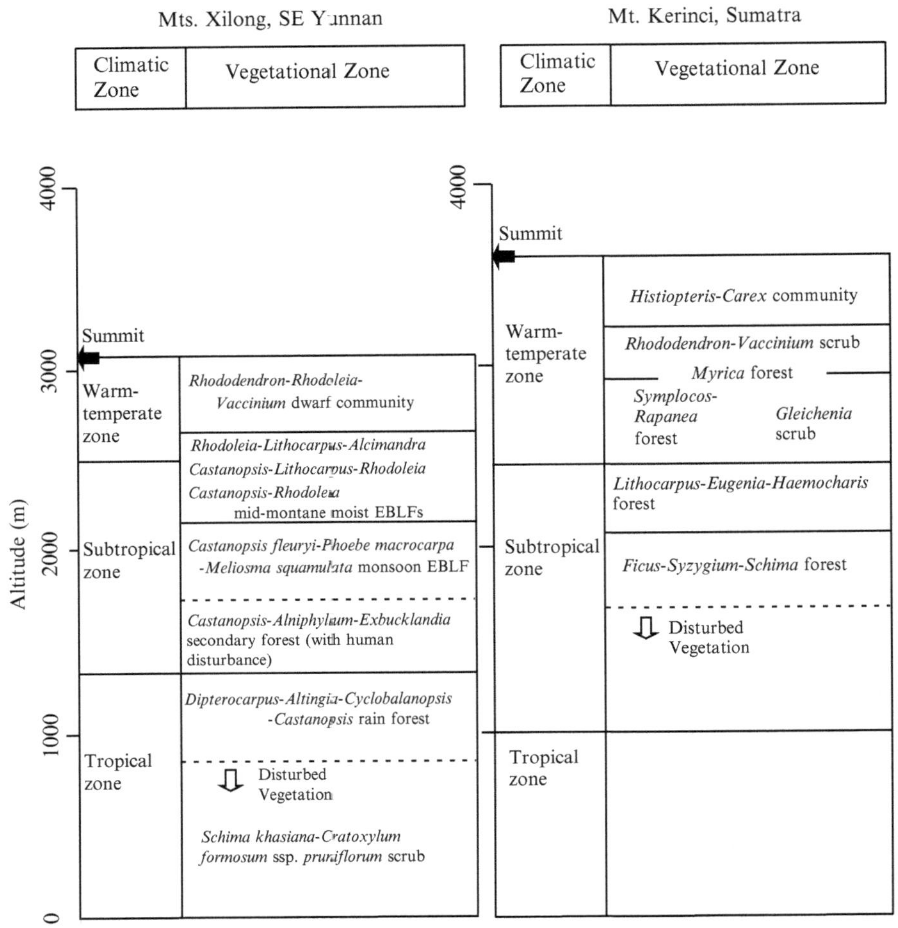

Fig. 8.22 Altitudinal zonation on Mts. Xilong, southeastern Yunnan and Mt. Kerinci, Sumatra. Partial data for Mts. Xilong from Zhou et al. (2002) (Data for Mt. Kerinci from Ohsawa (1985))

Alniphyllum fortunei and *Exbucklandia populnea*. The latter two deciduous species frequently appear in canopy gaps, at forest edges and at roadsides. The monsoon EBLF at 1,500–2,200 m (the middle part of the subtropical zone) has a wide distribution and is the main montane vegetation type in the region. The forest canopy, reaching 30 m in height, is mainly dominated by species of the Fagaceae and Lauraceae, such as *Castanopsis fleuryi*, *Castanopsis echinocarpa*, *Phoebe macrocarpa*, *Machilus salicina*, *Cinnamomum pittosporoides*, along with other evergreen *Meliosma squamulata* of Sabiaceae. The mid-montane moist EBLF, with a height of 25–30 m and at 2,200–2,700 m (subtropical/warm-temperate zone), is characterized by the species *Rhodoleia henryi*, *Castanopsis remotidenticulata*, *Castanopsis fleuryi*, *Castanopsis wattii*, *Lithocarpus xylocarpus*, *Lithocarpus hancei*, *Lithocarpus truncates*, *Alcimandra cathcartii*, *Michelia floribunda* and *Illicium petelotii*. This forest is wet, with many ferns, mosses and lianas on tree trunks and branches. The mossy dwarf forest, with a canopy height lower than 15 m and at 2,500 (2,700)–3,000 m (the warm-temperate zone), includes *Rhododendron irroratum*, *Rhodoleia henryi*, *Vaccinium mandarinorum*, *Lithocarpus pachyphylloides*, *Myrsine stolonifera*, *Symplocos* sp., *Ilex triflora* and *Pieris formosa*. From these, *Rhododendron*, *Vaccinium*, *Myrsine*, *Symplocos* and *Ilex* are common taxa in the understory of the EBLFs at lower altitudes.

On Mt. Kerinci, Sumatra, the tropical vegetation zone is heavily disturbed by human activity (Fig. 8.22). The subtropical EBLF at 1,000–1,750 m is dominated by *Ficus variegata*, *Bucklandia tricuspis*, *Schima wallichii* and *Syzygium lineatum*. In the mid-altitudes from 2,100 to 2,400 m, the wet forest is dominated by *Lithocarpus* sp., *Eugenia* sp. and *Haemocharis buxifolia*. This corresponds to the mid-montane moist EBLF on Mts. Xilong. In addition, the mossy dwarf forest at high altitude on Mts. Xilong is similar to Mt. Kerinci in that the forest zones at higher altitudes include many taxa of the understory of the forests at lower altitudes. Both the *Symplocos-Rapanea* zone and the *Rhododendron-Vaccinium* zone on Mt. Kerinci are examples of such extended understory elements. Therefore the zones at higher altitudes in tropical mountains seem to represent an impoverishment of the floristic pool of the lower zone, although some characteristic taxa are found in each zone (Ohsawa 1985). It demonstrates that the mountains situated in the transition from tropical to subtropical areas in Yunnan, have a vegetation zonation pattern similar to that of tropical mountains in SE Asia.

8.2.4 As compared with the Eastern Himalayas

8.2.4.1 Characteristics of Vegetation

The Sino-Himalayan Mountainous Region is made up of mid- and high-elevation forests, scrub and grasslands that cover the southern slopes of the Himalayas and the mountains of SW China and Indochina. SW China, as an extension of the eastern Himalayas, is a varied and unique physical and geographical region supporting a very rich flora. The characteristics of the vegetation of Yunnan as compared with those of the eastern Himalayas are summarized in Fig. 8.23a, b. The zonation is primarily due to climate differences along altitudinal gradients. In Yunnan, in the mountain gorges and deep river valleys (400–1,300 m), the vegetation appears as a semi-savanna characterized by *Indigofera cassioides*, *Phyllanthus emblica*, *Woodfordia fruticosa*, *Heteropogon contortus*, *Bothriochloa pertusa*, etc., and thorny succulent shrubland (e.g. *Euphorbia royleana*, *Opuntia monacantha*) (Fig. 8.23a). At low altitudes of 500–1,000 (1,300) m, the tropical rain forest is dominated by typical tropical elements such as *Parashorea chinensis*, *Dipterocarpus retusus*, *Pometia pinnata*, *Shorea assamica* and *Hopea chinensis*. In the tropical seasonal rain forest, some deciduous tree species such as *Bombax ceiba* and *Celtis wightii* are present. Tropical monsoon forest is often found in some local dry areas:

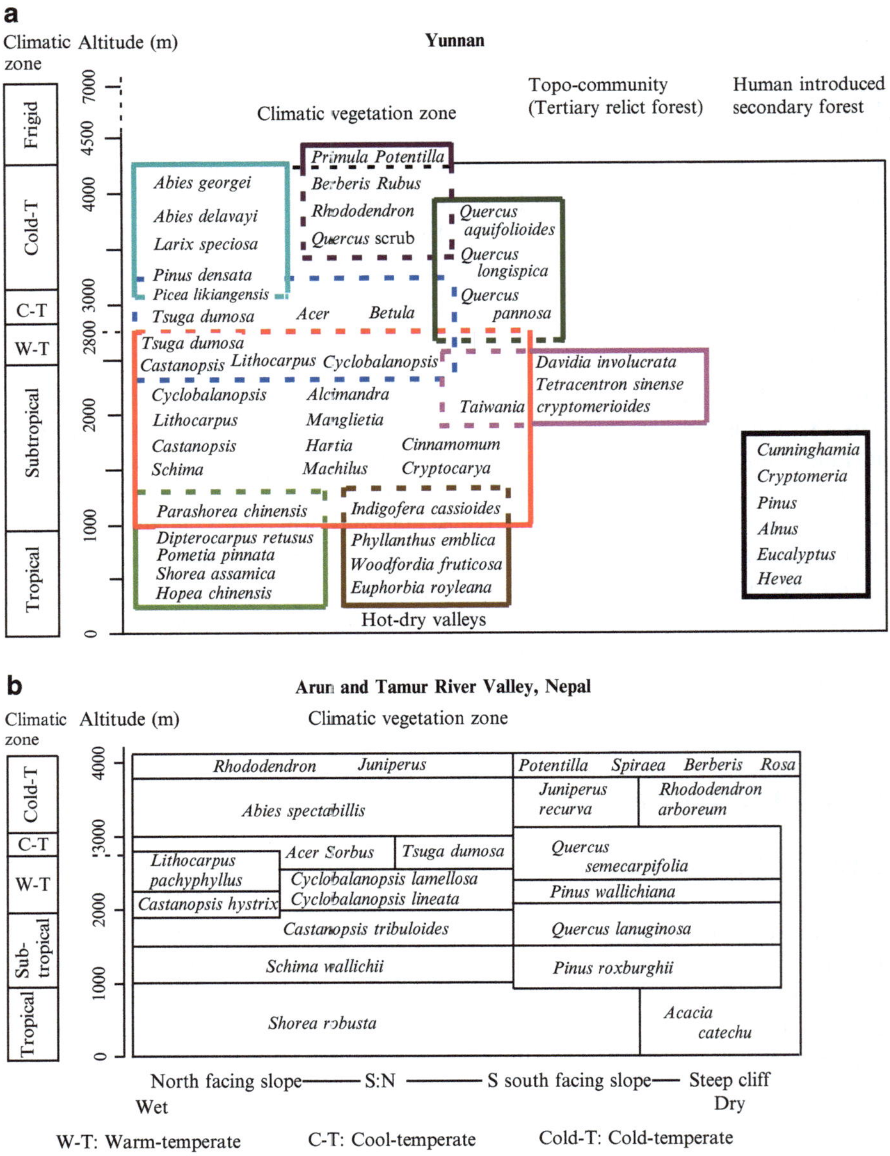

Fig. 8.23 Vegetation characteristics of Yunnan (**a**) and eastern Nepal (**b**) (Data source for b: Ohsawa et al. (1986))

a deciduous forest characterized by *Bombax ceiba*, *Chukrasia tabularis* var. *velutina*, *Erythrina stricta*, *Pterocarya tonkinensis* and *Albizia chinensis*. In this altitudinal belt, natural vegetation appears in mosaic patches, since human activities are frequent. The subtropical EBLF, with a wide distribution range from 1,000 to 2,800 m, is generally populated by many species of *Castanopsis*, *Lithocarpus*,

Cyclobalanopsis (e.g. *Lithocarpus pachyphyllus*, *Cyclobalanopsis lamellosa* and *Cyclobalanopsis oxyodon* at 2,100–2,700 m in the Dulongjiang and Gongshan areas, the northern Gaoligong Mountains, which are also found in the eastern Himalayas. see later paragraphs) and by *Schima wallichii*, *Schima villosa*, *Schima khasiana*, *Hartia sinensis*, *Machilus longipedicellata*, *Alcimandra cathcartii*, *Cinnamomum iners*, *Manglietia insignis* and *Cryptocarya depauperata*. Evergreen sclerophyllous *Quercus* (e.g. *Q. aquifolioides*, *Q. pannosa*, *Q. longispica*) forests can be found at 2,600–3,900 m. Coniferous *Tsuga dumosa* mixes with some evergreen broad-leaved trees (e.g. species of *Castanopsis*, *Lithocarpus* and *Cyclobalanopsis*) or deciduous broad-leaved trees (e.g. *Acer* and *Betula*) at 2,200–2,900 (or 3,200) m. In some mountain ranges such as the northern Gaoligong Mountians and eastern Ailao Mountains, *Taxus wallichiana* or *Taxus wallichiana* var. *mairei* can also be found in these altitudes. Finally, in the subalpine zone at 3,200–4,400 m *Picea likiangensis*, *Larix speciosa*, *Pinus densata*, *Abies delavayi* and *Abies georgei* are principal species. Deciduous species of *Berberis*, *Rubus*, *Spiraea* or conifer *Juniperus* are found at forest edges. *Abies georgei* reaches the treeline at 4,400 m. At 4,000 (3,500)–4,500 m evergreen sclerophyllous *Quercus* scrubs, *Rhododendron* scrubs, or meadows (with e.g. *Primula*, *Pedicularis* and *Potentilla*) or grasslands cover the land. As noted in Chap. 4, *Pinus yunnanensis* is widely distributed at altitudes of 400–3,200 m.

At altitudes below 1,800 m, human interference includes introduction of the tree species *Cunninghamia lanceolata*, *Cryptomeria japonica* var. *sinensis*, *Pinus yunnanensis*, *Pinus armandii*, *Alnus nepalensis* and *Eucalyptus smithii*.

Additionally, within the subtropical EBLF zone or in the transitional zone from EBLFs to deciduous broad-leaved forests, some important Tertiary relict forests (e.g. *Davidia involucrata* forests in NW and NE Yunnan, *Tetracentron sinense* forests in south-central and NE Yunnan, *Taiwania cryptomerioides* in NW Yunnan) survive in specific topo-habitats such as steep slopes, scree slopes, rocky cliffs, or mountain stream banks. The patterns of forest vegetation show how topography can be an important determinant of the floristic composition and structure of the vegetation. The karst topography in SE Yunnan is well developed and provides special habitats for the growth and reproduction of a number of unusual plants such as *Michelia coriacea* and *Talauma hodgsonii* of the Magnoliaceae. Again, in the Yangtze River valley on the boundary between Guizhou, Chongqing, Sichuan and Hubei, remarkable living fossil plants such as *Ginkgo biloba*, *Metasequoia glyptostroboides*, *Cathaya argyrophylla*, *Thuja sutchuenensis* and *Liriodendron chinense* persist in habitats that I described in Chap. 7.

In the eastern Himalayas, Ohsawa et al. (1986) studied the vegetation of the Arun and Tamur basins of Nepal in detail (Fig. 8.23b). The foothills up to 1,000 m are covered by the tropical species *Shorea robusta*. The upper limit of *Shorea* is at 1,300 m, where it is replaced by *Duabanga*, *Bombax* and *Lagerstroemia*. Above 1,000 m *Schima wallichii* becomes a dominant species, forming a belt above the *Shorea*. The lower part of this *Schima* belt contains mixed forests of *Schima*, *Albizia*, *Engelhardia spicata*, *Acer oblongum* and *Castanopsis indica*. Above 1,500 m, *Castanopsis tribuloides* replaces *Schima* as the dominant. However, it is rather difficult to situate the transition from the *Schima* to the *Castanopsis* belt because of severe disturbance of the natural vegetation by human activities. *Quercus lanuginosa* is scattered on rocky cliffs and grassy balds of the *Schima* and the *Castanopsis* belts. At around 2,000 m, *Castanopsis tribuloides* is replaced as the dominant species by *Cyclobalanopsis lamellosa*, *Cyclobalanopsis lineata*, *Litsea* spp. and *Machilus* spp. In some parts of eastern Nepal, a narrow transition belt dominated by *Castanopsis hystrix* appears between the *Castanopsis tribuloides* and the *Cyclobalanopsis* belts. In the *Cyclobalanopsis* belt the evergreen lauro-fagaceous trees often mix with deciduous trees, such as *Acer campbellii* and *Lyonia ovalifolia*. In the Singalila range bordering on Sikkim there is a typical East Himalayan type of forest dominated by *Lithocarpus pachyphyllus* (synonym: *Lithocarpus pachyphylla* in Ohsawa et al. (1986)) and *Symplocos* spp. at these altitudes. As was mentioned before, *Cyclobalanopsis lamellosa*, *Cyclobalanopsis oxyodon* and *Lithocarpus pachyphyllus* are also components of the mid-montane moist EBLF in the Dulongjiang and Gongshan areas, the northern Gaoligaong Mountains, Yunnan. The physiognomy of vegetation stands in the *Acer* belt can vary greatly as a result of the complicated topography. In the inner valleys or on north-facing slopes are *Tsuga* or *Lithocarpus* forests, while on south-facing slopes the evergreen sclerophyllous *Quercus semecarpifolia* forest or *Rhododendron arboreum* communities grow. Above the *Acer* belt is a belt dominated by *Abies spectabilis*. The upper limit of the *Abies* forests becomes the forest limit. At the upper margin of the forests, deciduous trees appear, such as *Betula utilis*, *Sorbus* spp., or the conifer *Juniperus indica*. Above the forest limit is an alpine scrub dominated by *Rhododendron* spp. There is a sharp contrast in physiognomy and floristic composition according to slope aspect in the alpine scrub: north-facing slopes have an almost pure community of evergreen *Rhododendron*, while south-facing slopes have a deciduous scrub community with high species diversity and dominated by *Potentilla fruticosa*, *Spiraea*, *Berberis* and *Rosa sericea*.

In dry habitats on south-facing slopes *Pinus roxburghii*, *Pinus wallichiana*, *Quercus semecarpifolia* and *Quercus lanuginosa* are found. These first three species form the principal zonal forests of the dry western Himalaya. Although eastern Nepal is the country's wettest part, some drier places are created by local topographic and climatic conditions. These afford habitats for western Himalayan forest types dominated by *Pinus roxburghii*, *Pinus wallichiana*, and *Quercus semecarpifolia*, and are considered to be relict patches within the wet eastern Himalayas (Ohsawa et al. 1986). *Quercus semecarpifolia* and *Pinus roxburghii* constitute relict fragments of the Tertiary flora which once had a wide distribution in south-western extra-tropical Eurasia (Meusel 1971).

Moreover, the EBLFs of the eastern Himalayas share major components with the mid-montane moist EBLFs of Yunnan, as exemplified by *Cyclobalanopsis oxyodon* and *Cyclobalanpsis lamellosa*. Ohsawa (1987, 1991) further studied the montane EBLFs in the Bhutan Himalaya, where *Cyclobalanopsis oxyodon* at 2,200–3,000 m and *Cyclobalanopis lamellosa* at 1,900–2,650 m are principal components, along with species of *Sloanea*, *Lindera*, *Machilus*, *Cinnamomum*, *Michelia*, and some species of Fagaceae. Around 1,500–2,100 m, *Altingia excelsa*, *Schima wallichii*,

Cyclobalanopsis glauca, *Castanopsis tribuloides*, *Castanopsis hystrix*, *Lithocarpus elegans*, *Betula alnoides*, *Toona ciliata*, *Alnus nepalensis*, and *Engelhardia spicata* are major trees of the EBLFs with various signs of natural or human disturbances. Between 2,600 and 3,000 m, Tertiary relict *Tetracentron sinense* is present on rocky slopes, in canopy gaps or in land sliding habitats. *Taxus wallichiana* (synonym: *Taxus baccata* subsp. *Wallichiana* in Ohsawa (1987, 1991)) and *Tsuga dumosa* are distributed at 2,600–3,000 m. In the Dulongjiang and Gongshan areas, NW Yunnan, *Cyclobalanopsis oxyodon* and *Cyclobalanopsis lamellosa* usually reach 20–25 m at canopy height, with various DBHs ranging from 20 to 90 cm at altitudes of 2,100–2,700 m. They are often chief components, accompanied by species of *Lithocarpus*, *Manglietia*, *Machilus*, *Schima*, *Cinnamomum*, and *Lindera* (see Fig. 2.4). Because of human activities at altitudes below 2,000 m, secondary forests prevail, represented by pioneer invaders *Alnus nepalensis*, *Betula cylindrostachya*, *Vernicia fordii*, *Populus yatungensis*, *Toona sinensis*, *Exbucklandia populnea* and *Pinus yunnanensis*. Around 2,400 m, Tertiary relict *Davidia involucrata* var. *vilmoriniana* along with *Meliosma oldhamii* thrive at wet sites by rivers. Two other Tertiary relict plants are present: *Taiwania cryptomerioides* is distributed on steep slopes and valley bottoms by rivers around 2,000–2,400 m in Qiqi and Bingzhongluo, Gongshan County, and *Tetracentron sinense* is present in canopy gaps or rocky soils or by rivers at 2,000–2,600 m. Between 2,400 and 2,800 m, sclerophyllous *Quercus gongshanensis*, or coniferous *Taxus wallichiana*, *Tsuga dumosa*, or *Picea brachytyla* var. *complanata* mix with *Cyclobalanopsis oxyodon* or *Lithocarpus pachyphyllus* and some other evergreen (e.g. *Illicium simonsii* and *Lithocarpus variolosus*) and deciduous broad-leaved trees (e.g. *Acer pectinatum* subsp. *taronense* and *Betula utilis*). *Cyclobalanopsis oxyodon*, *Cyclobalanopsis lamellosa* and *Lithocarpus pachyphyllus* reach their upper limit at about 2,700 (2,800) m. While *Larix speciosa* is found above 2,600 m, *Abies delavayi* appears above 3,000 m. As this brief comparison clearly shows, the vegetation patterns of Yunnan and the eastern Himalayas quite resemble each other.

References

Aiba S, David AH, Agetsuma N (2001) Comparison between old growth stands and secondary stands regenerating after clear felling in warm-temperate forests of Yakushima, southern Japan. For Ecol Manag 140:163–175

Angiosperm Phylogeny Group (2009) An update of the Angiosperm Phylogeny Group classification for the orders and families of flowering plants: APG III. Bot J Linn Soc 161(2):105–121

Auclair AN, Goff FG (1971) Diversity relations of upland forests in the western Great Lakes area. Am Nat 105:499–528

Axelrod DI (1983) Biogeography of oaks in the Arcto-Tertiary province. Ann Mo Bot Gard 70:629–657

Axelrod DI, Al-Shehbaz I, Raven PH (1998) History of the modern flora of China. In: Zhang AL, Wu SG (eds) Floristic characteristics and diversity of East Asian plants. China Higher Education Press, Beijing, pp 43–55

Bones TJ (1979) Atlas of fossil fruits and seeds from north-central Oregon, vol Occasional papers in natural history no. 1. Oregon Museum of Science and Industry, Portland, pp 1–23

Cannon CH, Manos PS (2003) Phylogeography of the Southeast Asian stone oaks (*Lithocarpus*). J Biogeogr 30(2):211–226

Chambers JC, Macmahon JA (1994) A day in the life of a seed: movements and fates of seeds and their implications for natural and managed systems. Annu Rev Ecol Syst 25:263–292

Chang Y, Zhong Q, Cheng D, Xu C, Bo H, Zhang Z (2013) Species diversity of ever-green broad-leaved mixed forest with different forest age in the north-west of Fujian province. Ecol Environ Sci 22(6):955–960

Chen W-L (1988) Forest. In: SEGQX (The scientific expedition group of the Qinghai-Xizang Plateau) (ed) Vegetation of Xizang (Tibet). Science Press, Beijing, pp 98–181

Chen B, Su Z (1995) Species diversity of evergreen broad-leaved forest in Babaoshan Nature Reserve, Guangdong. J South China Agric Univ 16(4):32–36

Da L-J, Yang Y-C, Song Y-C (2004) Population structure and regeneration types of dominant species in an evergreen broad-leaved forest in Tiantong National Forest Park, Zhejiang Province, China. Acta Phytoecol Sin 28(3):376–384

Deng T, Zhang D-G, Chen G-X (2008) Characteristics of evergreen broad-leaved forest in Xiaoxi National Natural Reserve, Hunan province and its ecological significance. J Jishou Univ (Natural Science Edition) 29(5):87–93

eFloras (2014). Flora of China. Published on the Internet http://www.efloras.org. Retrieved 22 Aug 2014. Missouri Botanical Garden, St. Louis & Harvard University Herbaria, Cambridge

Fukumoto H, Kajimura H (2001) Guild structure of seed insects in relation to acorn development in two oak species. Ecol Res 16:145–155

Gao G, Chen DL, Ren GY, Chen Y, Liao Y (2006) Spatial and temporal variations and controlling factors of potential evapotranspiration in China: 1956–2000. J Geogr Sci 16(1):3–12

Good R (1964) The geography of the flowering plants, 3rd edn. Longman, London

Guo K, Werger MJA (2010) Effect of prevailing monsoons on the distribution of beeches in continental East Asia. For Ecol Manag 259:2197–2203

Habeck JR (1968) Forest succession in the Glacier Park Cedar-Hemlock forests. Ecology 49:872–880

Halpern CB, Spies TA (1995) Plant species diversity in natural and managed forests of the Pacific Northwest. Ecol Appl 5:913–934

Hattori TH, Nakanishi S (1985) On the distributional limits of the lucidophyllous forest in the Japanese Archipelago. Botanical Magazine, Tokyo 98:317–333

Holdridge LR (1947) Determination of world plant formations from simple climatic data. Science 105:367–368

Holdridge LR (1967) Life zone ecology. Tropical Science Center, San Jose

Hsieh CF, Lai IL, Song GZ, Liao CC, Yang KC (1997) Biodiversity and conservation of the evergreen broad-leaved forests in Taiwan. Tropics 6:361–370

Hu Z, Yu M, Ding B, Fang T, Qiang H, Chen Q (2003) Types of evergreen broad-leaved forests and their species diversity in Gutian Mountain National Nature Reserve. Chin J Appl Environ Biol 9(4):341–345

Huang SF (2011) Historical biogeography of the flora of Taiwan. Natl Taiwan Mus 64(3):33–63

Huang Z, Kong G, Zhang Q, Liu S (1998) Structure, species diversity and population dynamics of the lower subtropical evergreen broad-leaved forest in Dinghushan Biosphere Reserve. In: Sun H-L (ed) The study of tropical and subtropical forest ecosystem, vol 8. Meteorology Press, Beijing, pp 64–75

Huang YH, Tang S-Q, Lu Z-Y, Liang X-T, Pan Q-D (2006) Characteristics of main vegetation types in Jinzhongshan Nature Reserve, Guangxi Province. J Trop Subtrop Bot 14(6):499–503

Itow S (1988) Species diversity of mainland and island forests in the Pacific area. Vegetatio 77:193–200

Jiang ZP (1993) Chinese oaks in the Tertiary. Acta Bot Sin 35(5):397–408

Kira T (1945) A new classification of climate in eastern Asia as the basis of agriculture geography. Horticultural Institute Kyoto University, Kyoto

Kizaki K, Oshiro I (1977) Paleogeography of the Ryukyu Islands. Mar Sci Mon 9:542–549

Kvaček Z, Walther H (1989) Paleobotanical studies in Fagaceae of European Tertiary. Plant Syst Evol 162:213–229

Li C (1997) The distribution of evergreen broad-leaved forest in East Asia. Nat Resour 2:37–46

Li HM, Zhou ZK (2007) Fossil nothofagaceous leaves from the Eocene of western Antarctica and their bearing on the origin, dispersal and systematics of *Nothofagus*. Sci China Earth Sci 50(10):1525–1535

Li JQ, Guo YS, Romane F (2000) Environmental heterogeneity and population variability of sclerophyllous oaks (*Quercus* Sec. suber) in East Himalayan Region. For Study China 2:1–15

Lin WL, Hsieh CF, Chen YK, Lo HC (2003) A community study of the *Keteleeria davidiana* var. *formosana* forest stands in Taiwan. J Exp For Natl Taiwan Univ 17:33–47

Liu M-Q, Zhou Z-K (2006) Modern and geological distribution of *Castanopsis* (Fagaceae). Acta Bot Yunnanica 28(3):223–235

López-Pujol J, Zhang F-M, Sun H-Q, Ying T-S, Ge S (2011) Centres of plant endemism in China: places for survival or for speciation? J Biogeogr 38:1267–1280

Luo Y, Zhou Z-K (2001) Phytogeography of *Quercus* subg. *Cyclobalanopsis*. Acta Bot Yunnanica 23(1):1–16

Manabe T, Nishimura N, Miura M, Yamamoto S (2000) Population structure and spatial patterns for trees in a temperate old-growth evergreen broad-leaved forest in Japan. Plant Ecol 151:181–197

Manchester SR (1994) Fruits and seeds of the middle Eocene nut beds flora, clarno formation, North Central Oregon. Palaeontogr Am 58:1–205

Manos PS, Stanford AM (2001) The historical biogeography of Fagaceae: tracking the tertiary history of temperate and subtropical forests of the Northern Hemisphere. Int J Plant Sci 162(Suppl 6):S77–S93

Metcalfe I (2005) Asia: south-east. In: Selley RC, Robin L, Cocks M, Plimer IR (eds) Encyclopedia of geology, vol 1. Elsevier, Oxford, pp 169–198

Metcalfe I (2006) Palaeozoic and Mesozoic tectonic evolution and palaeogeography of East Asian crustal fragments: the Korean Peninsula in context. Gondwana Res 9:24–46

Meusel H (1971) Mediterranean elements in the flora and vegetation of the West Himalayas. In: Davis PH, Harper PC, Hedge IC (eds) Plant life of South-West Asia. Botanical Society of Edinburgh, Edinburgh, pp 53–72

Miura M, Yamamoto S (2003) Effects of sprouting and canopy on the structure and dynamics of a *Castanopsis cuspidata* var. *sieboldii sapling* population in an old-growth evergreen broad-leaved forest. For Ecol Manag 183:387–400

Miura M, Manabe T, Nishimura N, Yamamoto S (2001) Forest canopy and community dynamics in a temperate old-growth evergreen broad-leaved forest, south-western Japan: a 7-year study of a 4-ha plot. J Ecol 89:841–849

Miyawaki A (ed) (1980) Vegetation of Japan, vol 1. Shibundo, Yakushima/Tokyo

Miyawaki A (ed) (1981) Vegetation of Japan, vol 2. Shibundo, Kyushu/Tokyo

Miyawaki A, Sasaki Y (1985) Floristic changes in the *Castanopsis cuspidata* var. *sieboldii* forest communities along the Pacific Ocean coast of the Japanese Islands. Vegetatio 59:225–234

Monk CD (1967) Tree species diversity in the eastern deciduous forest with particular reference to north central Florida. Am Nat 101:173–187

Morley RJ (1998) Palynological evidence for Tertiary plant dispersals in the SE Asian region in relation to plate tectonics and dispersal. In: Hall R, Holloway JD (eds) Biogeography and geological evolution of SE Asia. Backhuys, Leiden, pp 211–234

Negi SS, Naithani HB (1995) Oaks of India, Nepal and Bhutan. International Book Distributors, Dehra Dun

Numata M (ed) (1974) The flora and vegetation of Japan. Kodansha, Tokyo

Ohsawa M (1985) Altitudinal zonation of forest vegetation on Mount Kerinci, Sumatra: with comparisons to zonation in the temperate region of east Asia. J Trop Ecol 1:193–216

Ohsawa M (1987) Vegetation zones in the Bhutan Himalaya. In: Ohsawa M (ed) Life zone ecology of the Bhutan Himalaya. Chiba University, Chiba, pp 1–71

Ohsawa M (1991) Montane evergreen broad-leaved forests of the Bhutan Himalaya. In: Ohsawa M (ed) Life zone ecology of the Bhutan Himalaya. Chiba University, Chiba

Ohsawa M, Shakya PR, Numata M (1986) Distribution and succession of West Himalayan forest types in the Eastern part of the Nepal Himalaya. Mt Res Dev 6(2):143–157

Ohwi J, Meyer FG, Walker EH (1965) Flora of Japan: a combined, much revised, and extended translation by the author of his Flora of Japan (1953) and Flora of Japan, Pteridophyta (1957). Smithsonian Institution, Washington, DC

Peet RK (1978) Forest vegetation of the Colorado Front Range: patterns of species diversity. Vegetatio 43:131–140

Peng H, Du Y, Li J, Ouyang X, Huang Z (2006) Comparison of population distribution pattern of evergreen broad-leaved forests between Nanling Dadingshan Nature Reserve and Dinghushan. Ecol Environ 15(4):770–774

Qian H (2002) Floristic relationships between Eastern Asia and North America: test of Gray's hypothesis. Am Nat 160(3):317–332

Qian H, Wang S, He J-S, Zhang J, Wang L, Wang X, Guo K (2006) Phytogeographical analysis of seed plant genera in China. Ann Bot 98:1073–1084

Qiu Z-J, Liu P, Liu C-S, Gao J-G (2010) Community structure and regeneration types of dominant species in an evergreen broad-leaved forest in Mount Beishan of Jinhua, China. Guihaia 30(5):629–635

Quercus Portal (2014) Evolutionary biology. https://w3.pierroton.inra.fr/QuercusPortal/index.php?p=EVOLUTIONARY_BIOLOGY. Retrieved 16 Sept 2014

Richards PW (1952) The tropical rain forest. An ecological study. Cambridge University Press, Cambridge

Saito S (2002) Effects of a severe typhoon on forest dynamics in a warm-temperate evergreen broad-leaved forest in southwestern Japan. J For Res 7:137–143

Sang W, Ma K, Axmacher JC (2011) Securing a future for China's wild plant resources. Bioscience 61:720–725

Shi Y-F, Li T-Y, Li B-Y et al (1998) Uplift and environmental evolution of Qinghai-Xizang (Tibetan) Plateau. In: Su H, Zheng D (eds) Formation, evolution and development of Qinghai-Xizhang (Tibetan) Plateau. Guangdong Science and Technology Press, Guangzhou, pp 73–138

Soepadmo E (1972) Fagaceace. In: van Steenis CGGJ (ed) Flora Malesiana, Series 1, vol. 7 Part 2, Wolters-Noordhoff Publishing, Groningen, pp 265–403

Soepadmo E, Julia S, Rusea G (2000) Fagaceae. In: Soepadmo E, Saw LG (eds) Tree flora of Sabah and Sarawak, vol 3. Forest Research Institute Malaysia, Sabah Forestry Department and Sarawak Forestry Department, Kuala Lumpur

Song Y-C (ed) (2013) Evergreen broad-leaved forests in China: classification-ecology-conservation. Science Press, Beijing

Springer (2014) List of Korea evergreen plants. http://download.springer.com/static/pdf/928/bbm%253A978-94-011-16244%252F1.pdf?auth66=1410425161_897032acbd05d2f42261b4a4ae11ae8a&ext=.df. Retrieved 5 Aug 2014

Su HJ (1984) Studies on the climate and vegetation types of the natural forest in Taiwan. (II). Altitudinal vegetation zones in relation to temperature gradient. Q J Chin For 17:57–73

Suzuki K (1979) Vegetation of Ryukyu Islands, Japan. Bull Inst Environ Sci Technol Yokohama Natl Univ 5:87–160

Takhtajan A (1986) Floristic regions of the world. University of California Press, Berkeley

Tang Y-C (2000) On the affinities and the role of the Chinese flora. Acta Bot Yunnanica 22(1):1–26

Tang CQ (2006) Evergreen sclerophyllous *Quercus* forests in northwestern Yunnan, China as compared to the Mediterranean evergreen *Quercus* forest types in California, USA and northeastern Spain. Web Ecol 6:88–101

Tang CQ (2015) Distribution Patterns of the Subtropical Evergreen Broad-Leaved Forests of Southwestern China, as Compared with those of the Eastern Chinese subtropical regions. Collectanea Botanica (in press)

Tang CQ, Ohsawa M (2009) Ecology of subtropical evergreen broad-leaved forests of Yunnan, southwestern China as compared to those of southwestern Japan. J Plant Res 122:335–350

Tang CQ, Chiou C-R, Lin C-T, Lin J-R, Hsieh C-F, Tang J-W, Su W-H, Hou X (2013) Plant diversity patterns in subtropical evergreen broad-leaved forests of Yunnan and Taiwan. Ecol Res 28:81–92

Tanouchi H, Yamamoto S (1995) Structure and regeneration of canopy species in an old-growth evergreen broad-leaved forest in Aya district, southwestern Japan. Vegetatio 117:51–60

Tanouchi H, Sato T, Takeshita K (1994) Comparative studies on acorn and seedling dynamics of four *Quercus* species in an evergreen broad-leaved forest. J Plant Res 107:153–159

Terrestrial Biozones (2014) http://www.terrestrialbiozones.net/Endemic%20Floras/North%20 Indochinese%20Endemics.html. Retrieved 5 May 2014

Trelease W (1924) The American oaks. Mem Natl Acad Sci 20:1–255

Troll C (1948) Der asymmetrische Aufbau der Vegetationszonen und Vegetationsstufen auf der Nordund Sudhalbkugel. Jahresbericht Geobot Forschungsinstitutes Rubel i. Zurich f. 1947(Zurich):46–83

Turner IM, Wong YK, Chew PT, bin Ibrahim A (1997) Tree species richness in primary and old secondary tropical forest in Singapore. Biodivers Conserv 6:537–543

Wan H (2007) Plant diversity and its distribution pattern in the forest ecosystem of Lushan Mountain. Dissertation, Forestry University of Beijing, Beijing

Wang Y, Liu Y-Q, Yan Q-P, Deng Z-F, Wang B, Zeng B-S, Guo H (2009) A study on the community characteristics of evergreen broad-leaved forest in Dagangshan Mountains of Jiangxi Province. Acta Agric Univ Jiangxiensis 31(6):1055–1068

Wangda P, Ohsawa M (2006) Structure and regeneration dynamics of dominant tree species along altitudinal gradient in a dry valley slopes of the Bhutan Himalaya. For Ecol Manag 230:136–150

Whitmore TC (1975) Tropical rain forests of the Far East. Oxford University Press, Oxford

Wikipedia (2014a) http://en.wikipedia.org/wiki/Lithocarpus. Retrieved 10 Aug 2014

Wikipedia (2014b) http://en.wikipedia.org/wiki/List_of_Quercus_species. Retrieved 5 Aug 2014

Wolfe JA, Leopold EB (1967) Neogene and early quaternary vegetation of northwestern North America and northeastern Asia. In: Hopkins DM (ed) The bearing land bridge. Stanford University Press, Stanford, pp 193–206

Wu ZY (ed) (1980) Vegetation of China. Science Press, Beijing

Wu ZY (1983) On the significance of Pacific intercontinental discontinuity. Ann Mo Bot Gard 70:577–590

Wu C-Y (1988) Hengduan mountain flora and her significance. J Jpn Bot 63:297–311

Wu C-Y (1991) The areal-types of Chinese genera of seed plants. Acta Bot Yunnanica Suppl IV:1–139

Wu C-Y (1993) Addenda et corrigenda ad typi arealorum generorum spermatophytorum sinicarum. Acta Bot Yunnanica Suppl IV:141–178

Wu C-Y (1998) Delineation and unique features of the Sino-Japanese floristic region. In: Boufford DE, Ohba H (eds) Sino-Japanese Flora: its characteristics and diversification, University Museum, University of Tokyo, Bulletin No. 37. Tokyo University Press, Tokyo, pp 1–35

Wu C-Y, Wu SG (1998) A proposal for a new floristic kingdom (realm) – the Asiatic kingdom, its delineation and characteristics. In: Zhang A, Wu S (eds) Floristic characteristics and diversity of East Asian plants. China Higher Education Press, Beijing, pp 3–42

Wu ZY, Zhu YC, Jiang HC (eds) (1987) The vegetation of Yunnan. Science Press, Beijing

Xia A-M, Nie L-Q (2004) Discussion about the forest vegetation zonation in Anhui Province. J Wuhan Bot Res 22(6):523–528

Xiang QY, Soltis DE, Soltis PS (1998) The eastern Asia and eastern and western North American floristic disjunction: cogenetic patterns in seven diverse genera. Mol Phylogenet Evol 10:178–190

Yamamoto S (1992) Gap characteristics and gap regeneration in primary evergreen broad-leaved forests of western Japan. Botanical Magazine, Tokyo 105:29–45

Yamamoto S (1997) Some features of the forest dynamics of warm-temperate evergreen broad-leaved forests in Japan. Nat Hist Res, Special issue No. 4:127–131

Yang Y-F, Ding H, Xu H-G, Fang Y-M, Chen X, Yi X-G, Yang Q, Wen X-R, Xu X-J (2013) Point pattern analysis of four main plant populations in evergreen broad-leaved forests typical of Wuyi Mountains, Southeast China. J Ecol Rural Environ 29(2):184–190

Zheng GX, Lin ZL, Fang JY (2004) Tree species diversity along altitudinal gradient on southeastern and northwestern slopes of Mt. Huanggang, Wuyi Mountains, Fujian. China Biodivers Sci 12(1):63–74

Zhou ZK (1992) Origin, phylogeny and dispersal of *Quercus* from China. Acta Bot Yunnanica 14:227–236

Zhou ZK, Shui YM, Zhang GJ (2002) Plants and plant resources on Mt. Xilong, Jingping. In: Xu JC (ed) Scientific investigation in the Nature Reserve in the watershed on the boundary between Jingping and Fenshuilin, Yunnan. Science and Technology, Kunming, pp 151–162

Zhou ZK, Pu CX, Chen WY (2003) Relationships between the distributions of *Quercus* sect *heterobalanus* (Fagaceae) and uplift of Himalayas. Adv Earth Sci 18:884–890

Zhu H (2013a) Geographical elements of seed plants suggest the boundary of the tropical zone in China. Palaeogeogr Palaeoclimatol Palaeoecol 386:16–22

Zhu H (2013b) The floras of southern and tropical southeastern Yunnan have been shaped by divergent geological histories. PLoS One 8(5):e64213. doi:10.1371/journal.pone.0064213

Zhu H, Roos MC (2001) The tropical flora of southern China and its affinity to Indo-Malesian flora. Telopea 10(2):639–648

Zhu H, Yan L-C (2002) A discussion on biogeographical lines of the tropical-subtropical Yunnan. Chin Geogr Sci 12(1):90–96

Zhu H, Yan LC (2009) Biogeographical affinities of the flora of southeastern Yunnan, China. Bot Stud 50(4):467–475

Zhu B, Chan A-P, Liu Z-L, Li G-Z, Fang J-Y (2004) Changes in floristic composition, community structure, and tree species diversity of plant communities along altitudinal gradients on Mt. Maoer, Guangxi. China Biodivers Sci 12(1):44–52

Zhu H, Cao M, Hu H (2006) Geological history, flora, and vegetation of Xishuangbanna, Southern Yunnan, China. Biotropica 38(3):310–317

Index

C.Q. Tang, *The Subtropical Vegetation of Southwestern China*, Plant and Vegetation 11, DOI 10.1007/978-94-017-9741-2

MIX
Papier aus verantwortungsvollen Quellen
Paper from responsible sources
FSC® C105338
FSC
www.fsc.org

If you have any concerns about our products,
you can contact us on
ProductSafety@springernature.com

In case Publisher is established outside the EU,
the EU authorized representative is:
Springer Nature Customer Service Center GmbH
Europaplatz 3, 69115 Heidelberg, Germany

Printed by Libri Plureos GmbH
in Hamburg, Germany